Farmed Fish Quality

Farmed Fish Quality

Edited by

Steve C. Kestin and Paul D. Warriss
School of Veterinary Science
University of Bristol
Langford
Bristol, UK

Fishing News Books
An imprint of Blackwell Science

b

Blackwell
Science

Copyright © 2001
Fishing News Books
A division of Blackwell Science Ltd
Editorial Offices:
Osney Mead, Oxford OX2 0EL
25 John Street, London WC1N 2BS
23 Ainslie Place, Edinburgh EH3 6AJ
350 Main Street, Malden
  MA 02148 5018, USA
54 University Street, Carlton
  Victoria 3053, Australia
10, rue Casimir Delavigne
  75006 Paris, France

Other Editorial Offices:

Blackwell Wissenschafts-Verlag GmbH
Kurfürstendamm 57
10707 Berlin, Germany

Blackwell Science KK
MG Kodenmacho Building
7–10 Kodenmacho Nihombashi
Chuo-ku, Tokyo 104, Japan

First published 2001

Set in 10.5/12pt Garamond
by DP Photosetting, Aylesbury, Bucks
Printed and bound in Great Britain by
MPG Books Ltd, Bodmin, Cornwall

DISTRIBUTORS

Marston Book Services Ltd
PO Box 269
Abingdon
Oxon OX14 4YN
(*Orders:* Tel: 01865 206206
        Fax: 01865 721205
        Telex: 83355 MEDBOK G)

USA
Blackwell Science, Inc.
Commerce Place
350 Main Street
Malden, MA 02148 5018
(*Orders:* Tel: 800 759 6102
              781 388 8250
        Fax: 781 388 8255)

Canada
Login Brothers Book Company
324 Saulteaux Crescent
Winnipeg, Manitoba R3J 3T2
(*Orders:* Tel: 204 837 2987
        Fax: 204 837 3116)

Australia
Blackwell Science Pty Ltd
54 University Street
Carlton, Victoria 3053
(*Orders:* Tel: 03 9347 0300
        Fax: 03 9347 5001)

A catalogue record for this title
is available from the British Library

ISBN 0-85238-260-X

Library of Congress
Cataloging-in-Publication Data

Farmed fish quality/edited by S.C. Kestin and
  P.D. Warriss.
      p. cm.
    This book constitutes the results of an international
conference held on 7–9 April 1999.
    Includes bibliographical references (p.).
    ISBN 0-85238-260-X (hardcover)
    1. Fish-culture—Congresses.   2. Fishes—Quality—
Congresses.   I. Kestin, S.C.   II. Warriss, P.D.

SH151.F37 2000
639.3—dc21   00-062258

For further information on
Fishing News Books, visit our website:
http://www.blacksci.co.uk/fnb/

Contents

Contributors

Alderson, R.
Alderson Aquaculture, 3 Lumsdaine Drive,
Dalgety Bay, Fife, KY11 9YU, UK
(email: dick.alderson@btinternet.com)

Baker, R.T.M.
BASF AG, Global Marketing – Carotenoids,
67056 Ludwigshafen, Germany
(email: remi.baker@basf-ag.de)

Bell, J.G.
Institute of Aquaculture, University of
Stirling, Stirling, FK9 4LA, UK
(email: j.g.bell@stir.ac.uk)

Bencze Rørå, A.M.
Institute of Aquaculture Research
(AKVAFORSK), N-1432Ås, Norway

Branson, E.
Red House Farm, Llanvihangel, Monmouth,
Gwent, NP25 4HL, UK
(email: edward.branson@virgin.net)

Cooke, M.
Tesco House, P.O. Box 18, Delamare Road,
Cheshunt, Hertfordshire, EN8 9SL, UK

Day, B.P.F.
Campden and Chorleywood Food Research
Association, Chipping Campden,
Gloucestershire, GL55 6LD, UK

Einen, O.
Institute of Aquaculture Research
(AKVAFORSK), N-1432Ås, Norway

Erikson, U.
SINTEF Fisheries and Aquaculture, N-7465
Trondheim, Norway

Fjellanger, K.
Nutreco Aquaculture Research Centre,
Stavanger, Norway
(email: kurt.fjellanger@nutreco.com)

Greenhalgh, A.R.
'Woodlea', Hill Drive, Failand, Bristol, BS8
3UX, UK

Howgate, P.
26 Lavender Row, Stedham, Midhurst,
West Sussex, GU20 0NS, UK
(email: phowgate@rsc.co.uk)

Johansson, L.
Department of Domestic Sciences, Uppsala
University, Uppsala, S-75337, Sweden
(email: lisbeth.johansson@ihv.uu.se)

Johnston, I.A.
Fish Muscle Research Group, Gatty Marine
Laboratory, Division of Environmental
and Evolutionary Biology, School of
Biology, University of St Andrews, St
Andrews, Fife, Scotland KY16 8LB, UK
and Matre Aquaculture Station, Institute
of Marine Research, N-5984, Matredal,
Norway

Kestin, S.C.
School of Veterinary Science, University of
Bristol, Langford, Bristol BS40 5DU
(email: steve.kestin@bristol.ac.uk)

Kuhlmann, H.
Bundesforschungsanstalt für Fischerei,
Institut für Fischereiökologie,
Aussenstelle Ahrensburg, Wulffsdorfer
Weg 204, D-22926, Hamburg,
Germany
(email: kuhlmann_bfafisch_
hamburg@compuserve.com)

Laidler, L.A.
Marine Harvest (Scotland) Ltd, Lochailort,
Invernessshire, PH38 4LZ, UK

McEvoy, J.
Institute of Aquaculture, University of
Stirling, Stirling, FK9 4LA, UK

McGhee, F.
Institute of Aquaculture, University of
Stirling, Stirling, FK9 4LA, UK

Michie, I.
Marine Harvest (Scotland) Ltd., Craigcrook
Castle, Edinburgh, EH4 3TU, UK
(email: ian.michie@marine-
harvest.co.uk)

Mørkøre, T.
Institute of Aquaculture Research
(AKVAFORSK), N-1432Ås, Norway

Morris, P.C.
Trouw Aquaculture, Renfrew Mill, Wright
Street, Renfrew, PA4 8BF, UK
(email: paul.morris@nutreco.com)

Münkner, W.
Bundesforschungsanstalt für Fischerei,
Institut für Biochemie und Technologie,
Palmaille 9, D-22767 Hamburg,
Germany
(email: 100565.1223@compuserve.com)

Nickell, D.C.
Roche Products Limited, Heanor,
Derbyshire, DE75 7SG, UK
(email: david.nickell@roche.com)

Nute, G.R.
Department of Clinical Veterinary Science,
University of Bristol, Langford, Bristol,
BS40 5DU, UK
(email: geoff.nute@bris.ac.uk)

Obach, A.
Nutreco Aquaculture Research Centre,
Stavanger, Norway

Oehlenschläger, J.
Bundesforschungsanstalt für Fischerei,
Institut für Biochemie und Technologie,
Palmaille 9, D-22767 Hamburg,
Germany
(email: oehlenschlaeger.ibt@bfa-fisch.de)

Pottinger, T.G.
NERC Institute of Freshwater Ecology,
Windermere Laboratory, The Ferry
House, Far Sawrey, Ambleside, Cumbria,
LA22 0LP, UK

Robb, D.H.F.
Department of Clinical Veterinary Science,
University of Bristol, Langford, Bristol,
BS40 5DU, UK
(email: david.robb@bris.ac.uk)

Roberts, R.J.
9 Alexander Drive, Bridge of Allan, Stirling,
FK9 4QB, UK
(email: heronpisces@compuserve.com)

Rosenlund, G.
Nutreco Aquaculture Research Centre,
Stavanger, Norway

Sargent, J.R.
Institute of Aquaculture, University of
Stirling, Stirling, FK9 4LA, UK
(email: j.r.sargent@stir.ac.uk)

Schelvis-Smit, A.A.M.
Netherlands Institute for Fisheries Research
(RIVO), P.O. Box 68, 1970 AB
IJmuiden, The Netherlands

Shearer, K.D.
National Marine Fisheries Service, 2725
  Montlake Boulevard East, Seattle, WA
  98112 USA
  (email: kshearer@sci.nwfsc.noaa.gov)

Sigurgisladottir, S.
IceTec, Keldnaholt, Is-112 Reykjavik,
  Iceland

Sinnott, R.
Trouw Aquaculture, Wincham, Northwich,
  Cheshire, CW9 6DF, UK
  (email: rob.sinnott@nutreco.com)

Slinde, E.
Institute of Marine Research, P.O. Box 1870
  Nordnes, N-5817 Bergen, Norway

Smart, G.
Culmarex SA, Muelle del Hornillo, 30880
  Aguilas, Murcia, Spain

Southgate, P.
The Fish Veterinary Group, Rowandale,
  Dunscore, Dumfries, DG2 0UE, UK
  (email: pete@fishvet.co.uk)

Springate, J.R.C.
F. Hoffmann-La Roche Ltd, VMA,
  CH-4070 Basel, Switzerland
  (email: john.springate@roche.com)

Tejada, M.
Instituto del Frío (CSIC). C/ Ramiro de
  Maeztu, s/n. Ciudad Universitaria, 28040
  Madrid, Spain
  (email: mtejada@if.csic.es)

Torrissen, O. J.
Institute of Marine Research, P.O. Box 1870
  Nordnes, N-5817 Bergen, Norway
  (email: ole.torrissen@imr.no)

Undeland, I.
SIK – The Swedish Institute for Food and
  Biotechnology, P.O. Box 54 01, S-402 29
  Göteborg, Sweden
  (email: ummarine@tiac.net)

van de Vis, J.W.
Netherlands Institute for Fisheries Research
  (RIVO), P.O. Box 68, 1970 AB
  IJmuiden, The Netherlands
  (email: hans@rivo.dlo.nl)

Wall, A. J.
Fish Veterinary Group, 22 Carsegate Road,
  Inverness, IV3 8EX, UK
  (email: info@fishvet.co.uk)

Warriss, P.D.
School of Veterinary Science, University of
  Bristol, Langford, Bristol BS40 5DU
  (email: p.d.warriss@bris.ac.uk)

Webster, J.L.
Scottish Salmon Growers' Association
  Limited, Drummond House, Scott Street,
  Perth, PH1 5EJ, UK

Preface

This book constitutes the results of an international conference on Farmed Fish Quality held on 7–9 April 1999 and organised by the University of Bristol in collaboration with, and sponsored by, Trouw Aquaculture. Farmed fish quality is a developing area of concern. The issues are different from those associated with wild caught fish in that product quality is inherently amenable to influence or control throughout the whole production chain. The conference benefited from the quality and breadth of interest of the speakers, and the fact that they came from backgrounds ranging from academia to industry. This resulted in a complementary mixture of practical experience informed by basic science which generally seemed to find favour with participants. The scope of the conference was farmed fish quality in its widest sense and all aspects of quality were addressed. These were covered in seven sessions, starting with a consideration of the basis of fish quality and then addressing the influence of husbandry, handling, slaughter and post-harvest factors. Subsequent sessions were devoted to ways of measuring quality, value added products, and strategies to improve quality. The chapters in this book follow that format.

Much of the success of the conference was attributable to the quality of the speakers and the enthusiasm of the participants. However, it was undoubtedly a reflection of the large amount of effort that went into its organisation by the numerous individuals who contributed to it. Amongst these must be mentioned Rita Hinton, Christine Rawlings and Cassandra Peach, on whom the bulk of the organisation fell, and Steve Brown, Julie Edwards, Leisha Hewitt, Kate Hinton, Deirdre Lee, Justin McKinstry, Paul Riley, Paul Whittington and Steve Wotton who made the conference work. Kath McDonnell and Carol Hole were responsible for the collation of the final manuscript for this book. Our thanks go to all these people.

Steve C. Kestin
Paul D. Warriss

Chapter 1

Introduction

R.J. Roberts

Hagerman Laboratory, University of Idaho, Idaho, ID 83332, USA

Fish is an important component of the diet of most of the world's communities. In some areas, it may even comprise the only source of animal protein and for most societies it is the principal source of the essential polyunsaturated fatty acids. Fishing, the last survivor of the hunter–gatherer technologies, dependent as it is on natural production and sustainable capture strategies, has reached its maximal sustainable yield and can no longer supply the expanding markets. Increased population and excessive demand, allied to greater sophistication of harvesting, have led to over-fishing to the extent that, despite increased effort, annual world catches are actually diminishing.

Fortunately, the world is no longer dependent on wild fisheries. Aquaculture in all of its many forms is expanding exponentially to at least partially fill the supply deficit. Currently, the world *per capita* seafood consumption is of the order of 13.5 kg per annum. World population is expected to increase by 2 billion by 2025. If this population is to be able to consume similar levels of fish protein, given the fixed level of sustainable yield from fisheries, then aquaculture will have to provide all of the increase of 27 million metric tonnes (mmt) required. This represents a doubling of present production levels.

Already, over the decade from 1987 to 1997, global production from aquaculture has risen from 7.3 mmt. to 18.8 mmt. This is a remarkable increase and the growth is continuing. Whether it can continue at this level for the future, however, is critically dependent on the availability of feeds. Fish meal is a major component of most high value fish diets and since supply of that commodity is no more expandable than the wild fishery catches of which it is a by-product, alternative proteins capable of providing similar amino acid profiles to those of fish meal will have to be found.

Wild fish, while highly esteemed for quality and taste when perfectly preserved and processed, are notoriously variable in quality due to the vagaries of the capture, storage and processing technologies utilised. Farmed fish on the other hand can be harvested at will, and should in theory be able to be processed in the peak of condition to produce a superior product.

Unfortunately, even with the relatively good diets containing high levels of fish meal, which are available for most of the high value farmed species, this superior product of cultured fish technology is rarely produced. In the future, when diets will have to depend on substitute feed stocks, generally derived from modified vegetable origins rather than fish meal, quality of the end product may be even more difficult to assure.

Farmed fish product quality depends on a wide range of factors including the nature of the fish stock, husbandry and feeding procedures applied, slaughter techniques and methods of storage, processing and transport. Each of these will, if compromised, affect the quality of the product. Unfortunately, our current knowledge base in relation to the science underpinning these factors is inadequate. If we are to achieve the production levels expected of aquaculture over the next quarter century, and to ensure that farmed fish are both nutritionally and economically optimal, then this situation must be changed.

In April 1999, the School of Veterinary Science of the University of Bristol held an extremely successful international conference on the science behind farmed fish quality. They gathered a very wide range of speakers on aspects of the subject ranging from the basic structure of fish myotomes and myofibrils to methodologies for humane slaughter. It was clear to those of us who were fortunate enough to attend the conference that there was a need for a single source book of information covering all of the aspects of this multidisciplinary subject.

Normally, academic conferences generate volumes of papers, uncoordinated or unedited, and generally produced as symposium proceedings, which are of very limited value and are rarely used. The organisers of the Bristol conference, however, insisted that key contributors should produce definitive reviews of their whole subject area, for amalgamation as chapters into a proper reference book which would serve as the standard text on the subject for the next decade.

I believe they have succeeded well. In particular, the chapters at the beginning of the text on the fundamental aspects of fish biology relevant to fish quality will provide the basis for years of further, more applied, research and commercial development. Equally, the more practical aspects of quality assurance, slaughter technology and processing, and marketing require the same levels of attention to basic science if they are to move forward. The chapters devoted to these highlight the major deficiencies in present understanding as well as presenting the current situation and reference base.

If we are to meet the challenge of providing the fish protein needs of the twenty-first century, the multidisciplinary science base will need to be strengthened. In terms of the quality requirements of farmed fish, this book provides a balanced definition of our present understanding. I trust that the younger generations of scientist will rise to the challenges it presents and will ensure that the needs of mankind for more fish without the option of increasing the capture fisheries will be met in a rational, productive and humane fashion.

Chapter 2
The Nutritional Value of Fish

J.R. Sargent[1], J.G. Bell[1], F. McGhee[1], J. McEvoy[1] and J. L. Webster[2]

[1]*Institute of Aquaculture, University of Stirling, Stirling, FK9 4LA, UK*
[2]*Scottish Salmon Growers' Association Limited, Drummond House, Scott Street, Perth, PH1 5EJ, UK.*

Introduction

A briefing paper published in 1993 by the British Nutrition Foundation on nutritional aspects of fish, mostly gadoids such as cod, haddock and whiting, flatfish such as plaice and soles, and herring and mackerel, summarised the situation as follows (Anon 1993). Fish has an important role in our diet and its consumption should be strongly encouraged. Fish is a reliable source of high quality protein, iron, selenium and iodine. The livers of lean white fish, i.e. cod, haddock and whiting, are a very good source of vitamins A and D. The flesh of oil-rich fish such as herring and mackerel is an important source of the long-chain n-3 poly-unsaturated fatty acids (PUFA), eicosapentaenoic acid (EPA, $20:5n\text{-}3$) and docosahexaenoic acid (DHA, $22:6n\text{-}3$). However, the oil composition of fish varies with season and the type of feed available to the fish, and it can be altered by processing and cooking. The summary to the briefing paper highlighted the beneficial effects of long chain *n*-3 PUFA in a variety of human disorders and also emphasised the importance of DHA for development of the foetal brain and retina. It concluded:

> 'Consumption of oil-rich fish such as mackerel, herring, salmon, pilchards and sardines by UK adults should be substantially increased to about two portions per week. Intakes of fruits, vegetables and whole grain cereals should be increased at the same time to provide antioxidant substances which protect the vulnerable polyunsaturated fatty acids in fish oils from oxidation.'

That farmed fish were scarcely mentioned in the British Nutrition Foundation briefing paper is notable and regrettable but inevitable because, at that time, little or no data were available on the composition of farmed fish flesh and especially its oil content and fatty acid composition. Indeed unpublished information was circulating from the USA that farmed fish, specifically channel catfish, were relatively deficient in long chain *n*-3 PUFA but were rich instead in the *n*-6 PUFA linoleic acid (LA, $18:2n\text{-}6$). This reflected the extensive use of vegetable seed oils such as corn and soya oils in channel catfish farming in the USA, these oils being rich in $18:2n\text{-}6$. It was equally realised that salmon and trout farming in the UK and other European countries was based very largely on feeds derived mainly from fish meal and fish oil, so that the product was inevitably rich in long chain *n*-3 PUFA. However, doubts persisted that farmed salmonids might in some way be inferior compositionally and nutritionally to their wild counterparts.

Against that background, we have been involved in an ongoing study, sponsored by the Scottish Salmon Growers' Association. The study centres on the examination of certain important chemical and biochemical aspects of quality in Atlantic salmon produced in Scotland in accordance with the industry's Product Certification Scheme. Its major findings to date have been published (Bell *et al.* 1998). Here we summarise and update the salient findings of our study and assess their implications for the nutritional quality of farmed salmon and other farmed fish. We also consider the feasibility, and more so the desirability, of partially replacing fish meal and fish oil in farmed fish feeds with proteins and oils of vegetable origin, above all in relation to retaining the perceived health promoting properties of fish.

Fish meal and fish oils in farmed fish feeds

Fish meal and fish oils have traditionally formed the basis of farmed fish feeds for very sound reasons, both practically and scientifically. Meal and oil are derived simultaneously from industrial fisheries such as the anchovy and sardine fisheries in the southern hemisphere or the capelin, herring and sand eel fisheries in the northern hemisphere. Fish meal and fish oil have, until recently, been readily available and relatively cheap. Standard grade fish meal itself contains substantial amounts of fish oil, up to *c.* 10% of its dry weight, and is very readily accepted and easily digested by fish. The main protein components of fish meal are fish muscle proteins rich in essential amino acids, and the essential amino acid requirements of fish as determined experimentally can be met precisely by fish meal (Wilson 1989). This is not surprising since the species used for fish meal production are consumed by larger fish in the wild, e.g. cod and salmon consume substantial quantities of capelin, sand eels and herring in their natural diets. Thus, the amino acid compositions of the flesh of salmon, trout, channel catfish, a range of marine fish and fish meal are essentially the same (Wilson 1989; Sikorski *et al.* 1990 and references therein). Such amino acid compositions can, of course, be generated by various blends of animal (non-fish) and even vegetable proteins so that, providing the proteins in question are readily accepted and digested by fish, it is feasible both in theory and practice to substitute fish meal in farmed fish feeds with non-fish proteins. However, in practice vegetable proteins tend to be less digestible and to have a lower essential amino acid content than fish meal. Thus, in the high energy feeds currently favoured by salmon farmers the protein content of the diet is around 40% by weight which means that high quality, highly digestible protein must be used, limiting opportunities for replacing fish meal in the diets.

As fish meal is an ideal component for farmed fish feeds, so is fish oil which provides both energy yielding fatty acids, most notably saturated and monounsaturated fatty acids such as (in the case of typical 'northern hemisphere fish oils') 16 : 0, 18 : 1*n*-9, 20 : 1*n*-9 and 22 : 1*n*-11, and the long chain *n*-3 PUFA 20 : 5*n*-3 and 22 : 6*n*-3 that are necessary for forming cell membranes and, therefore, for growth of the fish (Sargent & Henderson 1995). The *n*-3 PUFA are dietary 'essential fatty acids' for all fish with marine fish having an absolute dietary requirement for the biologically active long chain n-3 PUFA, $20:5n$-3 and $22:6n$-3, because they cannot convert the shorter chain precursor linolenic acid (LNA, $18:3n$-3) to $20:5n$-3 and thence to $22:6n$-3 (Sargent *et al.* 1995). Many freshwater fish can perform this conversion but, not surprisingly, thrive on the end product $20:5n$-3 and $22:6n$-3 (Sargent *et al.* 1995). LNA is present in the glycolipids of chloroplasts, i.e. in green leaf plants which are not lipid-rich, but is not a major constituent of commonly available plant seed oils, linseed oil being a notable exception. There are few definitive studies of freshwater fish grown on diets containing $18:3n$-3 and the sole PUFA (see Sargent *et al.* 1995 for details) and few if any of the freshwater fish species currently being farmed intensively, most of which are carnivores and none of which are herbivores, would experience such a diet in nature. Not surprisingly, fish oil is used routinely in intensive culture of trout, salmon and marine fish including turbot, sea bream and sea bass, and halibut. There is currently no commercially available oil rich in $20:5n$-3 and $22:6n$-3 other than fish oil.

Thus, fish meal and fish oil are natural and ideal choices in feeds for farmed fish species that are naturally piscivorous and, as has often been stated, intensive fish farming is basically a means of converting relatively low value fish meal and fish oil from industrial fisheries to a relatively high value human food rich in high quality protein and long chain n-3 PUFA.

Health promoting effects of $20:5n$-3 and $22:6n$-3

The briefing paper by the British Nutrition Foundation on nutritional aspects of fish, referred to above, was prompted by an earlier report by the Foundation on the nutritional and physiological significance of unsaturated fatty acids (Anon 1992). This and other subsequent authoritative reports (Anon 1994a,b,c) highlighted the beneficial effects of fish oils and their constituent $20:5n$-3 and $22:6n$-3 in a range of human disorders including cardiovascular and inflammatory conditions and cancers (Various 1996).

One aspect of these beneficial effects which is relatively well understood concerns the actions of the eicosanoids which are a range of highly biologically active compounds produced in small amounts from the long chain n-6 PUFA, arachidonic acid (AA, $20:4n$-6). In contrast to fish, the main dietary essential fatty acid in terrestrial mammals including *Homo sapiens* is linoleic acid ($18:2n$-6) which is abundant in most commercially available vegetable seed oils and which is generally readily converted in most mammals to the biologically active $20:4n$-6. One of the major functions of $20:4n$-6 is to generate eicosanoids, broadly in response to physiologically stressful conditions. While this is a normal physiological response, excess production of eicosanoids is undesirable and is common in pathological conditions. This can occur when the diet contains an excess of n-6 relative to n-3 PUFA because $20:5n$-3, which is normally readily formed from $18:3n$-3 in *Homo sapiens* but is lacking under these conditions, attenuates eicosanoid levels by competitively inhibiting their formation from $20:4n$-6, whilst additionally inhibiting the formation of $20:4n$-6 itself from dietary $18:2n$-6. Thus, diets with very high ratios of $18:2n$-6 : $18:3n$-3, as are common in western societies including the UK, promote excess eicosanoid production and are associated with high incidences of cardiovascular and inflammatory disorders and cancers (Various 1996). It is against this background that directly supplementing the diet with $20:5n$-3, whether with fish oil preparations rich in this fatty acid or by eating oily fish, can be beneficial for health by damping down production of excess eicosanoids resulting from the conversion of excess dietary $18:2n$-6 to $20:4n$-6. In essence, the problem stems from a dietary imbalance due to ingesting an excess of $18:2n$-6, usually in the form of vegetable seed oils, and an insufficiency of green vegetables and fish. It should be noted that excess dietary intake of $18:2n$-6, as well as promoting excess eicosanoid production, also inhibits the conversion of $18:3n$-3 to $20:5n$-3 and $22:6n$-3 because $18:2n$-6 and $18:3n$-3 compete for the same fatty acid desaturases and elongases that convert the C18 PUFA ($18:2n$-6 and $18:3n$-3) to their C20 and C22 end products ($20:4n$-6, $20:5n$-3 and $22:6n$-3). Thus fish oil, by directly providing $20:5n$-3 and $22:6n$-3, can be beneficial even given an excess intake of $18:2n$-6. The alternative is to decrease the intake of $18:2n$-6, i.e. by eating less vegetable oils, and simultaneously increase the intake of $18:3n$-3, i.e. by eating more green leaf vegetables. The desired optimal dietary ratio of n-6 : n-3 PUFA in man is considered to be 4 : 1, with the n-6 PUFA being largely provided by the $18:2n$-6 in vegetable oils, and the n-3 PUFA being mainly provided by the $18:3n$-3 in green leaf vegetables supplemented by modest amounts of $20:5n$-3 and $22:6n$-3 provided directly by fish (Anon 1992, 1993).

Although $20:4n$-6 is the major PUFA present in mammalian cell membranes, $22:6n$-3 has a major role in neural tissues (the brain and the eye) where it is present in high concentrations. Brain development takes place in the uterus in the last trimester of pregnancy when there is a high demand for $22:6n$-3 by the foetus, provided by the mother across the placenta. Because an over high intake of $18:2n$-6 during pregnancy can restrict the production of $22:6n$-3 from $18:3n$-3 in the mother, it is deemed prudent to consume oily fish during pregnancy. More important, formulae feeds for premature infants are now routinely supplemented with $22:6n$-3, which is present in mother's milk, but until recently had not been added to premature infant feeds resulting in impaired development of visual and probably also cognitive functions. There is now also increasing evidence that dietary supplementation with long chain n-3 PUFA from fish oil may be beneficial in a range of mental disorders including schizophrenia (Peet 1997).

Much remains to be understood of the mechanisms whereby fish oils promote health and considerable controversy still exists in some areas of the field. However, it is certain that diets with an over high ratio of n-6 : n-3 PUFA, stemming from an over high intake of vegetable oils rich in $18:2n$-6, do not meet normal health criteria and that fish, and especially oily fish, have a unique role in improving the situation by directly providing $20:5n$-3 and $22:6n$-3 (see e.g. Various 1996; Okuyama *et al.* 1997).

Scottish farmed Atlantic salmon: a quality product rich in $20:5n$-3 and $22:6n$-3

Because of the importance of $20:5n$-3 and $22:6n$-3 in human nutrition and because little or no information was available prior to 1993 on the levels of these fatty acids in farmed salmon, we have studied Atlantic salmon farmed in Scotland over the past five years. Because fish meal and fish oil were and continue to be the major constituents of farmed salmon feeds, we anticipated that the fish would have high levels of $20:5n$-3 and $22:6n$-3 in their flesh. However, the early 1990s had seen the introduction of 'high energy' feeds in salmon farming, i.e. feeds with levels of oil up to and sometimes exceeding 33% of their dry weight, to increase the growth rate of the fish. We were interested in how the amount of oil deposited in the flesh related to the amount of the oil in the diet and, for very high oil diets, whether the flesh oil was adequately protected against peroxidation by antioxidant compounds.

Figure 2.1, based on analyses of more than 800 fish from five producers in Scotland and updated from Bell *et al.* (1998), establishes that the oil content of salmon flesh increases progressively with the oil content of the diet, up to somewhat less than 30% oil in the diet at which point the flesh contains around 10% oil. A notable feature in Fig. 2.1 is the large standard errors associated with individual points reflecting marked variations in flesh oil levels in groups of fish fed the same diet by the same producer, ostensibly under the same conditions. For example, the flesh of fish fed a diet containing 28% oil had levels of oil in their flesh ranging from 3% up to 17%. The reasons for such a large variation are almost certainly multifactorial but it is probable that genetic differences between individual fish contribute to the variation, in which case selective breeding of fish to standardise flesh oil levels may well be possible. Further analyses of the data in Fig. 2.1 showed a significant seasonal trend with highest levels of flesh oil occurring in spring and early summer, and lowest levels in late summer and autumn (Bell *et al.* 1998).

The mean flesh oil content of all the fish analysed in Fig. 2.1 was 10.1 ± 2.9%. Bell *et al.* (1998) analysed the oil in the flesh of 10 individual salmon from each of five producers, these five groups of fish each having a mean flesh oil content of 10.1%, and also that of wild

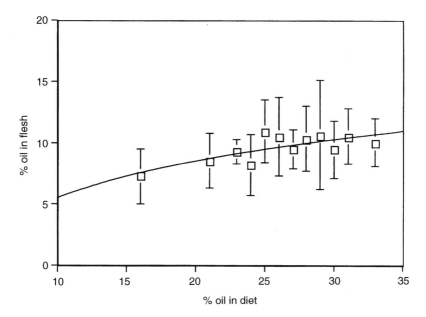

Fig. 2.1 Relationship between the % of oil in the flesh and the % of oil in the diet of farmed Atlantic salmon. Data are means ± standard deviations based on the analyses of 800 fish. Updated from Bell *et al.* (1998).

salmon caught by rod and line. Nearly 90% of the flesh oil was triacylglycerols, the remainder being mostly phospholipids in all cases. Free fatty acids were negligible. Flesh oils from fish from the five producers contained 10.4–12.4% of their fatty acids as 22 : 6n-3, only marginally less than wild fish with 12.9% 22 : 6n-3. The farmed fish also contained 5.0–5.8% of 20 : 5n-3, essentially the same as the wild fish. However, because the farmed fish had more flesh oil than the wild fish, they contained more than twice as much 22 : 6n-3 and 20 : 5n-3 on an absolute (g/kg) basis than the wild fish. The flesh oils of the farmed fish contained somewhat higher percentages of saturated fatty acids (20.7–22.5%) than the wild fish (19.3%), and corresponding lower percentages of monounsaturated fatty acids (43.6–45.3%) than the wild fish (49.8%). Interestingly, the percentage of 18 : 2n-6 in the farmed fish flesh oils, though low, was variable from 2.8–6.7% and notably higher than in wild fish (1.2%). The elevated 18 : 2n-6 in the farmed salmon reflects the presence of low levels of vegetable oils, derived from soya and wheat meals, rich in this fatty acid in at least some of the feeds used by the different producers. The elevated 18 : 2n-6 in the flesh oils of the farmed salmon resulted in an overall n-3/n-6 PUFA ratio ranging from 3.2–6.9, notably less than that of the wild fish (10.3) (Bell *et al.* 1998).

Table 2.1 shows trends in the measured parameters over the study period for all the fish analysed. After an initial rise, the flesh oil level has decreased somewhat over the past three years. The percentage of saturated fatty acids in the flesh oil has remained essentially constant whereas the percentage of monounsaturates appears to have increased. After an initial rise, the percentage of 18 : 2n-6 appears to have declined over the past three years. While the percentage of 20 : 5n-3 has varied somewhat, highest levels being recorded in the first and most recent years, the percentage of 22 : 6n-3 appears to have declined progressively and significantly, albeit modestly. This may reflect a change in sourcing the dietary fish oils, e.g. the use of less northern hemisphere oils and the use of more southern hemisphere oils, and/or natural trends in the composition of fish oils from either of these locations. Despite this apparent trend, the flesh oil remains rich in 22 : 6n-3. Table 2.1 also establishes that the flesh oil is well protected against peroxidation by vitamin E (α-tocopherol). The level of sup-

Table 2.1   Yearly trends in flesh parameters of Atlantic salmon farmed in Scotland.

Year	1994/95	1995/96	1996/97	1997/98
Weight of fish (kg)	2.9	3.1	3.2	3.1
Oil (% wet wt flesh)	10.1	11.5	10.0	9.4
Oil fatty acids (% of total)				
Sats	21.1	20.6	20.4	21.1
Monounsats	43.2	45.6	47.8	47.8
18 : 2n-6	4.3	5.2	4.7	4.1
20 : 5n-3	5.9	5.1	5.2	5.9
22 : 6n-3	12.1	11.1	10.0	9.5
Vitamin E (mg/kg)	25.8	29.1	27.9	25.3
Carotenoids (mg/kg)	5.6	8.3	8.4	8.3

plementary vitamin E in the feeds used in this study was *c*.250 mg vitamin E per kg feed, five times greater than the levels recommended by the US National Research Council (1993). The values of vitamin E in the flesh oil are more than ten times greater than found in previous studies with vitamin E deficient trout (Cowey *et al.* 1981) and vitamin E deficient salmon (Raynard *et al.* 1991). The different feeds used by the different producers contained various blends of the pigments astaxanthin and canthaxanthin, despite which the levels of pigment deposited in the flesh have remained constant over the past three years. Further details of the relationship between levels of astaxanthin and canthaxanthin in the diets and levels of these pigments deposited in the flesh are given by Bell *et al.* (1998). On the basis of flesh levels of both vitamin E and carotenoid pigments, the long chain *n*-3 PUFA in the flesh are clearly well protected against peroxidation. This conclusion is directly supported by the very low levels of peroxidation products such as 'thiobarbituric acid reacting substances' (TBARS), largely malonaldehyde, a product of lipid peroxide breakdown, detected in the flesh throughout the study.

More recently, we have investigated effects of smoking on the levels and composition of the flesh oils (unpublished observations). Our results have shown that oil levels can decrease by up to 45% after smoking and that carotenoid pigments can also decrease by up to 52%, the decreases being different in different regions of the fish, e.g. oil losses from the head region are greater than losses from the tail region. That peroxidation has occurred is confirmed by a substantial increase in TBARS (up to 70%). Interestingly, the vitamin E levels of the flesh are not reduced after smoking. Further studies are being conducted in this area.

These results establish that Atlantic salmon farmed in Scotland is a consistently high quality product in terms of its oil characteristics. The flesh is relatively but not excessively oil rich, the oil is rich in both DHA and EPA, it is well protected against peroxidation, being rich in vitamin E and carotenoid pigments, and shows no signs of degradation. The product conforms in all respects to the 'oily fish' recommended for consumption by nutritional and medical authorities (Anon 1992, 1993, 1994a,b,c).

The future: threats and opportunities

Clearly, farmed salmon is a high quality food providing both high quality protein and the nutritionally very important very long chain *n*-3 PUFA in the human diet. The same can be said of farmed fish in general, provided that fish meal and especially fish oil are the main basis of their feeds. Indeed, fish, whether farmed or wild, are virtually unique in directly providing EPA and DHA in the human diet and the importance of these nutrients for human well-being is beyond dispute. Given a rapidly growing world population and given that global fisheries are now stagnating if not declining, serious concerns are now emerging that the demand for EPA and DHA-rich foods, i.e. fish and sea foods in general, will soon outstrip supply. Intensive aquaculture clearly has an important role to play in providing EPA and DHA in the human diet, but intensive aquaculture has thus far been totally dependent on strictly limited global supplies of fish meal and fish oil. In 1996 aquaculture consumed some 576 000 tonnes of a global production of fish oil of some 1 400 000 tonnes (Tacon 2000). Of the 576 000 tonnes consumed, 36.3% was used for salmon farming and 21.6% for trout farming. Marine fish farming consumed 14.8%, carp farming 8.0%, eel farming 7.5%, marine shrimp farming 7.1%, milkfish farming 1.9% and catfish farming 1.8% (Tacon 2000). Given the now very rapid growth of freshwater fish farming in Asia, especially carp farming in China, it is clear there will be intense future competition for strictly limited global supplies of fish oil, even without seriously perturbing events such as El Niño. The same situation holds for fish meal but, while alternative protein sources for fish meal are available, the same is not true for fish oil. It is salutary to note that, on a global basis, fish oil production accounts for only 1.5% of total production of fats and oils with soya oil, palm oil and rapeseed oil accounting respectively for 22%, 18% and 12% of the total (O'Mara 1998). Production of these three oils is increasing rapidly and all are rich in $18:2n$-6. The only commercially available n-3 PUFA rich oil other than fish oil, linseed oil, is rich in $18:3n$-3 and accounts for a mere 0.7% of global production (O'Mara 1998). These data emphasise the current serious global imbalance between n-6 and n-3 PUFA and the resulting negative implications for human health. The question arises, how is intensive aquaculture to continue to deliver EPA and DHA to a rapidly growing human population faced with ever increasing supplies of $18:2n$-6 rich oils generated by an increasingly efficient agriculture and a static if not diminishing supply of fish oil?

Fish oils currently used in farmed fish feeds could be substituted at least partially with other oils. Castell *et al.* (1972) originally established that rainbow trout could be grown on diets containing $18:3n$-3 as the only PUFA, although serious pathologies occurred in trout grown on diets with $18:2n$-2 as the only PUFA, and many freshwater fish including carp and tilapia thrive on diets containing blends of $18:2n$-6 and $18:3n$-3 (references in Sargent *et al.* 1995). It has recently been authoritatively reported that marine fish can be grown successfully with 50% of the fish oil in their diets substituted with blends of palm oil and beef tallow (Watanabe & Cho 2000) and at least one fish feed manufacturer has successfully tested a commercial feed for salmon and trout containing a 50/50 blend of fish oil and vegetable oil (Anon 1998). Against this must be set the evidence of Bell *et al.* (1990) that salmon reared on a diet containing solely sunflower oil developed cardiac lesions, especially when subjected to stress. This raises concerns that fish may react in the same way as *Homo sapiens* to diets containing a high ratio of n-6:n-3 PUFA by being susceptible to stress-related disorders stemming, e.g., from overcrowding, increased temperature, low dissolved oxygen or infectious diseases. Irrespective of this, it is clearly undesirable from a human nutritional standpoint to substitute n-3 PUFA rich fish oils with n-6 PUFA rich vegetable oils in fish feeds, i.e. to use farmed fish as a vehicle for delivering yet more n-6 PUFA in the human diet when it is already undesirably high. Should substitution of fish oil in fish feeds become necessary, a better approach from the human nutritional perspective might be to use sparingly an $18:3n$-3 rich vegetable oil such as linseed oil or, in relation to 'high energy' feeds, oils low in PUFA but rich in $18:1n$-9, such as high $18:1$ sunflower oil, rapeseed oil or palm oil.

Obviously partial substitution of fish oils in fish feeds with vegetable oils will dilute the EPA and DHA levels in fish flesh, these being replaced with the dominant fatty acids in the vegetable oils, with implications for taste and consumer acceptance. These implications may not necessarily be wholly negative. Much research needs to be done in the area of partial substitution of fish oils in fish feeds but, a priori, the final outcome can only be a partial and transient solution to the fundamental problem of supplying EPA and DHA over and above that available from global fisheries.

A possible strategic solution to the problem may be to exploit the inherent ability of fresh water fish to convert $18:3n\text{-}3$ to $20:5n\text{-}3$ and thence to $22:6n\text{-}3$. Many such fish are omnivorous and naturally geared to converting $18:3n\text{-}3$ of plant origin to $22:6n\text{-}3$ and, unlike marine fish which are overwhelmingly carnivorous, are by no means ensured of a luxus of $20:5n\text{-}3$ and $22:6n\text{-}3$ in their natural diets. In this context it is perhaps regrettable that fish meal and fish oil are being increasingly used in feeds for carp and tilapia where traditional alternatives have long been available and where research in dietary oil substitution holds much promise. Such fish, of course, are not oily in the sense that they do not deposit oils in their flesh but, nonetheless, they contain valuable quantities of EPA and DHA rich phospholipids in their flesh. The ultimate solution to the forthcoming global deficit of EPA and DHA may be to maximise the inherent potential of freshwater fish to convert LNA to DHA, by modern genetic selection technology, to isolate the genetic systems and to introduce them into marine or anadromous fish. The ultimate challenge is to persuade the fish to deposit internally generated EPA and DHA in their flesh oils. Salmon and trout may yet hold the key to the problem.

References

Anon (1992) *Unsaturated Fatty Acids. Nutritional and Physiological Significance*. The Report of the British Nutrition Foundation's Task Force. 211pp. Chapman and Hall, London.

Anon (1993) *Nutritional aspects of fish. Briefing Paper 10, British Nutrition Foundation Publication*. 18pp. British Nutrition Foundation, London.

Anon (1994a) *Nutritional aspects of cardiovascular disease*. Report of the Cardiovascular Review Group Committee on Medical Aspects of Food Policy. Department of Health Report on Health and Social Subjects No. 46. 186pp. HMSO London.

Anon (1994b) Health message statement. *International Society for the Study of Fatty Acids and Oils News Letter*, 1, 3–4.

Anon (1994c) Recommendations for the essential fatty acid requirements for infant formulae. *International Society for the Study of Fatty Acids and Oils News Letter*, 1, 4–5.

Anon (1998) *Oils and Fats (Lipids) in Feeds for Salmon and Trout*. 4pp. Published by Trouw Aquaculture, Cheshire, UK.

Bell, J.G., McVicar, A.H., Park, M.T. & Sargent, J.R. (1990) Effects of high dietary linoleic acid on fatty acid compositions of individual phospholipids from tissues of Atlantic salmon (*Salmo salar*): association with a novel cardiac lesion. *Journal of Nutrition*, 121, 1163–72.

Bell, J.G., McEvoy, J., Webster, J.L., McGhee, F., Millar, R.M. & Sargent, J.R. (1998) Flesh lipid and carotenoid composition of Scottish farmed Atlantic salmon (*Salmo salar*). *Journal of Agricultural and Food Chemistry*, 46, 119–27.

Castell, J.D., Lee, D.J. & Sinnhuber, R.O. (1972) Essential fatty acids in the diet of rainbow trout; Lipid metabolism and fatty acid composition. *Journal of Nutrition*, 102, 93–9.

Cowey, C.B., Adron, J.W., Walton, M.J., Murray, J., Youngson, A. & Knox, D. (1981) Tissue distribution, uptake and requirement for α-tocopherol of rainbow trout (*Salmo gairdneri*) grown at natural, varying water temperatures. *Journal of Nutrition*, 111, 1556–7.

Okuyama, H., Kobayashi, T. & Watanabe, S. (1997) Dietary fatty acids – the $n\text{-}6/n\text{-}3$

balance and chronic elderly diseases. Excess linoleic acid and relative *n*-3 deficiency syndrome seen in Japan. *Progress in Lipid Research*, **35**, 409–57.

O'Mara, C.J. (1998) US oil seed industry looks at trade issues. *Inform*, **9**, 132–6. Published by Amer. Oil Chem. Soc. Press. Champaign, Ill., USA.

Peet, M. (1997) Schizophrenia and omega-3 fatty acids. *International Society for the Study of Fatty Acids and Oils News Letter*, **4**, 2–5.

Raynard, R.S., McVicar, A.H., Bell, J.G., Youngson, A., Knox, D. & Fraser, C.O. (1991) Nutritional aspects of pancreatic disease in Atlantic salmon: the effects of dietary vitamin E and polyunsaturated fatty acids. *Comparative Biochemistry and Physiology*, **98A**, 125–31.

Sargent, J.R. & Henderson, R.J. (1995) Marine (*n*-3) polyunsaturated fatty acids. In: *Developments in Oils and Fats*, (ed. R.J. Hamilton), pp. 32–65. Blackie Academic and Professional, London.

Sargent, J.R., Bell, J.G., Bell, M.V., Henderson, R.J. & Tocher, D.R. (1995) Requirement criteria for essential fatty acids. Symposium of European Inland Fisheries Advisory Commission. *Journal of Applied Ichthyology*, **11**, 183–98.

Sikorski, Z.E., Kolakowska, A. & Pan, B.S. (1990) The nutritional composition of the major groups of marine organisms. In: *Seafood: Resources, Nutritional Composition and Preservation*, (ed. Z.E. Sikorski), pp. 30–54. CRC Press, Boca Raton, Fl. USA.

Tacon, A.G.K. (2000) Aquaculture production and nutrient supply: a global perspective. *Proc. VIII Int. Symp. on Nutrition and Feeding of Fish. Recent Advances on Finfish and Crustacean Nutrition*, Las Palmas de Gran Canaria, 1–4 June 1998. *Aquaculture* (in press).

US National Research Council (1993) *Nutrient Requirements of Fish*. 114 pp. National Academy Press, Washington DC.

Various (1996) International Conference on Highly Unsaturated Fatty Acids in Nutrition and Disease Prevention, Bethesda, Maryland. *Lipids*, **31**, Suppl. S1–S32.

Watanabe, T. & Cho, C. Y. (2000) Development of fish feed: challenges and strategies for the 21st century. *Proc. VIII Int. Symp. on Nutrition and Feeding of Fish. Recent Advances on Finfish and Crustacean Nutrition*, Las Palmas de Gran Canaria, 1–4 June 1998. *Aquaculture* (in press).

Wilson, R.P. (1989) Amino acids and proteins. In: *Fish Nutrition*, (ed. J. Halver), 2nd edn, pp.111–151. Academic Press, New York.

Chapter 3

Implications of Muscle Growth Patterns for the Colour and Texture of Fish Flesh

I.A. Johnston

Fish Muscle Research Group, Gatty Marine Laboratory, Division of Environmental and Evolutionary Biology, School of Biology, University of St Andrews, St Andrews, Fife, KY16 8LB, UK and Matre Aquaculture Station, Institute of Marine Research, N-5984, Matredal, Norway

Introduction

Greene and Greene (1913) provided a detailed description of the complex architecture of the axial muscles in the king salmon (Fig. 3.1a). The muscle tissue is arranged in a series of blocks or myotomes on either side of the body. Individual muscle fibres are relatively short and are rarely more than 1–2 cm in length even in very large fish. They insert via short tendons into the sheets of collagen called myosepts that bound each myotome. The myotomes are cone-shaped and stacked in a metametric arrangement on either side of the medium septum. A transverse steak through the trunk will therefore section several myotomes at different levels (Fig. 3.1b).

Early studies identified different muscle fibre types on the basis of colour by a visual inspection of the fillet (Stirling 1885). 'Red muscle' fibres form a thin superficial layer thickening to a V-shaped wedge at the major horizontal septum (Fig. 3.1b). The red colour results from a well-developed blood supply and relatively high concentrations of myoglobin. Red fibres are relatively slow contracting and contain high volume densities of mitochondria (20–50%) (Johnston 1981). They function to power slow sustainable swimming using predominantly aerobic metabolic pathways (Bone 1978). The 'pale muscle' comprises the bulk of the myotome and it consists of faster contracting white fibres which appear pink or orange in Atlantic salmon because of the presence of carotenoid pigments. The white muscle fibres have a sparse capillary supply, contain relatively few mitochondria and are largely supported by anaerobic metabolic pathways (Black *et al.* 1966; Hochachka 1985). The white muscle powers fast swimming but is rapidly exhausted. A variety of other more minor muscle fibre types have been described in species of commercial importance such as turbot (Calvo and Johnston 1992) and gilthead sea bream (Mascarello *et al.* 1995); however, they are of little direct importance in the context of aquaculture.

The architecture and fine structure of the muscle determine its textural characteristics. The red muscle fibres run parallel to the longitudinal axis of the fish whereas the deeper white muscle fibres have more complex trajectories and are set at angles to the horizontal and median planes (Alexander 1969) (Fig. 3.1c), resulting in a gearing ratio of 2 to 3 during

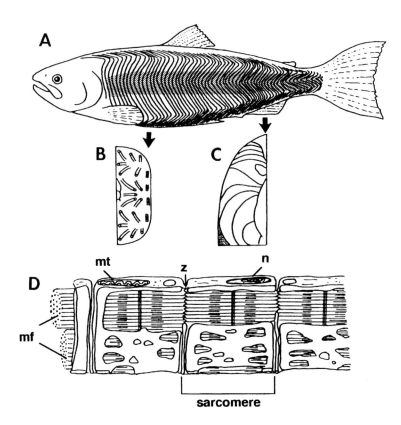

Fig. 3.1   The anatomy and structure of the lateral muscles in salmon species. (a) illustrates the chevron-shaped myotomes of the King salmon (based on Greene and Greene, 1913). (b) shows the distribution of the pale and dark muscle in a transverse section through the caudal myotomes (c) the orientation of muscle fibres in a transverse steak based on Alexander (1969) and (d) a diagrammatic representation of the fine structure of a white muscle fibre showing the repeating sarcomere structure and myofibrils (mf). In all the diagrams the red muscle is shaded. Other abbreviations: mt = mitochrondrion, n = myonucleus and z = z-line.

swimming (Wakeling & Johnston 1999). White muscle fibres contain tightly packed myofibrils around 1 µm diameter containing the contractile protein filaments of actin and myosin. The range of fibre diameters for the white muscle varies along the length of the fish, with the highest fibre density generally found in the more caudal myotomes (Love 1958).

Fish muscle contains less collagen which is significantly less cross-linked than in red meat (Sato *et al.* 1986). Collections of collagen fibres arise from the myosepts to form a connective tissue envelope around each muscle fibre called the perimysium. In the scanning electron microscope the muscle fibres therefore seem to fit into socket-like invaginations of the myocommata made of collagen (Bremner & Hallett 1985). Sheets of connective tissue also run between bundles of muscle fibres and are particularly numerous towards the tail. Both the connective tissue matrix and the muscle fibres contribute to the texture of raw fish. However, after cooking, the collagen fibres are thought to make relatively little contribution to the textural properties (Hatae *et al*. 1986).

The aim of fish farming is to provide muscle tissue for human consumption. It is therefore important to understand how genetics, production practices and diet can influence the quality of the flesh. Important quality attributes of the muscle include the nutritional value, freshness, storage properties, taste and texture of the fillet. The pattern of muscle growth is particularly important for texture and colour visualisation (Johnston *et al*. 2000a). This short review will focus on the growth of the white muscle fibres which constitute the main edible part of the myotomes, with an emphasis on the Atlantic salmon (*Salmo salar* L.).

Three phases of muscle growth can be distinguished in the Atlantic salmon (Johnston 1999). The first phase of myogenesis occurs in the embryo shortly after the formation of the somites on either side of the developing notochord. Following somitogenesis during the late embryo and early alevin stages further muscle fibres are formed from germinal zones at the dorsal and ventral extremities of the myotomes in salmon (Johnston & McLay 1997) and in several marine teleosts (sea bass: Veggetti *et al*. 1990; plaice: Brooks & Johnston 1993; gilt-head sea bream: Patruno *et al*. 1998). Just before first feeding, the final phase of myogenesis begins which persists throughout much of the life cycle. This third phase involves the activation of a population of muscle stem cells, termed satellite or myosatellite cells that provide nuclei for both new fibre recruitment and hypertrophic growth. As in other vertebrates satellite cells are also involved in muscle repair following injury (Rowlerson *et al*. 1997).

Early myogenesis

The somites are formed from the paraxial mesoderm in a rostral to caudal sequence. Studies with zebrafish (*Danio rerio*) have shown that a sub-set of cells express the myogenic transcription factor MyoD towards the end of gastrulation and become committed to a myogenic fate (Weinberg *et al*. 1996). A second phase of MyoD expression is observed following somitogenesis starting in a row of large cuboidal cells, termed 'adaxial cells' that flank the forming notochord (Devoto *et al*., 1996; Weinberg *et al*. 1996). A subset of these adaxial cells, termed the muscle pioneers, are the first to elongate and develop striations (Felsenfeld *et al*. 1991). The muscle pioneers are closely associated with the developing horizontal septum which divides the myotome into dorsal and ventral muscle masses. It has been suggested, but not proven, that the muscle pioneers have a role in the patterning of the characteristic chevron shape of the myotome that takes place at around the same time they begin to elongate (Halpern *et al*. 1993). The adaxial cells, including the muscle pioneers, migrate through the somite to form a superficial layer of muscle fibres that mature into the slow twitch red muscle fibres (Devoto *et al*. 1996).

The remaining cells, termed lateral presomitic cells express MyoD somewhat later and mature into the white muscle fibres (Devoto *et al*. 1996). Thus the red and white muscle fibres have distinct lineages that are specified even before the somites are formed. At hatch, Atlantic salmon alevins have around 300 red muscle fibres and 5000 white muscle fibres per myotome at the level of the pelvic fin insertions (Johnston & McLay 1997).

Germinal zone phase of myogenesis

Cell proliferation can be identified by *in vivo* labelling with the thymidine analogue bromodeoxyuridine or by staining tissue sections with proliferating cell nuclear antigen (PCNA). Active zones of myoblast proliferation are observed at the extremities of the myotomes after the early phase of embryonic myogenesis in several fish species (Rowlerson *et al.* 1995). In Atlantic salmon, the germinal zones become exhausted around first feeding (Johnston & McLay 1997).

Satellite cell phase of myogenesis

By first feeding in Atlantic salmon the number of white muscle fibres has increased to 9000 to 10 000 per myotome and very small diameter fibres are present throughout the myotome heralding the start of the satellite cell phase of myogenesis (Higgins & Thorpe 1990; Johnston & McLay 1997). The total number of nuclei per mm^2 white muscle in Atlantic salmon increases from around 2000 at hatching to 4500 at first feeding (Johnston & McLay 1997). At the electron microscope level a wide range of nuclear morphology is observed (Johnston *et al.* 1999). In addition to a small number of fibroblast nuclei and nuclei associated with capillary endothelial cells, there are a mixture of differentiated myonuclei, as well as quiescent and activated satellite cell nuclei.

Knowledge of the biology of satellite cells is relatively fragmented in fish. It has been suggested, although not proven, that the satellite cells originate from and proliferate within the mesenchymal tissue lining during late embryogenesis (Stoiber & Sänger 1996). Studies in mammals have shown that activated satellite cells undergo an asymmetric division to regenerate a replacement satellite cell and produce a daughter cell that is committed to terminal differentiation after a limited number of further divisions (Quinn *et al.* 1988; Schultz 1996). It is generally thought that the total number of satellite cells remains relatively constant throughout growth. The quiescent satellite cells are located between the basement membrane and plasma membrane of the muscle fibres (Mauro 1961). Satellite cells co-express MyoD and PCNA on entry to the cell cycle and subsequently express other myogenic regulatory factors (myogenin, myf-6) as they progress towards differentiation and the production of muscle-specific proteins such as desmin (Yablonka-Reuveni & Rivera 1994). Myogenic regulatory factors (MRFs) are a class of bhelix-loop-helix transcription factors that bind to a consensus region CANNTG (E box) present on the promoter regions of many muscle-specific genes. They have complex and redundant roles in controlling muscle commitment and differentiation and act in co-operation with the myocyte enhancer-2 family of muscle transcription factors (see Olson & Klein 1994 for a review).

A number of molecular markers for satellite cells have been identified in mammals including the c-met tyrosine kinase receptor (Cornelison & Wold 1997), Myocyte Nuclear Factor ß (MNF-ß) (Yang *et al.* 1997) and mitogen-activated kinases ERK1 and ERK2 (Yablonka-Reuveni *et al.* 1999). An antibody against c-met has been used to identify satellite cells in Atlantic salmon (Johnston *et al.* 1999). In addition, the division products of satellite cells and proliferating cells can be detected using antibodies to the myogenic regulatory factors (MyoD, myf-5, myf-6 and myogenin) and PCNA respectively (Johnston *et al.* 1999).

Cells expressing c-met represent 15% of total myonuclei in fry and smolts, and around 80% of these cells express one or more myogenic regulatory factor (MRFs) (Johnston *et al.* 1999, and unpublished results).

Tissue culture studies with mammalian cells have identified a number of growth factor families involved in controlling proliferation and differentiation of the satellite cells (Tatsumi *et al.*, 1998; Yablonka-Reuveni *et al.* 1999). For example, hepatocyte growth factor scatter factor (Tatsumi *et al.* 1998) and fibroblast growth factor 2 (FGF2) promotes recruitment of satellite cells in young and old rats (Yablonka-Reuveni *et al.* 1999). Insulin-like growth factors can modulate the proliferative activity of satellite cells (Florini *et al.* 1991).

In fish, the division products of satellite cells that are committed to differentiation have one of two fates. They are either absorbed into muscle fibres during hypertrophic growth to maintain the nuclear to cytoplasmic ratio within set limits as the fibre expands or they fuse together on the surface of existing fibres to form myotubes which mature into new fibres (fibre recruitment). A possible model for satellite cell behaviour in fish is presented in Fig. 3.2. A major uncertainty is whether there is a single class of satellite cell or alternatively distinct populations responsible for recruitment and hypertrophy (Koumans & Akster 1995;

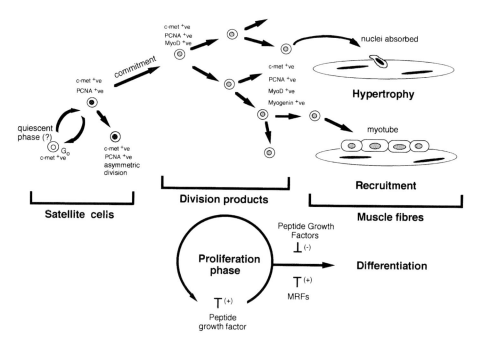

Fig. 3.2 A model illustrating the behaviour of muscle satellite cells in fish. Various molecular markers can be used to identify different cell types (see text). Satellite cells and their division products express c-met. All dividing cells express proliferating cell nuclear antigen whereas the division products of the satellite cells committed to differentiation express myogenic regulatory factors, including MyoD and myogenin. Also illustrated are the antagonistic signalling pathways promoting either proliferation or differentiation of the cells destined to become muscle nuclei. Adapted from Johnston (1999).

Johnston 1999). One possibility is that it is local signalling from the muscle fibres themselves that determines the ultimate fate of the division products of satellite cells.

The number of white muscle fibres per myotome continues to increase during freshwater growth reaching around 150 000 to 180 000 by the smolt stage (Higgins & Thorpe 1990; Johnston *et al.* 1999, 2000b). Hypertrophic growth can be investigated by determining the size distribution of muscle fibres at different stages from tissue sections. The use of frozen sections is preferable because it avoids problems associated with tissue shrinkage, making it easier to compare results from different stages and studies. Fibre diameters or cross-sectional areas can be plotted as histograms or better still a smoothed probability density function fitted to the data. The latter has the advantage that the value of realisation is preserved and subsequent statistical analysis is more straightforward (Johnston *et al.* 1999). In our laboratory we have used non-parametric kernel functions to fit smooth estimates of the probability density of diameter using measurements from 1000 muscle fibres per fish (Johnston *et al.* 1999). Bootstrap techniques can then be used to provide an estimate of the variability of the distributions and allow comparisons between different populations and/or treatments (Johnston *et al.* 1999). Different growth phases of the white muscle fibres have been distinguished histochemically on the basis of their myosin ATPase staining patterns in the sea bream (Mascarello *et al.* 1995) and sea bass (Ramírez-Zarosa *et al.* 1998).

In Atlantic salmon smolts, the distribution of muscle fibre diameters is essentially unimodal shortly after sea water transfer (Fig. 3.3a). In Scottish farmed salmon reared in cages the distribution of muscle fibre diameters was found to become bimodal in October at the end of the first summer at sea (Fig. 3.3b). Considerable variation in the size distribution of muscle fibres was observed between individual fish which represented six different families (Fig. 3.3b). The left-hand peak in the distribution was at about 15 μm diameter, whereas the broader right-hand peak was at around 70 μm. The 95th percentile of the distribution was at 145 μm and these presumably represent the oldest cohort of muscle fibres (Johnston *et al.* 2000b). By January the left peak in the distribution of diameters was at a higher density than the right peak (Fig. 3.3c, d). The winter months herald a decrease in day-length and water temperature and a period of slower growth that is particularly dramatic at high latitudes. It seems that during this period fibre recruitment slows less than does hypertrophy such that the proportion of very small diameter muscle fibres increases. A gap in present knowledge is how season affects the number and expression patterns of the satellite cells. The number of white muscle fibres per myotome reaches around 700 000 after 1-sea-winter fish and more than 1 million in 2-sea-winter fish (Johnston 1999; Johnston *et al.* 2000b).

Genetic variation in muscle growth patterns

Wild Atlantic salmon tend to home to their native streams to spawn, promoting reproductive isolation and genetic differentiation between stocks (Verspoor 1997). For example, fish from different river systems differ with respect to age and size at maturity, growth rate and body morphology (Taylor 1991). Many of these life history traits have been shown to be heritable and many of the major smolt producers have genetic improvement programs. Recently, we compared muscle growth patterns in two populations maintained by Marine Harvest McConnell in Scotland: a high grilsing stock of west coast Scottish origin (strain X)

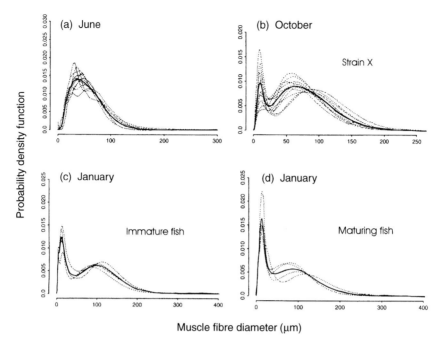

Fig. 3.3 The probability density of muscle fibre diameters in the white muscle of Atlantic salmon (*Salmo salar* L.) during seawater growth. The fish were reared in 5 × 5 × 5 m cages, were transferred to sea in April and sampled in (a) June, (b) October and (c, d) January. Anatomical evidence of sexual maturation in the January sample was used to sort the fish into immature (left) and maturing (right) individuals. The dotted lines represent the density distributions of individual fish and the solid lines the average density of all the fish in the sample. From Johnston *et al.* (2000b).

and a low grilsing stock (strain Y) of Norwegian origin (Johnston *et al.* 2000b). The fish, representing six families per strain, were PIT-tagged and transferred to a 5 m × 5 m × 5 m sea cage in April. Body mass, and the total cross-sectional area of white muscle at the level of the first dorsal fin ray, increased at a significantly faster rate after the first six months in seawater in the early maturing than in the late maturing population.

The contribution of fibre recruitment to growth varied between the strains at different stages of the production cycle. In June and July both populations had around 150 000 and 250 000 white muscle fibres per trunk cross-section respectively (Fig. 3.4a). However, between July and August the late maturing population (strain Y) recruited twice as many fibres for each mm² increase in white muscle cross-sectional area as fish from strain X (Fig. 3.4b). By January, strain X had caught up with strain Y and both populations had around 545 000 fibres per trunk cross-section (Fig. 3.4a). There was little further recruitment of fibres in strain Y until the time they were harvested in September, such that the last nine months' growth of the production cycle was entirely by fibre recruitment (Fig. 3.4a, b). In contrast, fibre number continued to increase in strain X, reaching 720 000 per cross-section by the time they were harvested in June. Around 30 fibres were added per mm² growth in

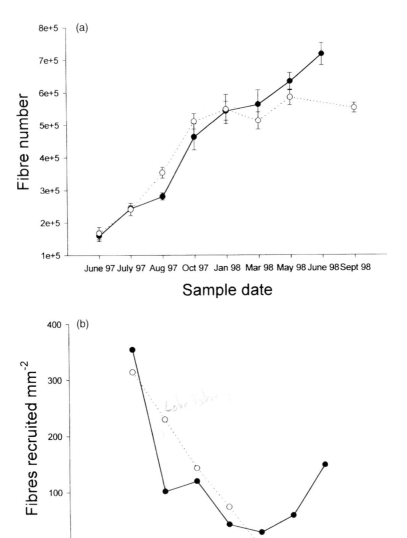

Fig. 3.4   (a) The number of white muscle fibres per cross-section of the trunk at the level of the first dorsal fin ray in Atlantic salmon (*Salmo salar* L.) from an early maturing (strain X; solid symbols/line) and a late maturing population (strain Y, open symbols/dotted line). (b) The number of white muscle fibres recruited for each 1 mm^2 increment of growth in white muscle cross-sectional area since the previous sample period for salmon from strain X (closed circles) and strain Y (open circles). For example, the July 1997 point represents the increase in average white muscle cross-sectional area between June and July divided by the average difference in the number of fibres. From Johnston *et al.* (2000b).

the winter (Jan–March), increasing to 150 fibres per mm^2 between May and June (Fig. 3.4b). All fish showed a bimodal distribution of muscle fibre diameters from the October sample and there were significant differences in fibre size distribution between the populations. The left-hand peak of the distribution was shifted to higher diameters in strain Y after March, reflecting the cessation of fibre recruitment (the May sample is illustrated in Fig. 3.5). On the other hand, the right-hand peak of the distribution was shifted to higher diameters in the early than in the late maturing population, indicating a faster rate of fibre hypertrophy. Thus the 95th percentile of the distribution was at higher diameters in strain X than strain Y even

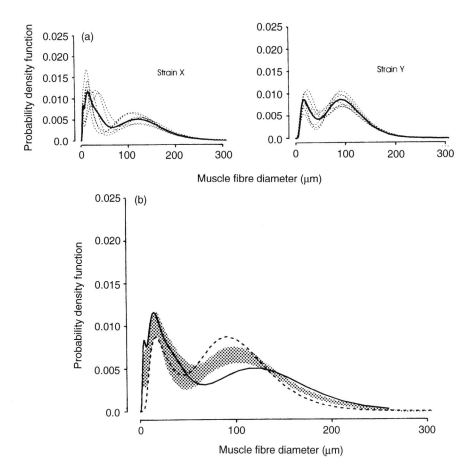

Fig. 3.5 The distribution of muscle fibre diameter in Atlantic salmon from early and late maturing populations sampled in May after 13 months growth in seawater. (a) The probability density function of muscle fibre diameter of individual fish (dotted lines) and the average for each population (solid lines). (b) The average probability density function for strain X (solid line) and strain Y (dotted line). The shaded area encloses 100 bootstrap estimates of the total population, strain X + strain Y. Areas where the solid and dotted lines lie outside the shaded area provide evidence for differences in the density distributions between populations. From Johnston *et al.* (200b).

for fish of equivalent fork-length (Fig. 3.6a, b). The superior growth performance of strain X over strain Y was therefore a function of a greater contribution of fibre recruitment to growth and a faster rate of fibre hypertrophy (Johnston *et al.* 2000b). Other studies have also noted a correlation between fibre recruitment and fast growth rate (Weatherley *et al.* 1988; Alami-Durante *et al.* 1997; Galloway *et al.* 1999).

An examination of immature and maturing fish in the January sample indicated that these differences in recruitment and fibre size between strains were not directly related to sexual maturation but to some other genetic difference between the populations (Johnston *et al.* 2000b). Thus in the immature fish within strain X the right-hand peak of the distribution was shifted to higher diameters relative to fish that had begun to sexually mature. Thus greater muscle fibre hypertrophy was observed in immature than in maturing fish of the same strain.

The genetic bases of these differences in muscle fibre recruitment in salmon are not known although studies with other animals do provide some important clues. In the Japanese quail (*Coturnix coturnix japonica*) selection for high body mass resulted in an increase in muscle length, fibre number and satellite cell density relative to control strains (Fowler *et al.* 1980; Campion *et al.* 1982). Similar selection for high body mass in the mouse also increased the number of satellite cells and muscle fibres present at birth, but not the ratio of satellite cells to myonuclei (Brown and Stickland 1993, 1994). Thus one possibility is that growth favouring fibre recruitment is associated with genetic differences in the density of satellite cells. Supporting this idea ploidy manipulation affects both fibre recruitment and the density of satellite cells in Atlantic salmon (Johnston *et al.* 1999). The density of satellite cells was around 24% higher in normal-sex-ratio and all-female diploid than triploid Atlantic salmon. Diploid fish recruited around 25–30% more fibres to reach a given muscle mass than triploids during sea-water growth. Thus, although fibre recruitment was compromised in triploids, there was a compensation with respect to hypertrophic growth, perhaps reflecting the larger size of satellite cells in triploids which may have enabled each nucleus to service a larger volume of fibre than in diploids (Johnston *et al.* 1999).

Although differences in satellite cell density might well be responsible for variation in muscle fibre number and recruitment potential between families, it is not a likely explanation for the differences observed between early and late maturing populations. The most plausible hypothesis to explain variations in the timing of muscle fibre recruitment is a difference in the proliferative capacity of the satellite cells. Indeed, satellite cells isolated from heavyweight strains of turkey show a higher proliferative capacity *in vitro* than cells from lightweight strains (Merly *et al.* 1998). This suggests that the proliferative capacity of satellite cells is under genetic as well as environmental control. Similar studies on the behaviour of satellite cells in fish showing different rates or patterns of growth have not yet been undertaken.

The hypertrophic growth of fibres also has a genetic component (Johnston *et al.* 2000b). Prime candidates for the regulation of fibre size are factors controlling the availability of nuclei and/or the stimulation of signalling pathways co-ordinating gene expression associated with fibre hypertrophy. Exercise is known to be a powerful stimulus for muscle fibre hypertrophy in fish (Johnston & Moon 1980; Totland *et al.* 1987). Population differences in fibre size might reflect differences in swimming behaviour and or genetic differences in the signalling pathways mediating exercise induced hypertrophy.

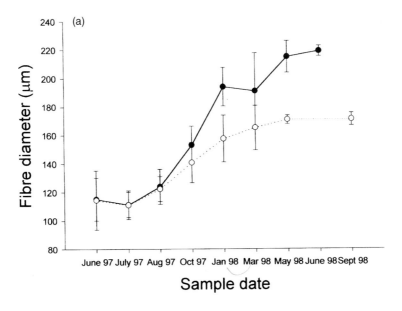

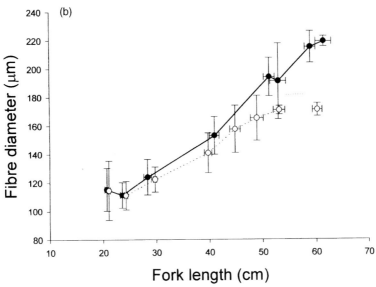

Fig. 3.6 The 95th percentiles of white muscle fibre diameter in an early (strain X, solid circles) and a late maturing (strain Y, open circles) population of Atlantic salmon (*Salmo salar* L. in relation to (a) sample date and (b) fork-length. The values represent mean ± SE. Ten fish from each population were analysed in the June, July, August and October 1997 samples, eight in January 1998, eight in March and May 1998, ten of strain X in June and eighteen of strain Y in September. From Johnston *et al.* (2000b).

Phenotypic plasticity of muscle fibre recruitment

Any factor that can affect the number and/or behaviour of the satellite cells has the potential to influence fibre recruitment and hypertrophy and hence muscle architecture. It is likely that there are ontogenetic windows, particularly during the embryo stages, when satellite cells are particularly sensitive to external factors. For example, studies with Atlantic herring (*Clupea harengus* L.) have shown that egg incubation temperature alters the number of presumptive satellite cells (Johnston 1993) and rates of muscle fibre recruitment if the hatched larvae are subsequently grown-on under similar conditions (Johnston *et al.* 1998). In Atlantic salmon, egg incubation temperature affects the number and diameter of muscle fibres formed during the early stage of myogenesis (Stickland *et al.* 1988; Johnston & McLay 1997) and the relative importance of fibre recruitment during the satellite cell phase of growth (Beattie *et al.* 2000). Beattie *et al.* (2000) incubated eggs from the Norwegian Namsen strain of Atlantic salmon in either heated water (median $7°C$) ($\pm 3°C$ range) or at ambient temperature (median $4°C$) ($\pm 2°C$ range) and monitored subsequent growth for 26 months up to the first winter in seawater. Fish reared at higher temperatures hatched earlier and had a growth advantage. The number of muscle fibres per myotome was almost four-fold higher in the heated group at the S1 parr stage compared to the ambient group. However, six months after seawater transfer there was no significant difference in fibre number between groups. For the majority of age classes the right-tail of the distribution of fibre sizes was shifted to higher diameters in the heated groups. Thus the average cross-sectional area of white muscle was higher whereas the fibre density was lower in the heated than in the ambient groups. The effects of temperature on satellite cell density were not examined in this study. However, one plausible hypothesis that could be tested is that the temperature around the time that the satellite cells are forming affects their number and hence fibre recruitment patterns at later stages in the life cycle.

The list of other factors that influence fibre recruitment and hypertrophy is substantial and includes photoperiod (C. Beattie, I.A. Johnston and T. Hansen, unpublished results), exercise (Johnston & Moon 1980; Totland *et al.* 1987), diet composition, and feeding regimes (Keissling *et al.* 1991). Other, as yet unstudied factors, may also be important and of potential relevance to fish farming. For example, starvation and re-feeding is associated with growth retardation and is followed by a compensatory 'catch-up' phase (Jobling 1994). Does dietary restriction reduce hypertrophy more than fibre recruitment, leading to an increase in the proportion of small diameter fibres that are subsequently expanded rapidly when energy becomes available? Do dietary components such as vitamins and polyunsaturated fatty acid composition affect prostaglandin biosynthesis and/or the concentrations of growth factors regulating satellite cell behaviour and fibre recruitment? These are just some of the key areas for future research.

The relationship of muscle architecture to flesh quality

Research on mammalian species has shown that both the connective tissue matrix and the muscle fibres themselves contribute to the intrinsic textural properties of meat (Offer *et al.* 1989). Because it is present in lower amounts, connective tissue is less important for texture in fish than in mammals and is probably only important for raw and smoked products. Following cooking, the muscle fibres themselves probably provide the main resistance to mastication (Dunajski 1979). A number of inter-specific comparisons have revealed significant correlations between average muscle fibre diameter and the 'firmness' of the flesh (Hatae *et al.* 1986; Hurling *et al.* 1996). For seven species of marine fish there was a negative correlation between firmness, as assessed by trained taste panels, and the average muscle fibre diameter in cooked fish. For example, the dab (*Limanda limanda* L.) had the highest sensory firmness and the smallest average diameter fibres, whilst flying fish (*Exocoetidae* sp) scored the lowest firmness and had the largest diameter fibres (Hurling *et al.* 1996).

In our recent study of muscle growth in two populations of farmed Atlantic salmon we produced fish which had a range of fibre densities reflecting differences in the relative importance of recruitment prior to harvest (Johnston *et al.* 2000b). The fish were sent to a commercial fish processor (Pinneys of Scotland Ltd) and the fillet from one side was salted and oak smoked. Following smoking, the fillet was trimmed, machine sliced in D slice format and vacuum packed, for evaluation of texture by a trained taste panel. The fresh fillet from the other side of the fish was used to determine muscle architecture, astaxanthin concentration, colour as determined with the Roche *Salmo*Fan™ and oil content.

Significant positive correlations were observed between muscle fibre density and all four measures of texture assessed by the taste panels, 'chewiness', 'firmness', 'mouthfeel' and 'dryness' (Johnston *et al.* 2000a). A high fibre density was associated with a firmer texture, with the best correlation ($r^2 = 0.4$) observed for the 'chewiness' score of the fillet (Fig. 3.7a, b). The oil content of the fillet was significantly higher for fish from strain X (11.2%) than for fish from strain Y (7.0%). However, there was no relationship between sensoric 'oiliness' score and the percentage oil content of the fillet.

An unexpected finding was a significant positive correlation between the Roche *Salmo*-Fan™ score and muscle fibre density, explaining between 26% and 44% of the variation in colour (Fig. 3.8). Astaxanthin concentration was independent of muscle fibre density. It

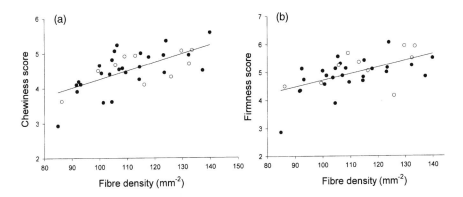

Fig. 3.7   The relationship between the density of fibres in the white muscle and the textural characteristics of the smoked fillet, as measured by trained taste panels, in Atlantic salmon from a high grilsing (closed circles) and low grilsing (open circles) population. The parameters illustrated are (a) 'chewiness' and (b) 'firmness'. From Johnston *et al.* (2000a).

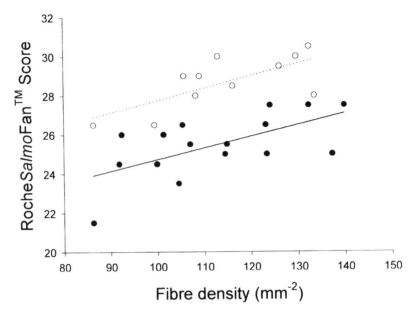

Fig. 3.8   The relationship between the density of fibres in the white muscle and the colour visualisation of the fillet as measured by the Roche *Salmo*Fan™ score in Atlantic salmon from a high grilsing (closed circles) and low grilsing (open circles) population. From Johnston *et al.* (2000a).

would seem that the higher the fibre density the better the colour visualisation of the fillet for a given level of pigment in the flesh which indicates that some component of the difference in muscle architecture was affecting light scattering properties. This important result indicates that broodstock selection and/or production regimes designed to maximise muscle fibre recruitment could result in the more efficient use of dietary pigments as well as bringing quality benefits in terms of the firm texture required for processing.

Conclusions

Research on muscle growth in fish is still in its infancy. In particular we know relatively little about the origin and behaviour of the satellite cells and factors controlling their proliferation and differentiation. However, there is strong evidence that fibre recruitment varies between families and populations of different genetic origin and can be modified by production practices. Furthermore, muscle architecture, particularly the number and size distribution of fibres and the amounts and arrangement of connective tissue, has been shown to affect texture and also tissue coloration in salmon. Firmness and strong colour – quality traits preferred by consumers – are both associated with a high fibre density. The economic endpoint of further research on fish muscle growth will be the development of strains, production practices and diets to give a high percentage of farmed fish of superior quality in order to reduce downgrading losses and retain the premium status of salmon products with the consumer.

Acknowledgements

The author's work in this area was supported by grants from the European Union and the Natural Environment Research Council of the UK under the Aquaculture Link Programme. It has involved a fruitful collaboration with Dick Alderson, Patrick Campbell (Biomar Ltd), David Mitchell, Remi Baker, Billy Robertson and Craig Selkirk (Marine Harvest McConnell Ltd), Alistair Dingwall (Pinneys of Scotland Ltd), David Nickell and John Springate (Roche Products Ltd), David Whyte (Aquasmart UK Ltd), Ray Johnstone (Fisheries Research Services, Scotland) and Tom Hansen (Matre Aquaculture Station, Institute of Marine Research, Bergen, Norway).

References

Alami-Durante, H., Fauconneau, B., Rouel, M., Escaffre, A.M. & Bergot, P. (1997) Growth and multiplication of white skeletal muscle fibres in relation to somatic growth rate. *Journal of Fish Biology*, **50**, 1285–302.

Alexander, R. McN. (1969) The orientation of muscle fibres in the myomeres of fishes. *Journal of the Marine Biological Association of the United Kingdom*, **49**, 263–90.

Beattie, C., Johnston, I.A., Bjørnevik, M., & Hansen, T. (2000) Egg incubation temperature influences muscle fibre recruitment during seawater stages of Atlantic salmon. *Journal of Experimental Biology* (in press).

Black, E.C., Robertson, A.C. & Parker, R.R. (1966) Combined effects of starvation and exercise on glycogen metabolism of rainbow trout, *Salmo gairdneri*. *Journal of the Fisheries Research Board of Canada*, **23**, 1461–3.

Bone, Q. (1978) Locomotor muscle. In: *Fish Physiology* (eds W.S. Hoar, D.J. Randall), pp.361–424, Academic Press, New York.

Bremner, H.A. & Hallett, I.C. (1985) Muscle fiber-connective tissue junction in the fish blue grenadier (*Macruronus novaezelandiae*). A scanning electron microscope study. *Journal of Food Science*, **50**, 975–80.

Brooks, S. & Johnston, I.A. (1993) Influence of development and rearing temperature on the distribution, ultrastructure and myosin sub-unit composition on myotomal muscle-fibre types in the plaice, *Pleuronectes platessa*. *Marine Biology*, **117**, 501–13.

Brown, S.C. & Stickland, N.C. (1993) Satellite cell content of muscles in muscles of large and small mice. *Journal of Anatomy*, **183**, 91–6.

Brown, S.C. & Stickland, N.C. (1994) Muscle at birth in mice selected for large and small body size. *Journal of Anatomy*, **184**, 371–80.

Calvo, J. & Johnston, I.A. (1992) Influence of rearing temperature on the distribution of muscle fibre types in the turbot *Scophthalmus maximus* at metamorphosis. *Journal of Experimental Marine Biology and Ecology*, **161**, 42–55.

Campion, D.R., Marks, H.L. & Richardson, L.R. (1982) An analysis of satellite cell content in the semimembranous muscle of Japanese quail (*Coturnix coturnix japonica*) selected for rapid growth. *Acta Anatomica*, **112**, 9–13.

Cornelison, D.D. W. & Wold, B.J. (1997) Single-cell analysis of regulatory gene expression in quiescent and activated mouse skeletal muscle satellite cells. *Development Biology*, 191, 270–83.

Devoto, S.H., Melançon, E., Eisen, J.S. & Westerfield, M. (1996) Identification of separate slow and fast muscle precursor cells in vivo, prior to somite formation. *Development,* 122, 3371–80.

Dunajski, E. (1979) Texture of fish muscle. *Journal of Texture Studies* 10, 301–18.

Felsenfeld, A.L., Curry, M. & Kimmel, C.B. (1991) The fub-1 mutation blocks initial myofibril formation in zebrafish muscle pioneer cells. *Development Biology,* 148, 23–30.

Florini, J.R., Ewton, D.Z. & Roof, S.L. (1991) Insulin-like growth factor-1 stimulates terminal differentiation by induction of myogenin gene expression. *Molecular Endocrinology*, 5, 718–24.

Fowler, S.P., Campion, D.R., Marks, H.L. & Reagan, J.O. (1980) An analysis of skeletal muscle response to selection for rapid growth in Japanese Quail (*Coturnix coturnix Japonica*). *Growth,* 44, 235–52.

Galloway, T.F., Kjørsvik, E., & Kryvi, H. (1999) Muscle growth and development in Atlantic cod larvae (*Gadus morhua* L.) related to different somatic growth rates. *Journal of Experimental Biology*, 202, 2111–20.

Greene, C.W. & Greene, C. H. (1913) The skeletal musculature of the king salmon. *Bulletin of the Bureau of Fisheries, Washington,* 33, 25–59.

Halpern, M.E., Ho, R.K., Walker, C. & Kimmel, C.B. (1993) The induction of muscle pioneers and floor plate is distinguished by the zebrafish mutation *no tail*. *Cell*, 75, 99–111.

Hatae, K., Tobimatsu, A., Takeyama, M. & Matsumoto, J.J. (1986) Contribution of the connective tissues on the texture differences of various fish species. *Bulletin of the Japanese Society of Scientific Fisheries,* 52, 2001–2008.

Higgins, P.J. & Thorpe, J.E. (1990) Hyperplasia and hypertrophy in the growth of skeletal muscle in juvenile Atlantic salmon, *Salmo salar* L. *Journal of Fish Biology,* 37, 505–19.

Hurling, R., Rodell, J.B. & Hunt, H.D. (1996) Fibre diameter and fish texture. *Journal of Texture Studies*, 27, 679–85.

Hochachka, P.W. (1985) Fuels and pathways as designed systems for support of muscle work. *Journal of Experimental Biology* 115, 149–64.

Jobling, M. (1994) *Fish Bioenergetics*. Chapman and Hall, London.

Johnston, I.A. (1981) Structure and function of fish muscles. In: Vertebrate Locomotion (ed. M.H. Day). *Symposia of Zoological Society of London*, 48, 71–113, Academic Press: London.

Johnston, I.A. (1993) Temperature influences muscle differentiation and the relative timing of organogenesis in herring (*Clupeus harengus*) larvae. *Marine Biology*, 116, 363–79.

Johnston, I.A. (1999) Muscle development and growth: potential implications for flesh quality in fish. *Aquaculture*, 177, 99–115.

Johnston, I.A. & McLay, H.A. (1997) Temperature and family effects on muscle cellularity at hatch and first feeding in Atlantic salmon (*Salmo salar* L.). *Canadian Journal of Zoology*, 75, 79–91.

Johnston, I.A. & Moon, T.W. (1980) Exercise training in skeletal muscle of brook trout (*Salvelinus fontinalis*). *Journal of Experimental Biology,* 87, 177–94.

Johnston, I.A., Cole, N.J., Abercromby, M. & Vieira, V.L.A. (1998) Embryonic temperature modulates muscle growth characteristics in larval and juvenile herring. *Journal of Experimental Biology*, 201, 623–46.

Johnston, I.A., Strugnell, G., McCraken, M.L. & Johnstone, R. (1999) Muscle growth and development in normal-sex-ratio and all-female diploid and triploid Atlantic salmon. *Journal of Experimental Biology*, **202**, 1991–2016.

Johnston, I.A., Alderson, D., Sandham, C., Dingwall, A., Mitchell, D., Selkirk, C., Nickell, D., Baker, R., Robertson, B., Whyte, D. & Springate, J. (2000a) Muscle fibre density in relation to the colour and texture of smoked Atlantic salmon (*Salmo salar* L.). *Aquaculture*, **189**, 335–49.

Johnston, I.A., Alderson, D., Sandham, C., Mitchell, D., Selkirk, C., Dingwall, A., Nickell, D., Baker, R., Robertson, B., Whyte, D. & Springate, J. (2000b) Patterns of muscle growth in early and late maturing populations of Atlantic salmon (*Salmo salar* L.). *Aquaculture*, **189**, 307–33.

Kiessling, A., Storebakken, T., Åsgård, T. & Kiessling, K.-H. (1991) Changes in the structure and function of the epaxial muscle of rainbow trout (*Oncorhynus mykis*) in relation to ration and age: I. Growth dynamics. *Aquaculture*, **93**, 335–6.

Koumans J.T.M. & Akster, H.A. (1995) Myogenic cells in development and growth of fish. *Comparative Biochemistry and Physiology*, **110A**, 3–20.

Love, R.M. (1958) Studies of the North Sea cod. I. Muscle cell dimensions. *Journal of the Science of Food and Agriculture*, **9**, 195–8.

Mascarello, F., Rowlerson, A., Radaelli, G., Scaolo, P.-A. & Veggetti, G. (1995) Differentiation and growth of muscle in the fish *Sparus aurata* (L). I. Myosin expression and organisation of fibre types in lateral muscle from hatching to adult. *Journal of Muscle Research and Cell Motility*, **16**, 213–22.

Mauro, A. (1961) Satellite cells of skeletal muscle fibres. *Journal of Biophysics, Biochemistry and Cytology*, **9**, 493–5.

Merly, F., Magras-Resch, C., Rouaud, T., Fontaine-Perus, J., & Gardahaut, M.F. (1998) Comparative analysis of satellite cell properties in heavy- and lightweight strains of turkey. *Journal of Muscle Research and Cell Motility*, **19**, 257–70.

Offer, G., Knight, P., Jeacocke, R., Almond, R., Cousins, T., Elsey, J., Parsons, N., Sharp, A., Starr, R. & Purslow, P. (1989) The structural basis of the water-holding, appearance and toughness of meat and meat products. *Food Microstructure*, **8**, 151–70.

Olson, E.N. & Klein, W.H. (1994) BHLH factors in muscle development: dead lines and commitments, what to leave in and what to leave out. *Genes Dev.*, **8**, 1–8.

Patruno, M., Radaelli, G., Mascarello, F. & Candia Carnevali, M.D. (1998) Muscle growth in response to changing demands of functions in the teleost *Sparus aurata* (L.) during development from hatching to juvenile. *Anatomical Embryology*, **198**, 487–504.

Quinn, L.S., Norwood, T.H. & Nameroff, M. (1988) Myogenic stem cell commitment probability remains constant as a function of organismal and mitotic age. *Journal of Cell Physiology*, **134**, 324–36.

Ramírez-Zarosa, G., Gil, F., Vázquez, J.M., Arenciba, R., Latorre, O., López-Albors, A., Ortega, A. & Moreno, F. (1998). The post-larval development of lateral musculature in gilthead sea bream (*Sparus aurata* L.) and sea bass (*Dicentrarchus labrax* L.). *Anatomy, Histology and Embryology*, **27**, 21–9.

Rowlerson, A., Mascarello, F., Radaelli, G. & Veggetti, A. (1995) Differentiation and growth of muscle in the fish *Sparus aurata* (L): II. Hyperplastic and hypertrophic growth of lateral muscle from hatching to adult. *Journal of Muscle Research and Cell Motility*, **16**, 223–36.

Rowlerson, A., Radaelli, G., Mascarello, F. & Veggetti, A. (1997) Egeneration of skeletal muscle in two teleost fish: *Sparus aurata* and *Brachydanio* rerio. *Cell and Tissue Research*, 289, 311–22.

Sato, K., Yoshinaka, R., Sato, M. & Shimizu, Y. (1986) Collagen content in the muscle of fishes in association with their swimming movements and meat texture. *Bulletin of the Japanese Society of Scientific Fisheries*, 52, 1595–600.

Schultz, E. (1996) Satellite cell proliferative compartments in growing skeletal muscles. *Developmental Biology*, 175, 84–94.

Stickland, N.C., White, R.N., Mescall, P.E., Crook, A.R. & Thorpe, J.E. (1988) The effect of temperature on myogenesis in embryonic development of the Atlantic salmon (*Salmo salar* L.). *Anatomical Embryology*, 178, 253–7.

Stirling, W. (1885) On the red and pale muscles in fishes. *Appendix to the 4th Annual Report of the Fisheries Board of Scotland*, pp.166–71.

Stoiber, W. & Sänger, A.M. (1996) An electron microscopic investigation into the possible source of new muscle fibres in teleost fish. *Anatomical Embryology* 194, 569–79.

Tatsumi, R., Anderson, J.E., Nevoret, C.J., Halevy, O. & Allen, R.E. (1998) HGF/SF is present in normal adult skeletal muscle and is capable of activating satellite cells. *Developmental Biology*, 194, 114–28.

Taylor, E.B. (1991) A review of local adaptation in Salmonidae, with special reference to Pacific and Atlantic salmon. *Aquaculture*, 98, 185–207.

Totland, G.K., Kyrivi, H., Jødestol, K.A., Christiansen, E.N., Tangerås, A. & Slinde, E. (1987) Growth and composition of the swimming muscle of adult Atlantic salmon (*Salmo salar* L.) during long-term sustained swimming. *Aquaculture*, 66, 299–313.

Veggetti, A., Mascarello, F., Scapolo, P.A., & Rowlerson, A. (1990) Hyperplastic and hypertrophic growth of lateral muscle in *Dicentrarchus labrax* (L.). *Anatomical Embryology*, 182, 1–10.

Verspoor, E. (1997) Genetic diversity among Atlantic salmon (*Salmo salar* L.) populations. *ICES Journal of Marine Science*, 54, 965–73.

Wakeling, J.M. & Johnston, I.A. (1999) White muscle strain in the common carp and red to white muscle gearing ratios in fish. *Journal of Experimental Biology*, 202, 521–8.

Weatherley, A.H., Gill, H.S. & Lobo, A.F. (1988) Recruitment and maximal diameter of axial muscle fibres in teleosts and their relationship to somatic growth and ultimate size. *Journal of Fish Biology*, 33, 851–9.

Weinberg, E.S., Allende, M.L., Kelly, C.S., Abdelhamid, A., Murakami, T., Anderman, P., Doerre, O.G., Grunwald, D.J. & Riggleman, B. (1996) Developmental regulation of zebrafish MyoD in wild-type, no tail and spadetail embryos. *Development* 122, 271–80.

Yablonka-Reuveni, Z., & Rivera, A.J. (1994) Temporal expression of regulatory and structural muscle proteins during myogenesis of satellite cells of isolated adult rat fibers. *Developmental Biology*, 164, 588–603.

Yablonka-Reuveni, Z., Seger, R. & Rivera, A.J. (1999) Fibroblast growth factor promotes recruitment of skeletal muscle cells in young and old rats. *Journal of Histochemistry and Cytochemistry*, 47, 23–42.

Yang, Q., Bassel-Duby, R., & Williams, R.S. (1997) Transient expression of winged-helix protein, M NF-ß, during myogenesis. *Molecular and Cell Biology*, 17, 5236–43.

Chapter 4

The Effect of Diet Composition and Feeding Regime on the Proximate Composition of Farmed Fishes

K.D. Shearer

National Marine Fisheries Service, 2725 Montlake Boulevard East, Seattle, WA 98112, USA

Introduction

The primary objective of commercial fish production is to produce a marketable product at the maximum profit. This involves a continuing effort to reduce feed cost, which is a major portion of the cost of production, while trying to maintain fish health, feed efficiency and rapid growth. In an expanding market relatively little emphasis may be placed on product quality, but as the market matures, competition between producers for market share and competition from other meat sources will put more emphasis on product quality. One quality parameter that is receiving increased attention is fish composition. There have been several previous reviews of the effects of diet on fish composition (Love 1970, 1988; Buckley & Groves 1979; Spinelli 1979; Haard 1992; Shearer 1994). The purpose of the current review is to provide some tools to aid interpretation of published studies, which often report conflicting results, and to summarise briefly what these studies tell us about the effect of diet composition and feeding regime on fish proximate composition. Space limitations prevent discussion in depth of the papers cited, so readers should consult the original references for more details and also Chapter 16 by Paul Morris in this text.

Interpretation of published studies

There is considerable conflict in the literature, concerning the effects of diet and feeding regime on fish composition. It is therefore worth examining some of the causes of these conflicting results. First, in many studies more than one dietary component is varied at one time, as one component replaces another. The observed effects may therefore be attributed to a change in either or both components. This is often the case when protein is replaced with fat, as the diets are neither isonitrogenous nor isoenergetic (Table 4.1). Although it is possible to design an experiment where the diets are isonitrogenous, a preferred, fully factorial design will allow examination of the effects of each component and their interaction on body composition (Table 4.1).

Table 4.1   Three experimental designs for examining the effects of various protein/fat ratios on the composition of fish. Design A is neither isonitrogenous or isoenergetic. Design B is iso-nitrogenous but not isoenergetic. Design C allows the effect of each dietary component and their interaction to be determined.

Design		%	%	%	%
A	Protein	40	45	50	55
	Fat	20	15	10	5
B	Protein	45	45	45	45
	Fat	20	15	10	5
	Inert	0	5	10	15
C	Protein	Fat (%)			
	40	20	15	10	5
	45	20	15	10	5
	50	20	15	10	5
	55	20	15	10	5

A second problem, which has often led to misinterpretation of results, is that the proximate composition of the fish or a tissue is often expressed on a dry weight basis. If the percentage of one component changes, the others must change since the total must remain at 100%. Comparing results on a dry weight basis frequently leads to an effect being found that is not there, or in some cases to an effect that is the opposite to the one reported (Britz & Hecht 1997). The interdependence of the percentages of the components, when they are reported on a dry weight basis, can also be shown graphically (Fig. 4.1).

A third problem is that small differences between treatments (for example, tenths of a percentage unit in a component) are sometimes reported as significantly different, but if the results of the proximate analysis for each group are summed, differences of several per cent (totals of 97–102%) are found. This means that the differences may be due to analytical error rather than the dietary treatments. One should also be concerned when the sum of a large number of samples is consistently 100% (Berger & Halver 1987). This often means that one of the components was determined by difference so there is no way to determine the accuracy of the other analysis. When interpreting results, one should be most critical of protein values as protein is the most difficult of the components to determine.

A fourth problem is that fish are genetically programmed to change composition as they grow and sexually mature (reviewed by Shearer 1994). A general trend, under culture conditions for most species, is for protein to increase during the juvenile phase up to a relative constant level in the adult. Ash will also increase during the juvenile period as bones calcify but then it decreases since weight increases faster than length (muscle mass increases faster than skeletal mass). The situation for fat and moisture is a well known reciprocal relationship

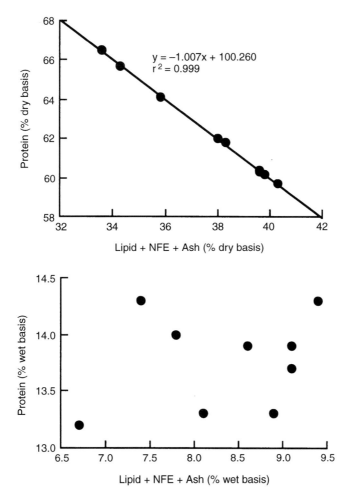

Fig. 4.1 The relationship between protein and the sum of lipid, nitrogen free extract (NFE) and ash on a dry basis (upper) and a wet basis (lower). On a dry basis the percentage of protein is dependent on the percentages of the other components. On a wet basis it is independent. Data from Britz and Hecht (1997).

of fat increasing faster than weight (fat increases in percentage) and moisture decreasing in percentage as fat increases. The significance of these changes is that if treatments produce fish which differ in size, an adjustment must be made to normalise size before treatment effects can be interpreted. Differences in body fat between treatment groups often disappear when size is used as a covariate (Weatherup *et al.* 1997; Johansen & Jobling 1998). The effect of size, at least for protein content, is much greater in juveniles, while the largest difference in fat attributable to size usually occurs in larger fish (Shearer *et al.* 1994). An additional factor that may need to be taken into account is the effect of sexual maturation, since maturation normally leads to increased fat retention (Tveranger 1985; Asknes *et al.* 1986). This can produce differences in the percentage of fat between males and females or differences among fish in treatments if the frequency of maturation differs.

A final problem is that there has been little effort to standardise the portion of the fish being sampled. There is little problem comparing studies where the whole fish is analysed but results are often reported for carcass, fillet, cutlet or mince. The term carcass is poorly defined; sometimes this represents the whole fish, while in other cases it may refer to the eviscerated fish with or without the head or gills. It also is important for the location of the sample to be specified. Nortvedt and Tuene (1998) showed that the fat content of Atlantic halibut fillets ranges from 1.9% at the tail to 17.5% near the head. The Norwegian quality cut is an attempt to use a standardized sample for Atlantic salmon.

The effect of dietary composition on fish composition

Protein

The protein content of the whole fish, the carcass or the muscle appears to be fixed for each species at a given size regardless of the diet fed if the fish is growing (Shearer 1994). This also appears to be the case for the amino acid profile (Schwarz & Kirchgessner 1988). Helland and Grisdale-Helland (1998) showed that increased levels of dietary protein are used less efficiently for growth in juvenile Atlantic salmon but that higher protein lower fat diets produced fish with similar levels of protein but lower carcass and visceral fat. Shyong *et al.* (1998) reported that muscle protein was increased by feeding higher levels of dietary protein in juvenile *Zacco babata* but fish weight also increased so that the higher muscle protein levels appeared to be due to fish size alone. A similar muscle protein effect was reported by Asknes *et al.* (1996) in Atlantic halibut (*Hippoglossus hippoglossus*) which also showed increased weight gain. In contrast, Webster *et al.* (1997) working with *Lepomis cyanellus*, and van der Meer *et al.* (1995) with *Colossoma macropomum* have reported that high levels of dietary protein resulted in elevated body protein. Additional studies may clarify these conflicting results.

Increasing dietary protein generally increases growth up to the point where protein intake exceeds the ability of the fish to synthesize additional protein. At this point any additional protein is deaminated and stored as fat. The source of the protein appears to have little effect on body protein content. Smith *et al.* (1988) fed isonitrogenous and isoenergetic diets containing plant or animal protein to rainbow trout and reported no differences in carcass composition, dressing-out percentage or organoleptic properties attributed to diet. Similar results were reported by Kaushik *et al.* (1995) for rainbow trout fed soy protein. High levels of protein may suppress growth (van der Meer *et al.* 1995). Increasing dietary fat generally spares protein for growth. A problem, however, is that an increasing portion of the dietary fat appears to be stored. A consequence of adding fat to the diet therefore, is that by improving protein efficiency, additional fat is retained in the carcass and the viscera. The amino acid profile of the whole body or the muscle appears to be relatively unaffected by the amino acid content of the diet (Schwarz & Kirchgessner 1988), although the free amino acid pool can be influenced.

Fat

In animals in general, increasing dietary fat tends to increase whole body fat (West & York 1998). In fish, the location of the deposition of the additional fat appears to differ depending on the species and fish size (Sheridan 1988; Navarro & Gutiérrez 1995). Einen and Roem (1997) showed that fat deposition, when a given diet was fed, is fish size dependent (Fig. 4.2). Hillestad *et al.* (1998) reported that ration affected fillet fat in Atlantic salmon but that increasing dietary fat affected visceral but not fillet fat. They also indicated that the dressing-out percentage of fish fed high fat diets was reduced. This contrasts somewhat with many other studies that have shown that increasing dietary fat increases muscle fat (Hemre & Sandness 1995; Bjerkeng *et al.* 1997; Einen & Skrede 1998; Jobling *et al.* 1998). Weatherup *et al.* (1997) fed diets containing graded levels of protein and fat at different feeding rates so that fish received equivalent amounts of energy supplied by either protein or fat. They concluded that increasing dietary fat significantly increased both fillet and offal fat. Nortvedt and Tuene (1998) showed that the fat content of Atlantic halibut increased with both fish size and with the level of dietary fat.

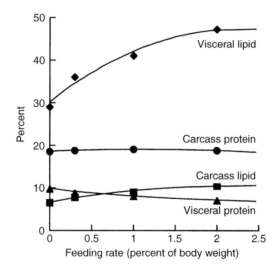

Fig. 4.2 The effect of feeding rate on carcass and visceral fat and protein of juvenile Atlantic salmon fed at ration levels from 0 to 2.0% of body weight. Data from Storebakken *et al.* (1991).

Fatty acids

The fatty acid profile of fish reflects the fatty acid profile of their diet. Kirsch *et al.* (1998) reported that the fatty acid profile of Atlantic cod (*Gadus morhua*) changed to the fatty acid profile of their new diet after three weeks. Nettleton (1990) also states that the fatty acid profile of cultured fishes can be modified a few weeks prior to harvest. The fatty acid profile of Atlantic salmon has also been shown to reflect that of their diet, and may modify the organoleptic qualities of the flesh (Bergström 1989; Thomassen & Røsjø (1989). Fauconneau *et al.* (1997) reported that adiposite size was affected by the type of fatty acids fed, being increased by *n*-6 fatty acids.

Carbohydrate

High levels of available carbohydrate increased visceral fat but appear to have little effect on carcass composition (Bergot 1979; Berger & Halver 1987). Dietary fibre appears to have a minimal effect on body composition. High levels of fibre will, however, reduce yields by increasing the visceral/somatic index since stomach size increases as dietary energy density decreases (Hilton *et al.* 1983; Bromley & Adkins 1984).

Ash

The ash level of fish fed a mineral complete diet appears to be fixed for fish of a given size (Shearer 1994). However, whole body fat levels have been shown to be elevated when phosphorus deficient diets are fed (Skonberg *et al.* 1997).

Dietary supplements and biochemical regulators

Dietary additions of several additives have been shown to alter carcass or muscle composition. Stowell and Gatlin (1992) showed that pantethine (a synthetic coenzyme) modified the fatty acid profile of channel catfish muscle, while Zoccarato *et al.* (1995) showed a fatty acid modifying effect from avoparcin (an antibiotic) in rainbow trout. Increased whole body protein has been reported when carnitine was fed to channel catfish (Burtle & Liu 1994.)

The effects of feeding regime (feeding rate, meal frequency, time of feeding, fasting) on fish composition

Composition effects attributed to feeding regime appear to be due primarily to differences in feed intake. Kiessling *et al.* (1991) reported that the levels of protein in red and white muscle increased in rainbow trout with size and age up to 1.4 years, then remained relatively constant regardless of ration level. Fat levels in all tissues examined increased with age and increasing ration. The rate of fat increase in red and white muscle declined in older fish. Much larger changes were seen in visceral fat in relation to ration level. Increasing the feeding rate increased carcass and visceral fat and reduced visceral protein (Fig. 4.2) in rainbow trout (Storebakken *et al.* 1991). Johansen and Jobling (1998) also reported that fish fed to satiation had higher levels of muscle fat than fish fed at a lower predetermined ration, but indicated that this could be attributed to differences in fish size. Company *et al.* (1999) reported that the percentage of body fat was increased by increasing ration in gilthead sea bream (*Sparus aurata*). They also reported an increase in viscerosomatic index.

Grayton and Beamish (1977) reported increases in weight gain and fat deposition in rainbow trout when they increased the number of daily feedings (one, three or six times to satiation). They attributed this to greater daily feed intake in the fish fed more frequently.

When they varied the number of daily feedings and restricted the ration so that all groups had the same daily intake, no differences were reported.

The time of day a meal is fed has been shown to affect feed intake and therefore growth. Boujard *et al.* (1995) examined the effects of feeding time on the composition of rainbow trout, but the effect of meal time is unclear despite the authors' claim that both protein and fat were reduced in fish fed at midnight versus those fed at dawn. Composition was reported on a dry weight basis and fish differed in size at the end of the experiment.

Azzaydi *et al.* (1998) reported that European sea bass (*Dicentrarchus labrax* L.) showed daily feeding rhythms when fed using demand feeders. When fed *ad libitum* or when feeding time was restricted, growth and feed efficiency were reduced but body composition was not affected.

Several studies have examined the effects of fasting for a period prior to harvest on body or product composition. Lie and Huse (1992) starved Atlantic salmon for 78 days and observed that weight loss and reduction of fillet fat slowed progressively with time. The largest loss of fat occurred in the viscera and fillet fat decrease was marginal. This was confirmed by Wathne (1995). Einen and Thomassen (1998) starved 5 kg fish for 86 days and concluded that starvation was a relatively poor way to improve fillet quality.

Future studies

Smith *et al.* (1988) reported differences in growth, dressing-out percentage and carcass composition attributable to rainbow trout strain. There appears to be some scope for selection for a desired body composition (Gjerde & Schaffer 1989; Rye 1991). These differences may, however, be due solely to differences in growth rate (Ayles *et al.* 1979). In their recent review of the role of body composition in meat production, Wray-Cahen *et al.* (1998) suggest that genetic manipulation (both selection and engineering) hold the greatest potential for modifying the carcass composition of meat animals. They suggest that control of composition lies in a better understanding of the relationships between nutrients, hormones and genes. Pérez-Sánchez *et al.* (1995) have recently shown that ration size and protein level affect plasma growth hormone levels. It has also been shown that proliferation of adipose tissue mass is dependent on the type of fat fed to rats (Shillabeer & Lau 1994).

References

Asknes, A., Gjerde, B. & Roald, S. (1986) Biological, chemical and organoleptic changes during maturation of farmed Atlantic salmon, *Salmo salar*. *Aquaculture*, **53**, 7–20.

Asknes, A., Hjertnes, T. & Opstvedt, J. (1996) Effects of dietary protein level on growth and carcass composition in Atlantic halibut (*Hippoglossus hippoglossus* L.). *Aquaculture*, **145**, 225–33.

Ayles, G.B., Bernard, D. & Hendzel, M. (1979) Genetic differences in lipid and dry matter content between strains of rainbow trout (*Salmo gairdneri*) and their hybrids. *Aquaculture*, **18**, 253–62.

Azzaydi, M., Madrid, J.A., Zamora, S., Sánchez-Vázquez, F.J. & Martínez, F.J. (1998) Effect of three strategies (automatic, *ad libitum* demand-feeding and time-restricted demand-feeding) on feeding rhythms and growth in European sea bass (*Dicentrarchus labrax* L.). *Aquaculture*, 163, 283–94.

Berger, A. & Halver, J.E. (1987) Effect of dietary protein, lipid and carbohydrate content on growth, feed efficiency and carcass composition of striped bass, *Morone saxatilis* (Walbaum), fingerlings. *Aquaculture and Fisheries Management*, 18, 345–56.

Bergot, F. (1979) Carbohydrate in rainbow trout diets: effects of the level and source of carbohydrate and the number of meals on growth and body composition. *Aquaculture*, 18, 157–67.

Bergström, E. (1989) Effect of natural and artificial diets on seasonal changes in fatty acid composition and total body lipid content of wild and hatchery-reared Atlantic salmon (*Salmo salar* L.) parr-smolt. *Aquaculture*, 82, 205–17.

Bjerkeng, B., Refstie, S., Fjalestad, K.T., Storebakken, T., Rødbotten, M. & Roem, A.J. (1997) Quality parameters of the flesh of Atlantic salmon (*Salmo salar*) as affected by dietary fat content and full-fat soybean meal as a partial substitute for fish meal in the diet. *Aquaculture*, 157, 297–309.

Boujard, T., Gelineau, A. & Corraze, G. (1995) Time of a single daily meal influences growth performance in rainbow trout, *Oncorhynchus mykiss* (Walbaum*). Aquaculture Research*, 26, 341–9.

Britz, P.J. & Hecht, T. (1997) Effect of dietary protein and energy level on growth and body composition of South African abalone, *Haliotis midae*. *Aquaculture*, 156, 195–210.

Bromley, P.J. & Adkins, T.C. (1984) The influence of filler on feeding, growth and utilization of protein and energy in rainbow trout, *Salmo gairdnerii* Richardson. *Journal of Fisheries Biology*, 24, 235–44.

Buckley, J.T. & Groves, T.D.D. (1979) Influence of feed on the body composition of finfish. In: *Fish Nutrition and Fish Feed Technology*, (eds J.E. Halver & K. Tiews). Proceedings of a Symposium, 20–23 June 1978 at Hamburg, Germany. Heenemann, Berlin, Vol.II, pp. 335–43.

Burtle, G.J. & Liu, Q. (1994) Dietary carnitine and lysine affect channel catfish lipid and protein content. *Journal of the World Aquaculture Society*, 25, 169–74.

Company, R., Calduch-Giner, J.A., Kaushik, S. & Pérez-Sánchez, J. (1999) Growth performance and adiposity in gilthead sea bream (*Sparus aurata*): risks and benefits of high energy diets. *Aquaculture*, 171, 279–92.

Einen, O. & Roem, A.J. (1997) Dietary protein/energy ratios for Atlantic salmon in relation to fish size: growth, feed utilization and slaughter quality. *Aquaculture Nutrition*, 3, 115–26.

Einen, O. & Skrede, G. (1998) Quality characteristics in raw and smoked fillets of Atlantic salmon, *Salmo salar*, fed high energy diets. *Aquaculture Nutrition*, 4, 99–108.

Einen, O. & Thomassen, M.S. (1998) Starvation prior to slaughter in Atlantic salmon (*Salmo salar*) II. White muscle composition and evaluation of freshness, texture and color characteristics in raw and cooked fillets. *Aquaculture*, 169, 37–53.

Fauconneau, B., Andre, S., Chmaitilly, J., Le Bail, P.-Y., Krieg, F. & Kaushik, S.J. (1997) Control of skeletal muscle fibres and adipose cell size in the flesh of rainbow trout. *Journal of Fisheries Biology*, 50, 296–314.

Gjerde, B. & Schaffer, L.R. (1989) Body traits in rainbow trout II. Estimation of heritabilities and phenotypic and genetic correlations. *Aquaculture*, 80, 25–44.

Grayton, B.D. & Beamish, F.W.H. (1977) Effects of feeding frequency on food intake, growth and body composition of rainbow trout (*Salmo gairdneri*). *Aquaculture,* 11, 159–72.

Haard, N.F. (1992) Control of chemical composition and food quality attributes of cultured fish. *Food Research International,* 25, 289–307.

Helland, S.J. & Grisdale-Helland, B. (1998) The influence of replacing fish meal in the diet with fish oil on growth, feed utilization and body composition of Atlantic salmon (*Salmo salar*) during the smoltification period. *Aquaculture,* 162, 1–10.

Hemre, G.I. & Sandness, K. (1995) Effect of dietary lipid level on muscle composition in Atlantic salmon *Salmo salar*. *Aquaculture Nutrition,* 5, 9–16.

Hillestad, M., Johnsen, F., Austreng, E. & Åsgård, T. (1998) Long-term effects of dietary fat level and feeding rate on growth, feed utilization and carcass quality of Atlantic salmon. *Aquaculture Nutrition,* 4, 89–97.

Hilton, J.W., Atkinson, J.L. & Slinger, S.J. (1983) Effect of increased dietary fiber on growth of rainbow trout (*Salmo gairdneri*). *Canadian Journal of Fisheries and Aquatic Sciences,* 40, 81–5.

Jobling, M., Koskela, J. & Savolinen, R. (1998) Influence of dietary fat level and increased adiposity on growth and fat deposition in rainbow trout, *Oncorhynchus mykiss* (Walbaum). *Aquaculture Research,* 29, 601–7.

Johansen, S.J.S. & Jobling, M. (1998) The influence of feeding regime on growth and slaughter traits of cage-reared Atlantic salmon. *Aquaculture International,* 6, 1–17.

Kaushik, S.J., Cravedi, J.P., Lalles, J.P., Sumpter, J. Fauconneau, B. & Laroche, M. (1995) Partial or total replacement of fish meal by soybean protein on growth, protein utilization, potential estrogenic or antigenic effects, cholesterolemia and flesh quality in rainbow trout, *Oncorhynchus mykiss*. *Aquaculture,* 133, 257–74.

Kiessling, A., Åsgård, T., Storebakken, T. Johansson, L. & Kiessling, K.-H. (1991) Changes in the structure and function of the epaxial muscle of rainbow trout (*Oncorhynchus mykiss*) in relation to ration and age. III. Chemical composition. *Aquaculture,* 93, 373–87.

Kirsch, P.E., Iverson, S.J., Bowen, W.D., Kerr, S.R. & Ackman, R.G. (1998) Dietary effects on the fatty acid signature of whole Atlantic cod (*Gadus morhua*). *Canadian Journal of Fisheries and Aquatic Sciences,* 55, 1378–86.

Lie, Ø. A. & Huse, I. (1992) The effect of starvation on the composition of Atlantic salmon (*Salmo salar*). *Fiskerdirektoratets Skrifter, Ernæring,* 5, 11–16.

Love, R.M. (1970) *The Chemical Biology of Fishes.* 547 pp. Academic Press, New York.

Love, R.M. (1988) *The Food Fishes; Their Intrinsic Variation and Practical Implications.* 276 pp. Farrand Press, London.

van der Meer, M.B., Machiels, M.A.M. & Verdegem, M.C.J. (1995) The effect of dietary protein level on growth, protein utilization and body composition of *Colossoma macropomum* (Curvier). *Aquaculture Research,* 26, 901–9.

Navarro, I. & Gutiérrez, J. (1995) Fasting and starvation. In: *Metabolic Biochemistry,* (eds T.P. Mommsen & P.W. Hochachka) pp.339–434. Elsevier, Amsterdam, The Netherlands.

Nettleton, J.A. (1990) Comparing nutrients in wild and farmed fish. *Aquaculture Magazine,* Jan–Feb, 34–41.

Nortvedt, R. & Tuene, S. (1998) Body composition and sensory assessment of three weight groups of Atlantic halibut (*Hippoglossus hippoglossus*) fed three pellet sizes and three dietary fat levels. *Aquaculture,* 161, 295–313.

Pérez-Sánchez, J. Martí-Planca, H. & Kaushik, S.J. (1995) Ration size and protein intake affect circulating growth hormone concentration, hepatic growth hormone binding and plasma insulin-like growth factor-1 immunoreactivity in a marine teleost, the gilthead sea bream, (*Sparus aurata*). *Journal of Nutrition*, **125**, 546–552.

Rye, M. (1991) Prediction of carcass composition in Atlantic salmon by computerized tomography. *Aquaculture*, **99**, 35–48.

Schwarz, F.J. & Kirchgessner, M. (1988) Amino acid composition of carp (*Cyprinus carpio* L.) with varying protein and energy supplies. *Aquaculture*, **72**, 307–17.

Shearer, K.D. (1994) Factors affecting the proximate composition of cultured fishes with emphasis on salmonids. *Aquaculture*, **119**, 63–88.

Shearer, K.D., Åsgård, T., Andorddóttir, G. & Aas, G.H. (1994) Whole body elemental and proximate composition of Atlantic salmon (*Salmo salar*) during the life cycle. *Journal of Fisheries Biology*, **44**, 785–97.

Sheridan, M.A. (1988) Lipid dynamics in fish: aspects of absorption, transformation, deposition and mobilization. *Comparative Biochemistry and Physiology*, **90**, 679–90.

Shillabeer, G. & Lau, D.C.W. (1994) Regulation of new fat cell formation in rats: the role of dietary fats. *Journal of Lipid Research*, **35**, 592–600.

Shyong, W.-J., Huang, C.-H. & Chen, H.-C. (1998) Effects of dietary protein concentration on growth and muscle composition of juvenile *Zacco barbata*. *Aquaculture*, **167**, 35–42.

Skonberg, D.L., Yogev, L., Hardy. R.W. & Dong, F.M. (1997) Metabolic response to dietary phosphorus intake in rainbow trout (*Oncorhynchus mykiss*). *Aquaculture*, **157**, 11–24.

Smith, R.R., Kincaid, H.L., Regenstein, J.M. & Rumsey, G.L. (1988) Growth, carcass composition, and taste of rainbow trout of different strains fed diets containing primarily plant or animal protein. *Aquaculture*, **70**, 309–21.

Spinneli, J. (1979) Influence of feed on finfish quality. In: *Fish Nutrition and Fishfeed Technology* (eds J.E. Halver & K. Tiews). Proceedings of a Symposium, 20–23 June 1978 at Hamburg, Germany. Heenemann, Berlin, Vol.II, pp. 345–52.

Storebakken, T., Hung, S.S.O., Calvert, C.C. & Plisetskaya, E.M. (1991) Nutrient partitioning in rainbow trout at different feeding rates. *Aquaculture*, **96**, 191–203.

Stowell, S.L. & Gatlin, D.M. III. (1992) Effects of dietary pantethine and lipid levels on growth and body composition of channel catfish, *Ictalurus punctatus*. *Aquaculture*, **108**, 177–88.

Thomassen, M.S. & Røsjø, C. (1989) Different fats in feed for salmon: Influence on sensory parameters, growth rate and fatty acids. I. muscle and heart. *Aquaculture*, **79**, 129–35.

Tveranger, B. (1985) Variation in growth rate, liver weight and body composition at first sexual maturation in rainbow trout. *Aquaculture*, **49**, 89–99.

Wathne, E. (1995) *Strategies for directing slaughter quality of farmed Atlantic salmon* (Salmo salar) with emphasis on diet composition and fat deposition. Dr. Sci. thesis, Agricultural University of Norway, Ås.

Weatherup, R.N., McCracken, K.J., Foy, R., Rice, D., McKendry, J., Mairs, R.J. & Hoey, R. (1997) The effects of dietary fat content on performance and body composition of farmed rainbow trout (*Oncorhynchus mykiss*). *Aquaculture*, **151**, 173–84.

Webster, C.D., Tiu, L.G. & Tidwell, J.H. (1997) Growth and body composition of juvenile hybrid bluegill *Lepomis cyanellus* x *L. macrochirus* fed practical diets containing various percentages of protein. *Journal of the World Aquaculture Society*, **28**, 230–40.

West, D.B. & York, B. (1998) Dietary fat, genetic predisposition, and obesity: a lesson from animal models. *American Journal of Clinical Nutrition*, 67 (suppl.), 505S–512S.

Wray-Cahen, C.D., Kerr, D.E., Evock-Clover, C.M. & Steele, N.C. (1998) Redefining body composition: nutrients, hormones, and genes in meat production. *Annual Review of Nutrition*, 18, 63–92.

Zoccarato, I., Gasco, L., Leveroni Calvi, S., Fortina, R., Bianchini, M.L. & Rollin, X. (1995) Effect of dietary avoparcin on performance and carcass composition in rainbow trout, *Oncorhynchus mykiss* (Walbaum). *Aquaculture Research*, 26, 361–6.

Chapter 5

Texture and Technological Properties of Fish

O.J. Torrissen[1], S. Sigurgisladottir[2] and E. Slinde[1]

[1]*Institute of Marine Research, P.O. Box 1870 Nordnes, N-5817 Bergen, Norway*
[2]*IceTec, Keldnaholt, Is-112 Reykjavik, Iceland*

Introduction

The consumer judges quality of fish on the basis of assumed freshness, fat content, colour and texture. In addition, the production yields are of great importance for the processing industry. All these quality factors interact, and it is known that the consumer judges pale and soft salmon to be fat. Fish tend to lose texture during storage and consumers may classify them as fatter fish. Similarly, the salmon smoking industry expects lower yields of fat fish than lean fish, but it is known that increased fat levels in the fish improve the yield during smoking (Rora *et al.* 1998; Torrissen *et al.* 2000). It is a problem for the fishery and aquaculture industry that misinterpretations and wrong conclusions are drawn by the market, as this takes the focus away from the real problems.

The production of Atlantic salmon (*Salmo salar*) in 1997 was approximately 650 000 tons worldwide (Norske Fiskeoppdretteres Forening, Trondheim). About 40% of this quantity, or more than 250 000 tons, is smoked (The Norwegian Export Council for Fish, Tromsø, Norway). Technological properties of farmed salmon can therefore be regarded as those contributing to fitness for smoking or filleting. As in most other productions systems, yield is a factor of crucial importance. A 1% improvement in the yield in salmon smoking represents a quantity of 2500 tons of smoked salmon, but in economic value it represents three to four times the value of this quantity.

Texture of raw fish is commonly tested in the industry by the 'finger method'. A finger is pressed on the skin or the fillet and the firmness is evaluated as a combination of the hardness when pressed on the fillet and the mark or hole left in the fillet after pressing. This method depends to a large extent on the subjective assessment of the person who is performing the measurement (Sigurgisladottir *et al.* 1997).

Texture can, however, be measured objectively by mechanical testing equipment but because of the heterogeneity of fish fillets it is difficult to get reproducibility in texture measurements (Børresen 1986; Botta 1991a; Reid & Durance 1992).

In this paper we will discuss technological properties of salmon in relation to cold smoking and measurement of texture and textural properties of fish, especially salmon.

Technological properties

Yield

There is a general impression that an increased amount of fat in fish feed will give a higher lipid level in the fish and a decreased processing yield. The relationship between increased dietary lipid level from 31% up to 48% and the processing yields during the cold smoking process of Atlantic salmon were studied by Torrissen *et al.* (2000), who found small but significant differences in the processing yield. The total yield was highest in the groups fed the highest dietary lipid level (Fig. 5.1). Similar results are presented by Einen and Roem (1997) and Rora *et al.* (1998). The process losses during the individual processing stages of cold smoking of salmon are shown in Table 5.1 which shows higher values for gutting but lower values for salting and smoking losses by increasing dietary fat level. This indicates a higher adipose deposit in fish fed high fat levels in the feed compared to the one fed lower levels.

By multiple linear regression an inverse relationship was found between the process losses in relation to fillet lipid level and loss during filleting, salting and smoking, and a direct relationship between fillet lipid level and gutting and trimming. The decreased loss during salting and smoking can be explained by less dehydration in fat fish compared to leaner fish. Increased fish weight marginally increased the yields at all individual stages of processing (Table 5.2)

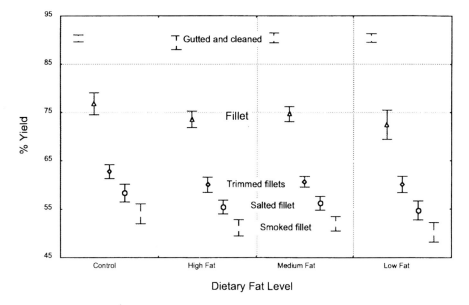

Fig. 5.1 Yield in smoked salmon in relation to fat level in the diets. Control is a standard commercial diet containing 31% fat, high fat is a microwave produced diet of 48% fat, medium fat a microwave diet containing 38% fat and low fat a microwave diet containing 32% fat.

Table 5.1   Process losses during the individual stages of cold smoking of salmon in groups fed a standard commercial extruded diet of 31% fat and three experimental groups containing 32%, 38% and 48% fat (Torrissen *et al.* 2000).

	Gutting	Filleting	Trimming	Salting	Smoking
Control	9.7 ± 0.7	13.3 ± 1.9	14.1 ± 1.8	4.4 ± 1.1	4.2 ± 0.5
High lipid level	10.4 ± 1.4	16.0 ± 1.9	13.3 ± 2.2	4.7 ± 1.2	4.3 ± 0.5
Medium lipid level	9.7 ± 1.1	16.0 ± 1.8	13.6 ± 1.8	4.5 ± 1.1	4.3 ± 0.4
Low lipid level	9.6 ± 0.9	17.8 ± 3.0	12.5 ± 2.3	5.2 ± 1.1	4.6 ± 0.5

Table 5.2   Multiple linear regression data. Slope for fillet lipid level (%) and body weight (g) for losses during gutting, filleting, trimming, salting, smoking and for the total losses during the whole process of cold smoking of Atlantic salmon (Torrisen *et al.* 2000).

	Gutting	Filleting	Trimming	Salting	Smoking	Total
Fillet lipid (%)	0.05	−0.60	0.24	−0.14	−0.10*	−0.56***
Body weight (g)	−0.0002	−0.0003	−0.00008	−0.0005*	−0.0004***	−0.0015***

Levels of significance: $* = p < 0.05$, $*** = p < 0.001$

Gaping

Gaping is a serious problem for the salmon farming industry. The reasons for the development of gaping or fillet rupture may be multifactorial. The strength and duration of *rigor mortis*, muscle fibre sizes, fat level in the fillet or, possibly more importantly, the location of fat in the muscle are probably important factors in addition to the pH in the muscle. Sheehan *et al.* (1996) reported pronounced gaping in smoked flesh from salmon fed diets containing 30% lipid. However, in our work (Torrissen *et al.* 2000) we could not observe any gaping of significance. But this experiment was terminated the day after smoking while Sheehan *et al.* (1996) measured an increased gaping through storage time, and at 24 days after smoking more than twice the severity of gaping in the high dietary lipid groups compared to the low dietary lipid groups. They also reported fillets from fish fed the low lipid diet (20%) to be significantly softer than those from fish fed the medium (25%) and high lipid diets (30%). The occurrence of gaping in fatter fish may be due to infiltration of lipid droplets in the myocommata, and as a consequence decreased strength of the collagen.

Colour

The carotenoid astaxanthin is the natural pigment for salmon flesh, and is also the most common pigment used for pigmentation of farmed Atlantic salmon (*Salmo salar*) (Torrissen *at al.* 1989; Storebakken & No 1992), but another carotenoid, canthaxanthin, is also commonly used either alone or in combination with astaxanthin. Carotenoids are found throughout the plant and animal kingdoms, but can only be biosynthesised *de novo* by fungi, algae and higher plants, while other organisms obtain carotenoids solely from their dietary sources. Commercially farmed Atlantic salmon are fed pigment-fortified diets from smoltification (50–150 g) until harvest (> 2 kg).

The flesh colour of salmon is one of the most important quality parameters, and salmon with insufficient pigmentation are not accepted by the market. For smoked salmon, not only an adequate pigment level is required, but the individual variation in pigmentation, as well as variation within the fillet, should be minimal. Farmed Atlantic salmon are competing in the market with the naturally redder Pacific salmon species (Torrissen *et al.* 1989). Therefore it is a scientific challenge to increase the colour above the normal level for this species. The wide availability of smoked salmonid products has also led the consumer to expect the flesh to be consistent and uniformly pigmented. A panel of consumers held in the departure lounge at Dublin airport indicated a preference for smoked salmon which had a light pink colour. This effect was independent of sex and nationality, but frequent buyers preferred a more orange product (Gormley 1992).

The reflectance spectra of smoked and fresh Atlantic salmon are shown in Fig. 5.2. These are relatively similar, but with decreased values in smoked salmon compared to fresh salmon. The change in colour from fresh to smoked salmon can be characterised by decreased lightness ($\Delta L^* = 8 \pm 1.8$), slightly decreased redness ($\Delta a^* = 1 \pm 1.2$) and increased yellowness ($\Delta b^* = -6 \pm 1.4$) (Torrissen *et al.* 2000). On a group basis the differences in lightness were not significant between fish fed different dietary lipid levels. However, there was a significant interaction between dietary lipid level and astaxanthin source/level. The changes in redness ($p < 0.05$) and yellowness were influenced by dietary lipid level ($p < 0.001$) and by a significant interaction between dietary lipid level and astaxanthin source ($p < 0.01$). The astaxanthin losses during smoking are generally low and insignificant,

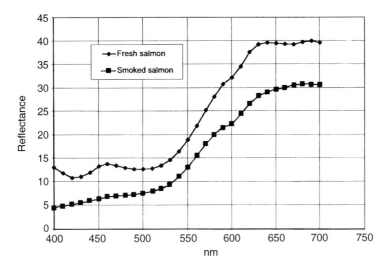

Fig. 5.2 Reflectance of fresh and smoked Atlantic salmon fillet in the visual spectrum.

but the colour losses of canthaxanthin-pigmented fish can be significant after frozen storage. It has been speculated that this can be due to crystallisation of canthaxanthin in the frozen flesh.

Fat

Torrissen *et al.* (2000) reported average lipid levels in the fillet of Atlantic salmon fed dietary lipid levels of 48%, 38% and 31% to be 17.2%, 17.0% and 16.4% respectively, where only the low lipid group gave a significantly lower lipid level in the fillet. However, the lipid levels in the fillets increased with increasing fish weight (Fig. 5.3).

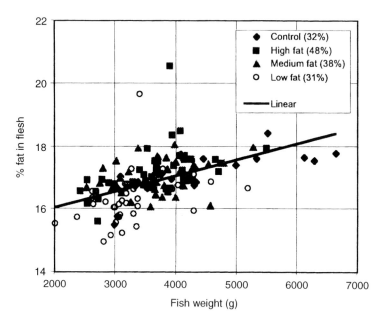

Fig. 5.3   Lipid content in the flesh of Atlantic salmon fed diets ranging in lipid level from 31% to 48%. Lipid (%) = 15.0 + 0.5* fish weight (kg). Correlation $r = 0.50$. Control is a group fed a commercial diet; the other groups are fed diets prepared by a microwave technique. (Torrissen *et al.* 2000).

Fat in relation to processing of salmon is primarily a problem in relation to lipid and water losses in vacuum packed sliced or whole smoked fillets. It is also reported that fillets with a high lipid level have a marbled appearance due to wide myocommata. A higher tendency for gaping may also occur in fillets with very high lipid levels.

Texture

The connective tissue forms a supporting network through the whole fish muscle. The content of connective tissue is lower and more evenly distributed in fish muscle compared to tissue of warm blooded animals. The content varies in different parts of the musculature because of the size and distribution of the myotomes (Dunajski 1979). Dunajski stated that water and lipids reduced the structural factors of muscle tissue and lowered its mechanical strength. Factors such as fat content, fatty acid profile and distribution of muscle fat may influence meat texture (Haard 1992). Hatae *et al.* (1990) stated that textural differences observed among different fish species are due to differences in the content of sarcoplasmic proteins and differences in the sizes of the muscle fibres. They claim that the firmness is higher in muscle with a large number of smaller muscle fibres compared to muscle with larger and relatively fewer fibres.

The post-mortem tenderisation of fish muscle has been demonstrated in several histological studies to be closely related to the degradation of collagen fibrils of the endomysium and perimysium. In the case of severe collagen breakdown, the myotomes are separated from the myocommata and gaping occurs (Bremner & Hallet 1985, 1986; Hallet & Bremner 1988; Ando *et al.* 1991). Although little evidence exists on the exact mechanism of this degradation, these same studies suggest that endogenous proteinases are involved. Such evidence has also been found in studies on the solubility of collagen from fish muscle (Sato *et al.* 1987; 1991; Montero & Borderias 1990), and in studies on sexually maturing salmon (Konagaya 1982; Ando *et al.* 1986; Yamashita and Konagaya 1990; Reid & Durance 1992; Reid *et al.* 1993).

The proteolytic degradation may be facilitated by the post mortem accumulation of lactic acid and reduced pH, as these conditions can induce leakage of proteolytic enzymes from lysosomes (Whitting *et al.* 1975) and break collagen crosslinks, thus making the collagen more available as a substrate for the proteolytic enzymes (Asghar & Yeates 1978; Montero & Borderias 1990).

Measuring texture

Different instruments have been used to determine the texture properties of food materials throughout the years, and Breene (1975) reviewed the instrumental evaluation methods on texture profile analysis (TPA). Sigurgisladottir *et al.* (1997) reviewed available instruments as well as alternative measuring procedures and evaluations of the results.

Puncture

Several different methods have been used for measuring the force needed to puncture samples by 'tooth like' objects. Ando *et al.* (1991) evaluated toughness of fish muscle by simulation of a molar tooth by using a cylindrical plunger of 8 mm diameter. The maximum force was recorded after piercing into a slice of fish muscle parallel to the orientation of the muscle fibres and was recorded as the breaking strength. The evaluation of muscle toughness showed good correlation with the measured breaking strength.

Botta (1991a,b) developed a rapid method to measure the texture of cod fillets. The method is based on measuring the sample deformation by a standard force and the rebound when the pressure is released. The firmness was calculated as rebound to deformation ratio. The method of Botta is similar to the TPA described above.

Shear force

The shear force is the force needed to cut a sample in two (Fig. 5.4), and shearing has been the most popular method for evaluation of food texture for many years (Sigurgisladottir *et al.* 1997). The Kramer shearing device is reported to give values which significantly correlate with sensory evaluations both for raw, cooked and canned fish (Gill *et al.* 1979; Karl & Schreiber 1985). The Warner-Bratzler shear apparatus measures the breaking resistance. Sigurgisladottir *et al.* (1999) concluded that shear force measurements were more sensitive than those from compression methods.

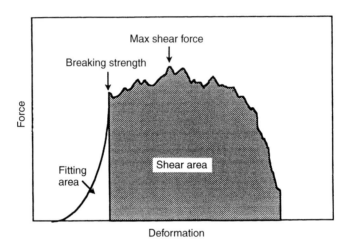

Fig. 5.4   Schematic illustration of a force versus time plot from a shear test.

Compression

Double compression (Fig. 5.5) makes it possible to calculate the TPA parameters of Bourne (1978) from the force versus time curves. TPA has been used for evaluation of textural changes of canned chum salmon in relation to sexual maturation (Durance & Collins 1991; Reid & Durance 1992). They reported that TPA correlated well with sensory evaluation of acceptability and firmness.

Tensile strength

Weinberg (1983) determined the binding properties of fish muscle by estimating the tensile force required to tear a sample of fish muscle apart by an Instron instrument after it had been secured at both ends. Tensile strength tests seem to agree fairly well with the subjective assessments of the binding properties of fresh fish fillets.

Sampling and sample preparation

The instrumental evaluation of textural properties by shearing, tensile strength, compression

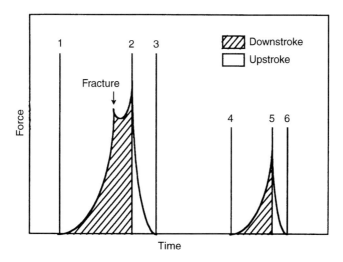

Fig. 5.5   Schematic illustration of the force–time plot of a TPA test. The events of zero or maximum compression are marked with vertical lines numbered 1 to 6, where 1–2 represents the first down stroke, 2–3 the first up-stroke, 3–4 the rest time, 4–5 the second down stroke and 5–6 the second up stroke (Veland & Torrissen 1999).

and puncture are all reported to correlate with sensory assessments of textural properties. However, this refers to evaluations made on the same fish species taken at the same position in the fish fillet and under standardised conditions, including time and storage conditions after harvest. The problem is far more complicated if the intention is to get an overall figure representing the whole fillet, and if it is to be used for describing the textural properties of a population, or of a group varying in live weight.

Veland and Torrissen (1999) measured the hardness of Atlantic salmon fillets of different sizes by pressing a sphere 6 or 10 mm into the fillet. The highest hardness was measured by 10 mm compression, but both for 6 and 10 mm the hardness decreased with increasing fillet thickness. At fillet thickness above 12 and 18 mm for 6 and 10 mm penetration respectively, this decrease was less significant (Fig. 5.6).

Sigurgisladottir *et al.* (1999) investigated differences in both shear force measured with a blade with thickness 3.0 mm (knife angle 60°) and hardness measured by a sphere with the diameter of 2.54 cm and puncture by a flat ended cylinder of 2.5 cm diameter. They measured the textural properties of Atlantic salmon at seven locations from head to tail. The conclusion from this investigation was that the hardness and shear force increased from the head to the tail in all measurements (Figs 5.8, 5.9 and 5.10). Sigurgisladottir *et al.* (1999) measured the hardness by sphere and puncture both on fillets at their natural thickness and also on samples of adjusted thickness. Sample thickness had a clear effect on the measured values (Figs 5.7, 5.8, 5.10 and 5.11). The differences in the measurements from head to tail, and also the force involved, were largest in the shear force measurements followed by the puncture with the flat ended cylinder and the sphere. The shear force measured on a standard piece of the muscle 4 cm in diameter and 2 cm thick, and the values are thus independent of fillet thickness.

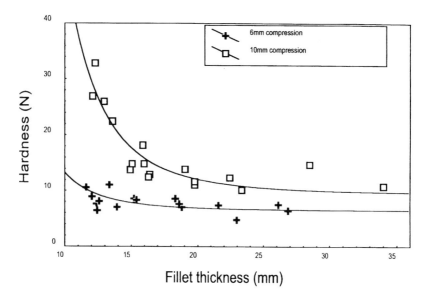

Fig. 5.6 Hardness of Atlantic salmon fillets measured by 6 and 10 mm penetration (compression) versus thickness of the fillet. (Veland & Torrissen, 1999).

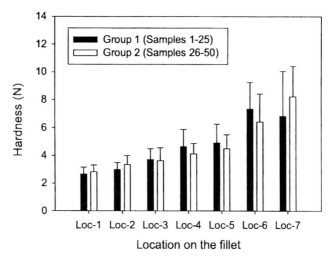

Fig. 5.7 Hardness of Atlantic salmon fillets measured by puncture with a flat ended cylinder at seven locations from the head to the tail. The samples had natural thickness, and the thickness decreased from head to tail. The data are means and standard deviations of 25 fish in two independent samplings. Locations 6 and 7 are significantly different from the other locations (Sigurgisladottir *et al.* 1999).

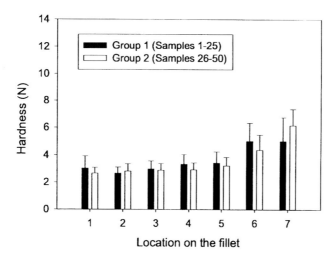

Fig. 5.8 Hardness of Atlantic salmon fillets measured by compression with a sphere at seven locations from the head to the tail. The samples had natural thickness, and the thickness decreased from head to tail. The data are means and standard deviations of 25 fish in two independent samplings. Locations 6 and 7 are significantly different from the other locations (Sigurgisladottir *et al.* 1999).

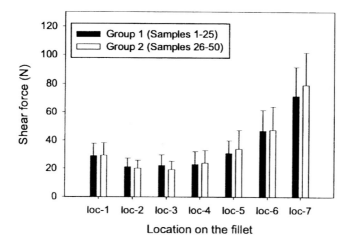

Fig. 5.9 Shear force of salmon fillets measured by cutting with a blade from the front to the back. Data are means and standards deviation of 25 fish from two independent samplings. Locations 6 and 7 are significantly different from the rest (Sigurgisladottir *et al.* 1999).

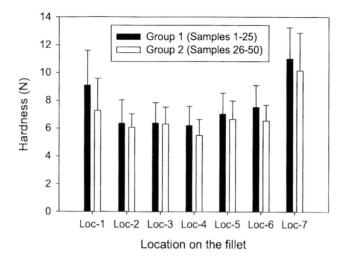

Fig. 5.10 Hardness of salmon fillets measured by puncture with a flat ended cylinder from head to tail. The thicknesses of the samples were adjusted. The data are means and standard deviations of 25 fish in two independent samplings. Values at location 7 are significantly different from the other locations (Sigurgisladottir *et al.* 1999).

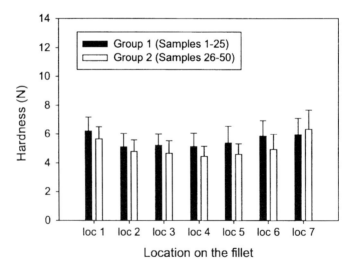

Fig. 5.11 Hardness of salmon fillets measured by compression with a sphere at seven locations from head to tail. The thicknesses of the samples were adjusted. The data are means and standard deviations of 25 fish in two independent samplings. Location 7 is significantly different from locations 2 to 6 (Sigurgisladottir *et al.* 1999).

Based on the results from Sigurgisladottir *et al.* (1999), the shear force should be measured below or in front of the dorsal fin. The shear force increases from the dorsal fin to the tail, and the location will thus influence the results. TPA measurements by sphere and puncture by a flat ended cylinder are very sensitive to differences in fillet thickness, and, based on the results from Veland and Torrissen (1999) and Sigurgisladottir *et al.* (1999), the measurements should be made on samples of the same thickness and in the central 2/4 of the fillet. Generally shear force gives the most sensitive measurements, but the drawback is that it is destructive to the fillets. However, in order to get comparable results by puncture, or compression by sphere, the thickness needs to be adjusted.

Factors determining textural properties of fish

Veland and Torrissen (1999) studied the effect of sample temperature on the hardness and work performed during a shear test (Fig. 5.12). The study was conducted by adjusting samples from the same fish to four different temperatures: 0°, 4°, 10° and 20°C, and then measuring the hardness by a sphere and shear by Warner-Bratzler knife (Figs 5.4 and 5.5). The results indicated that the sample temperature should be between 4° and 10°C.

Veland and Torrissen (1999) also studied the effect of storage of Atlantic salmon on the textural properties. They measured TPA and shear force development in fish stored on ice from 0 to 24 days after harvest. The hardness and maximum shear force are shown in Figs 5.13 and 5.14. Both the shear force and the hardness decreased from harvest until approximately four days later; thereafter the decrease in both hardness and shear force were insignificant. The relatively high firmness of the fillet during the first four days are due to *rigor mortis*.

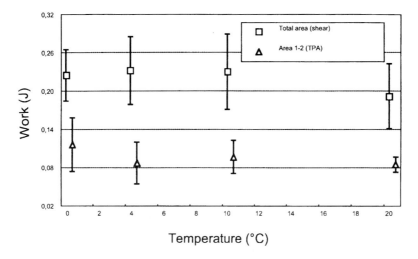

Fig. 5.12 Hardness (TPA) and total area (Shear) measured in samples from the same fish but where the sample temperatures were adjusted to 0°, 4°, and 10° and 20°C (Veland & Torrissen 1999).

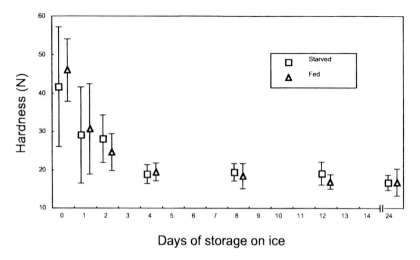

Fig. 5.13   Hardness (TPA) measured by 12.5 mm compression of Atlantic salmon fillets stored on ice (Veland & Torrissen 1999).

Veland and Torrissen (1999) compared two groups, one group fed until harvest and one group starved for 14 days prior to harvest, the latter according to industry standards. Small, non-significant, higher shear force values and hardness measurements were found in the fed group compared to the starved group at the first two sample times; thereafter the measured values were the same. The standard deviations in the first sample times were also high. This could be due to differences in onset and strength of *rigor mortis*.

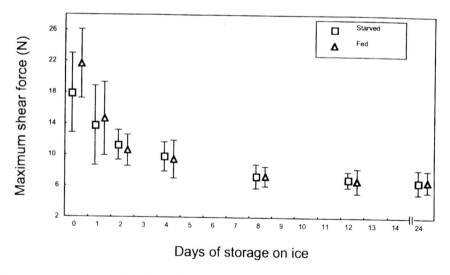

Fig. 5.14   Maximum shear force (*N*) measured in cylindrical samples (diameter 25 mm) from Atlantic salmon stored on ice (Veland & Torrissen 1999).

Sigurgisladottir *et al.* (1999) studied differences in texture properties of Atlantic salmon from different origins: sea cage reared fish, fish farmed in land-based tanks, and ocean ranched fish (comparable to wild Atlantic salmon). They found that ocean ranched fish had a higher shear force than sea cage or tank reared fish. Between the latter there were no significant differences. Hardness values measured by sphere were not significantly different between any treatment.

References

Ando, M., Toyohara, H. & Sakaguchi, M. (1991) Post-mortem tenderization of rainbow trout muscle caused by the gradual disintegration of collagen fibres in the pericellular connective tissue. *Nippon Suisan Gakkaishi,* **58**, 567–70.

Ando, S., Hatano, M. & Zama, K. (1986) Protein degradation and protease activity of chum salmon (*Oncorhynchus keta*) muscle during spawning migration. *Fish Physiology and Biochemistry*, **1**, 17–26.

Asghar, A. & Yeates, N.T.M. (1978) The mechanism for promotion of tenderness in meat during the post-mortem process: a review. *CRC Critical Reviews in Food Science and Nutrition*, **8**, 115–45.

Botta, J.R. (1991a) Instrument for nondestructive texture measurement of raw Atlantic cod (*Gadus morhua*) fillets. *Journal of Food Science*, **56**, 962–8.

Botta, J.R. (1991b) Method of measuring the firmness of meat. Patent, International patent no. WO 91/08479.

Bourne, M.C. (1978) Texture profile analysis. *Food Technology*, **7**, 62–6.

Breene, W.M. (1975) Application of texture profile analysis to instrumental food texture evaluation. *Journal of Texture Studies*, **6**, 53–82.

Bremner, H.A. & Hallet, I.C. (1985) Muscle fiber-connective tissue junctions in the fish Blue Grenadier (*Macruronus noevazelandiae*). A scanning electron microscope study. *Journal of Food Science*, **50**, 975–80.

Bremner, H.A. & Hallet, I.C. (1986) Muscle fibre-connective tissue junctions in the Spotted Trevalla (*Seriolella punctata*) examined by Scanning Electron Microscopy. *Journal of Food Science and Agriculture*, **37**, 1011–18.

Børresen, T. (1986) Fish texture. A discussion paper submitted to WEFTA working group on analytical methods for fish products. pp.1–14. *Fiskeriministeriets forsøkslaboratorium*, Lyngby, Denmark.

Dunajski, E. (1979) Texture of fish muscle. *Journal of Texture Studies*, **10**, 301–18.

Durance, T.D. & Collins, L.S. (1991) Quality enhancement of sexually mature chum salmon *Oncorhynchus keta* in retort pouches. *Journal of Food Science*, **56**, 1281–6.

Einen, O. & Roem, A.J. (1997) Dietary protein/energy ratios for Atlantic salmon in relation to fish size: growth, feed utilisation and slaughter quality. *Aquaculture Nutrition*, **3**, 115–26.

Gill, T.A., Keith, R.A. & Lall, B.S. (1979) Textural deterioration of red hake and haddock muscle in frozen storage as related to chemical parameters and changes in the myofibrillar proteins. *Journal of Food Science*, **44**, 661–7.

Gormley. T.R. (1992) A note on consumer preference of smoked salmon colour. *Irish Journal of Agricultural and Food Research,* **31**(2), 199–202.

Hallet, I.C. & Bremner, H.A. (1988) Fine structure of myocommata-muscle fibre junction in hoki (*Macruronus novazelandiae*). *Journal of the Science of Food and Agriculture,* 44, 245–61.

Hatae, K., Yoshimatsu, F. & Matsumoto, J.J. (1990) The role of muscle fibres in contributing firmness of cooked fish. *Journal of Food Science,* 55, 693–6.

Haard, N.F. (1992) Control of chemical composition and food quality attributes of cultured fish. *Food Research International,* 25, 1–19.

Karl, H. & Schreiber, W. (1985) Textural analysis of canned fish. *Journal of Texture Studies,* 16, 271–80.

Konagaya, S. (1982) Enhanced protease activity in muscle of chum salmon (*Oncorhynchus keta*) during spawning migration. *Bulletin of the Japanese Society of Scientific Fisheries,* 48, 1503.

Montero, P. & Borderias, J. (1990) Effect of *rigor mortis* and ageing on collagen in trout (*Salmo irrideus*) muscle. *Journal of the Science of Food and Agriculture,* 52, 141–6.

Reid, R.A. & Durance, T.D. (1992) Textural changes of canned chum salmon related to sexual maturity. *Journal of Food Science,* 57, 1340–2.

Reid, R.A., Durance, T.D., Walker, D.C. & Reid, P.E. (1993) Structural and chemical changes in muscle of chum salmon (*Oncorhynchus keta*) during spawning migration. *Food Research International,* 26, 1–9.

Rora, A.M.B., Kvale, A., Morkore, T., Rorvik, K.A., Steien, S.H. & Thomassen, M.S. (1998) Process yield, colour and sensory quality of smoked Atlantic salmon (*Salmo salar*) in relation to raw material characteristics. *Food Research International,* 31, 601–609.

Sato, K., Yoshinaka, R., Sato, M. & Shimizu, Y. (1987) Isolation of native acid-soluble collagen from fish muscle. *Nippon Suisan Gakkaishi,* 53, 1431–6.

Sato, K., Ohashi, C., Ohashi, C., Ohtsuki, K. & Kawabata, M. (1991) Type V collagen in trout (*Salmo gairdneri*) muscle and its solubility change during chilled storage of muscle. *Journal of Agricultural and Food Chem*istry, 39, 1222–5.

Sheehan, E.M.. O'Connor, T.P., Sheehy, P.J.A., Buckley, D.J. & Fitzgerald. R. (1996) Effect of dietary fat intake on the quality of raw and smoked salmon. *Irish Journal of Agricultural and Food Research,* 35, 37–42.

Sigurgisladottir, S., Torrissen, O., Lie, Ø., Thomassen, M. & Hafsteinsson, H. (1997) Salmon quality: methods to determine the quality parameters. *Reviews in Fisheries Science,* 5, 223–52.

Sigurgisladottir, S., Hafsteinsson, H., Jonsson, A., Lie, Ø., Nortvedt, R., Thomassen, M. & Torrissen, O. (1999) Textural properties of raw salmon fillets as related to sampling method. *Journal of Food Science,* 64, 99–104.

Storebakken, T. & No, H.K. (1992) Pigmentation of rainbow trout. *Aquaculture,* 100, 209–29.

Torrissen, O.J., Hardy, R.W. & Shearer, K.D. (1989) Pigmentation of Salmonids – carotenoid deposition and metabolism. *CRC Critical Reviews in Aquatic Sciences,* 1, 209–25.

Torrissen, O.J., Hemre, G.-I. & Sandnes, K. (2000) High energy diets – effect of flesh pigmentation and quality of smoked Atlantic smoked fillets. *Aquaculture Nutrition* (submitted).

Veland, J.O. & Torrissen, O.J. (1999) The texture of Atlantic salmon (*Salmo salar*) muscle as measured instrumentally using TPA and Warner-Bratzler shear test. *Journal of the Science of Food and Agriculture,* 79, 1737–46.

Weinberg, Z.G. (1983) A comparison of the binding properties of fish flesh. *Journal of Food Technology*, **18**, 441–51.

Whitting, R.C., Montgomery, M.W. & Anglemeir, A.F. (1975) Stability of rainbow trout muscle lysosomes and their relationship to rigor mortis. *Journal of Food Science*, **40**, 854–7.

Yamashita, M. & Konagaya, S. (1990) Participation of Cathespin L. into extensive softening of the muscle of chum salmon caught during spawning migration. *Nippon Suisan Gakkaishi*, **56**, 1271–7.

Chapter 6

Pigmentation of Farmed Salmonids

D.C. Nickell[1] and J.R.C. Springate[2]

[1] *Roche Products Ltd, Heanor, Derbyshire, DE75 7SG, UK*
[2] *F. Hoffmann-La Roche Ltd, VMA, CH-4070 Basel, Switzerland*

Introduction

Intensive aquaculture by definition requires farmed animals to be fed a complete diet that includes all the macro and micro-ingredients that are essential to growth, health and appearance. For end-product quality and marketing, carotenoids are among the most important micro-ingredients used in many intensive aquaculture systems. This group of compounds have been linked to many important biological functions associated with their strong antioxidant properties, but here their important role as biological colorants only will be discussed.

Carotenoids are a group of naturally occurring organic pigments that are responsible for the red, orange and yellow colours in the skin, flesh, shell and exoskeleton of aquatic animals. The word pigment originates from the Latin word *pigmentum,* which refers to painting materials and cosmetics, and implies a notion of colour (Shahidi *et al.*1998). Any substance which can impart colour to the tissues or cells of animals or plants can be called a pigment. The name of this group of pigments comes from the carrot (*Daucus carota* L.) from which the yellow-orange pigment β-carotene was isolated by Wackenroder in a crystalline form as long ago as 1831.

Carotenoids are widespread in plants and animals and over 600 have been characterised to date. They are responsible for the colours of many fruits (tomatoes, paprika, chillies, citrus fruits), vegetables (carrots, potatoes), flowers (water lily, rose hips, tagetes), many birds (flamingo, ibis, oriole, canary) insects (ladybird, Colorado beetle), fungi (chanterelle mushrooms) and aquatic animals (salmon, trout, shrimps, lobster, goldfish). Although carotenoids occur in low concentrations, the total carotenoid production in nature has been estimated to be over 100 million tonnes per year (Isler *et al.* 1967), with the majority being synthesised via the photosynthetic pathway and subsequently being stored in leaves, algae and zooplankton. However, higher organisms, including salmonids, cannot synthesise carotenoids and therefore rely on a dietary intake. Under intensive culture conditions, the relevant carotenoid is included in the complete feed in order to produce the desired flesh colour. This is essential if the farmed fish product is to mimic its 'wild' counterpart and have maximum consumer appeal.

The major carotenoid in the aquatic system is astaxanthin (3,3'-dihydroxy-β,β-carotene-4,4'-dione). Attempts to isolate the molecule and map its structure in the 1930s failed because of the common practice of treating carotenoid extracts with alkali (Bjerkeng, 1997). Kuhn and Sørensen finally elucidated the structural constitution of the pigment in an isolate from the lobster, *Astacus gamarus*, in 1938 and thus named it astaxanthin after the lobster's Latin name. With the exception of birds, in which canthaxanthin and astaxanthin are almost equally common, astaxanthin is the predominant red carotenoid in all animal groups (Shahidi *et al.*, 1998) and is the major pigment in salmon flesh (Schiedt *et al.* 1981) (Fig. 6.1).

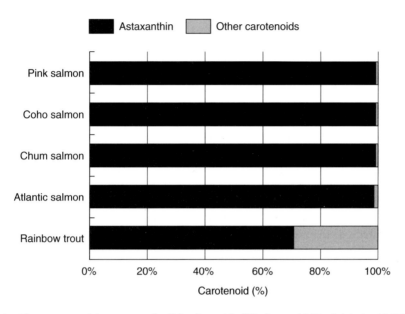

Fig. 6.1 The carotenoid content of wild salmonids (Kitahara, 1983; Schiedt, 1987).

The other carotenoid that is sometimes used for pigmentation in aquaculture is canthaxanthin. It occurs in small quantities in the aquatic environment, although is rarely encountered in wild Atlantic salmon (*Salmo salar*) (Bjerkeng 1997). Canthaxanthin was first detected as the major colouring principle in the edible chanterelle mushroom, *Cantharellus cinnabarinus*, from which it derived its name (Haxo 1950). Canthaxanthin has been found in algae and also in low levels in shrimp and fish. It is generally found in low concentrations since it is often an intermediate in the metabolism of ß-carotene to astaxanthin.

The distinctive pink-red colour of salmon and trout flesh, produced by the muscle retention of astaxanthin, provides an immediate indication of product quality and is, therefore, an essential factor in product marketing. It has long been recognised that the colour in fish is much more than a cosmetic effect and that the consumer associates natural coloration of a product with health and high quality (Schiedt 1998). The wide availability of salmonid products has led to an increase in the demands of the processors and end consumers for the flesh to be strongly and uniformly pigmented. It is, therefore, important that during the farming process the chosen carotenoid and its form is effectively used to produce fish with the desired flesh colour that will satisfy customers' demands. Intensive aquaculture systems allow the farmer complete control over the animal's feed intake, thereby making it possible to produce uniform and optimal colour that will meet the consumer's expectations.

The increased importance placed on flesh colour means that farmers must endeavour to pigment all their fish to the minimum level that is acceptable to the market. However, throughout the growth of the fish, the pigmentation of the flesh is influenced by a variety of exogenous factors including dietary pigment concentration, dietary source, carotenoid type and feeding period. Moreover, endogenous factors such as growth rate, genetics and stage of maturation also influence flesh pigmentation. The combination of these exogenous and endogenous factors collectively results in high variation in flesh pigmentation in any one population and a subsequent need for careful management of pigment regimes.

The biological limits in the retention efficiency of carotenoids in salmonids and the inherent inter and intra-stock variability in flesh colour, influence the ability of the farmer to pigment any given population of fish to an acceptable level without incurring economic loss due to product rejection. Understanding the pigmentation process is an important area of research and has an influence on the development and implementation of specific pigmentation regimes designed to achieve target carotenoid levels in the flesh of farmed salmonids at the most economical rate.

Carotenoid chemistry

Most carotenoids are polyunsaturated C40 hydrocarbons, comprising two terminal ring systems. Carotenoids that are composed entirely of carbon and hydrogen are known as carotenes (such as β-carotene), while those that contain oxygen are termed xanthophylls (such as astaxanthin). It is the long chain of several conjugated double bonds that is of special interest, as it constitutes the chromophore. It is this long chain of conjugated double bonds that shifts the wavelength of maximum absorption into the visible range of 400–600 nm (Davies *et al*. 1960). Consequently, carotenoids which have 7–15 conjugated double bonds appear yellow to red in colour. Furthermore, the chromophore system can be influenced by additional double bonds and various other functional groups (Latscha 1990).

Since carotenoids consist of several conjugated double bonds and cyclic end groups, there is a loss of free rotation about certain C-C bonds. This, along with the presence of various asymmetric (chiral) C-atoms, causes carotenoids to adopt a variety of configurations and stereoisomers that have different chemical and physical properties. The most important carotenoid forms are the geometric (cis/trans or Z/E) and the optical isomers. In nature carotenoids are usually present as the all-E isomer; however isomerisation is a common occurrence especially when the carotenoid is subjected to heat or light. The two most common geometric isomers of astaxanthin are 9Z and 13Z.

The three possible optical isomers of astaxanthin are designated (3S,3'S), (3R,3'S) and (3R,3'R). Optical isomerisation is a result of the orientation plane of the hydroxyl groups on the third carbon of the terminal rings. These small differences cause subtle colour changes due to changes in light refraction. These small differences in light refraction are also seen in geometric isomers. In salmonids the predominant geometrical and optical isomers are the trans (E) and 3S,3'S forms, respectively (Schiedt *et al*. 1981). The bioavailability and apparent digestibility of the all-E-astaxanthin isomer has been shown to be higher than that of the 9Z and 13Z-astaxanthin isomers in rainbow trout (Bjerkeng *et al.* 1997; Østerlie *et al.* 1999). Moreover, this difference between geometric isomers resulted in significantly redder flesh in fish fed all-E-astaxanthin (Bjerkeng *et al.* 1997).

Carotenoid esters are also found in nature (Renstrom *et al.* 1981), but often have reduced bioavailability compared to the free carotenoid form (Storebakken *et al.* 1987). Of importance is that astaxanthin is usually deposited in the free form in the muscle and fatty tissue of salmon and trout, although in the skin and eggs it is usually deposited as mono- and diesters (Storebakken & No 1992). Furthermore, the attachment of protein molecules to the carotenoid structure can also result in the visualisation of a variety of colours. This process is most frequently observed in crustacea. In the lobster, astaxanthin is attached to a protein to produce the carotenoprotein, crustacyanin. This carotenoprotein imparts a blue colour in the living lobster, but when heated the carotenoprotein molecule is cleaved which results in the characteristic red colour of cooked lobsters.

Although free carotenoids are often more bioavailable they are nevertheless very unstable and are highly sensitive to oxygen, light, acids, bases and heat. When subjected to adverse conditions, carotenoids are highly susceptible to a variety of oxidative conversion and degradation processes, or isomerisation. Commercially available forms of free carotenoids are formulated to improve the stability of these highly oxidative compounds.

Carotenoid sources

The main carotenoid used throughout the world in intensive aquaculture is astaxanthin. There are a number of alternative sources of astaxanthin, although the major source (over 95%) is the nature-identical, synthesised, non-esterified, free form. These forms contain a high content of all-E astaxanthin and are uniformly formulated into stable, highly useable and proven products that currently contain a minimum 8–10% astaxanthin activity.

Alternative astaxanthin sources are krill, shrimp, crawfish, basidiomycetes, yeast and algae (Fig. 6.2). However, these sources have a highly variable astaxanthin content and reduced bioavailability (Torrissen *et al.* 1989; Sommer *et al.* 1991; Gouveia *et al.* 1996). Shrimp meals and algae have been used as astaxanthin sources for salmonid pigmentation (Torrissen *et al.* 1981; Choubert & Heinrich 1993). These sources also contain a high level of esterified astaxanthin with resultant decreased bioavailability (Johnson & An 1991; Aas 1998). In order to attain adequate flesh pigmentation using material of reduced bioavailability and low activity a high dietary inclusion level is required (Torrissen *et al.* 1989; Johnson & An 1991). In contrast, the majority of astaxanthin found in *Phaffia rhodozyma* yeast is non-esterified, although it appears as the 3R,3'R isomer (Andrewes & Starr 1976). However, the bioavailability of yeast astaxanthin compared to that derived from synthetic sources is often reduced due to the indigestible cell walls. Lysing of these cell walls by enzymatic treatment or mechanical rupture can increase the astaxanthin bioavailability (Gentles & Haard 1991).

In addition to existing commercial astaxanthin sources, earlier studies have examined the use of some higher plants (*Adonis aestivalis*) for salmonid pigmentation, but have also found that these sources contain a large proportion of esterified astaxanthin and anti-nutritional factors (Kamata *et al.* 1990) and are therefore of limited commercial value. In contrast to

(a)

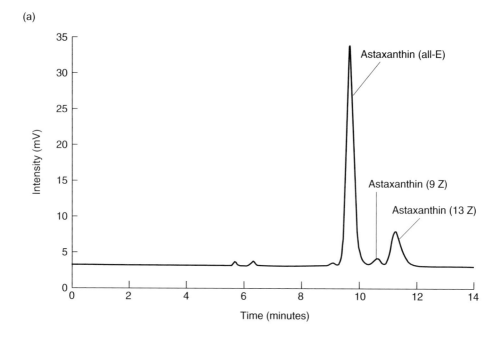

(b)

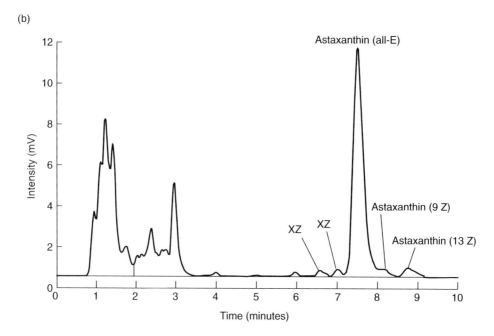

Fig. 6.2 (a–d)   Chromatograms of the carotenoid content of different pigment sources.

(c)

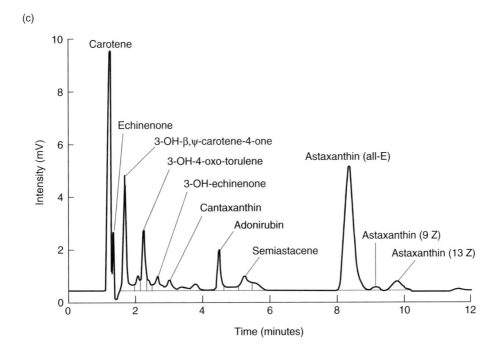

(d)

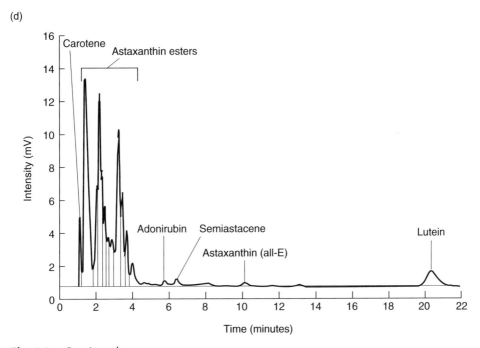

Fig. 6.2   *Continued.*

astaxanthin, there is only one source of canthaxanthin used for intensive aquaculture; this is a commercially synthesised, non-esterified product.

The importance of flesh colour

Visual appearance, especially colour, is the most important characteristic of foods in determining their selection prior to actual consumption (Chichester 1981). Buying decisions are frequently based on the appearance of the product. Given two food products of equal price, and perceived nutritional equality, the consumer will choose the most visually appealing product. Packaging and/or display can enhance appearance but often it is the appearance of the product itself that leads to the purchase. This is particularly true of some of the economically important salmonids. Ostrander *et al.* (1976) used a trained consumer panel to assess samples of Pacific salmon and rainbow trout (*Oncorhynchus mykiss*). Colour was considered to be of prime importance to both panels in determining their preference for salmon and trout. Farmed salmon, which did not have a typical salmon colour, were not considered as being different from non-pigmented trout by sensory panels. The authors concluded that, in the future, aquaculturists would have to pay more attention to the sensory attributes of the end product.

The intensive culture of fish has so far been limited to the high value, so-called 'luxury fish', but as culture methods have become successful and supply exceeds demand then prices of some farmed species, such as Atlantic salmon and rainbow trout, have dropped. The reduction in market price has focused the attention of fish farmers on investigating all possible methods of improving production efficiency and reducing costs. Therefore, effective feed and pigment management is vitally important to the economics of intensive aquaculture. While ensuring economical feed and pigment use, optimal flesh quality characteristics must not be ignored. Flesh quality parameters such as freshness, colour and fat levels are particularly important to consumers and must be monitored.

Colour assessment of fish stocks is a destructive and therefore expensive process and often quality control samples are not large enough to assess accurately the colour variation within one stock. Poor sampling and an inadequate pigment regime can lead to severe problems at harvest time when it is too late to return sub-optimal coloured fish for pigmentation improvement. It is not possible to pigment dead fish. Therefore, understanding the processes that affect flesh quality and being able to predict the general quality of any one harvest of fish, is vitally important to the salmonid farming industry.

One of the most important flesh quality characteristics is flesh colour and the problems associated with the large inter and intra-stock variability in pigmentation. Such variations in pigmentation are present within and between different fish stocks on individual farms and between farms growing the same genetic stock, consequently making it difficult to manage pigmentation regimes effectively. Because of the requirement for all-year round supplies of product of consistent size and quality, these seasonal, inter-farm and intra-stock variations in flesh colour can cause the farmer problems, with significant proportions of product being downgraded or rejected by processors and retailers. This problem has become more of an issue since greater quantities of farmed salmonids are now being processed compared to the beginning of the decade. Consequently, minimising downgrading and rejection of fish due to insufficient pigmentation, while attempting to increase the efficiency of carotenoid utilisation through the implementation of specific pigmentation regimes, is an area of interest to the researcher and farmer alike.

Utilisation of carotenoids

The deposition of carotenoids in the flesh of salmonids is influenced by several endogenous factors. These include the digestibility of the carotenoid (Foss *et al.* 1987; Choubert *et al.* 1995), the absorption from the intestine (Choubert *et al.* 1987; Hardy *et al.* 1990; Torrissen *et al.* 1990), transport in the blood by lipoprotein (Ando *et al.* 1985; Nakamura *et al.* 1985; Choubert *et al.* 1994a), its metabolism (Kitahara 1983; Schiedt *et al.* 1985, 1986; Ando *et al.* 1989; Guillou *et al.* 1992) and attachment to the muscle fibre (Henmi *et al.* 1987, 1989, 1990). However, these separate processes can each significantly influence flesh carotenoid concentrations and colour visualisation and therefore any limitation in one part of the process may result in insufficient flesh pigmentation.

The carotenoid form and raw material matrix (feed) in which it is presented have an effect on the digestibility of the pigments and their subsequent efficacy in flesh pigmentation. The digestibility of carotenoids will have an influence on the necessary dietary inclusion rate and the regime employed for flesh pigmentation.

The pigment form and diet composition has been shown to have an effect on carotenoid digestibility (Choubert *et al.* 1991). The astaxanthin apparent digestibility coefficient (ADC) in rainbow trout (*Oncorhynchus mykiss*) has been reported to be greater than 70% and shown to be higher than that of canthaxanthin (Choubert *et al.* 1995; Choubert & Storebakken 1996). Values obtained for canthaxanthin have been reported to range from 35–70% in rainbow trout (Foss *et al.* 1987; Torrissen *et al.* 1990; Choubert *et al.* 1991), which agrees with those reported for Atlantic salmon (*Salmo salar*) (Storebakken *et al.* 1987). Moreover, the ADC of synthetic astaxanthin, astaxanthin dipalmitate and canthaxanthin fed to rainbow trout, Atlantic salmon and sea trout (*Salmo trutta*) were shown to exhibit large variation (Foss *et al.* 1987; Storebakken *et al.* 1987). These variations in digestibility results have been linked to destruction of carotenoids in the gut, or during feed storage and the incomplete extraction of faeces from experimental fish (No & Storebakken 1991). However, differences in ADC and pigmenting efficiency are not only limited to carotenoid type, but also between geometric isomers of the same carotenoid (Bjerkeng *et al.* 1997; Østerlie *et al.* 1999). Nevertheless, of importance is that these differences in carotenoid digestibility can influence the inclusion rate of specific carotenoids in most diets.

The digestibility of the raw material in which the carotenoid is found can also influence carotenoid bioavailability. This was clearly demonstrated by Johnson *et al.* (1980) when rainbow trout were fed whole or lysed *Phaffia rhodozyma* yeast cells. Similar results were obtained by Torrissen *et al.* (1981), when studying astaxanthin digestibility from different shrimp waste preparations. Increased astaxanthin digestibility was achieved using ensiled shrimp waste meal compared to fresh material. Therefore, the pigment form and its formulation are vitally important to carotenoid digestibility. As a consequence the aquaculture industry uses products that are specifically formulated to ensure optimal stability and bioavailability. Because of the low astaxanthin content in yeast, algal and shrimp products, high dietary inclusion levels have to be used. Such high inclusion levels can affect the formulation and fabrication of modern high-energy diets. Furthermore, pigment sources that contain a mixture of esterified forms of astaxanthin may increase the feeding period and carotenoid dose necessary to achieve a desired flesh colour (Choubert *et al.* 1995).

An increase in dietary lipid has been shown to lead to improved pigmentation in Atlantic salmon (Einen & Roem 1997; Einen & Skrede 1998). Similarly, studies on rainbow trout have demonstrated that an increase in dietary fat concentration results in increased carotenoid digestibility (Choubert *et al.* 1991), absorption (Torrissen *et al.* 1990) and retention efficiency (Torrissen 1985; Nickell & Bromage 1998a). As a consequence the use of high oil diets can influence pigmentation regimes. Such effects on carotenoid utilisation will ultimately affect the amount of carotenoid that has to be fed in order to ensure the required amount is available for absorption and subsequent deposition in the flesh.

The efficiency of pigment retention has been reported to vary according to carotenoid type. The free form of astaxanthin has been shown to be 1.3 to 2.5-fold more efficient than free canthaxanthin in rainbow trout flesh pigmentation (Foss *et al.* 1984; Choubert & Storebakken 1989; Bjerkeng *et al.* 1990; Torrissen & Christiansen 1991; Choubert *et al.* 1995). This increased pigmenting efficiency of astaxanthin could be partly explained by differences in the absorption and plasma kinetics of the two carotenoids (Schiedt *et al.* 1985; Guillou *et al.* 1992; Choubert *et al.* 1994b; Gobantes *et al.* 1997). The work of Gobantes *et al.* (1997) demonstrated in rainbow trout that ingestion of astaxanthin resulted in greater absorption, higher peak plasma concentrations and a 2.4 fold mean plasma retention time compared to the same dose of canthaxanthin. Subsequently, achieving and maintaining higher plasma carotenoid concentrations may result in greater flesh pigmentation (Storebakken & Goswami 1996).

The main area for carotenoid absorption in rainbow trout is in the pyloric caeca, although significant amounts are absorbed in the posterior intestine (Torrissen *et al.* 1990). However, the actual process of absorption has not been elucidated, although active, selective and passive uptake mechanisms have been proposed (Bird & Savage 1989; Bjerkeng *et al.* 1997; Choubert *et al.* 1994b; Kiessling *et al.* 1995; Storebakken & Goswami 1996). Nevertheless, the deposition of free astaxanthin in the flesh does indicate that the carotenoid is absorbed in its free form (Torrissen & Braekkan 1979). Understanding the absorption mechanism and determining the factors that have an influence may enable manipulation of the process and greatly improve the amount of carotenoid that can be absorbed; this could ultimately improve the efficiency of salmonid flesh pigmentation. However, within any population there is considerable variation in food intake (McCarthy *et al.* 1992) and differences in the ability of individual fish to absorb astaxanthin from the intestine (Choubert *et al.* 1994b; Gobantes *et al.* 1997; Nickell 1997). As such, these factors are unavoidable and could be a possible cause of the observed variation in flesh colour within and between salmonid stocks.

Following digestion, absorbed carotenoids and triacylglycerols are packaged into chylomicrons in the enterocytes and eventually released into the blood. The chylomicrons are metabolised and over time the carotenoids become incorporated into a variety of different density lipoproteins, which vary in function depending on the physiological state of the fish (Fremont & Marion 1982; Babin & Vernier 1989). As such, this dynamic lipoprotein system could be a limiting factor in the transport and deposition of carotenoids in the flesh of salmonids.

Several studies have examined the lipoprotein transport of astaxanthin in salmonids (Ando *et al*. 1986; Choubert *et al*. 1994a; Nickell 1997), but little is known about the effect of diet and season on the distribution of astaxanthin within the different lipoprotein classes. However, work by Wallaert and Babin (1994) reported a circannial variation in the circulating low-density lipoprotein concentration of rainbow trout, which was unrelated to water temperature but related to the annual photocycle. Any rhythmical variation in lipoprotein levels could ultimately affect the amount of astaxanthin in circulation and its delivery to peripheral tissues, which may directly influence flesh pigmentation. Other studies have also demonstrated a seasonal variation in lipoprotein levels in rainbow trout (Nickell, 1997). However, other factors such as dietary lipid level appear to have little influence on the lipoprotein transport of astaxanthin (Nickell 1997). Similar results have been found for α-tocopherol (Lie *et al*. 1994) and would suggest that there is a non-specific association of astaxanthin with the different lipoproteins (Ando *et al*. 1986; Choubert *et al*. 1994a) or that specific plasma proteins are involved in the blood transport of carotenoids (Aas 1998). Nevertheless, a greater understanding of the blood transport of astaxanthin throughout the life of the fish may provide important information concerning the utilisation of carotenoids and how differences in transport could affect pigmentation regimes and the resulting flesh colour.

It has been demonstrated that carotenoids bind non-specifically to actomyosin hydrophobic binding sites by weak hydrophobic bonds (Henmi *et al*. 1990). Furthermore, salmonid muscle has the ability to bind various carotenoids due to the many hydrophobic binding sites on the actomyosin (Henmi *et al*. 1990). However, *in vivo* salmonid muscle shows a preferential accumulation of free astaxanthin (Schiedt *et al*. 1985; Bjerkeng *et al*. 1992) which could indicate a preferential transfer of the xanthophyll across the muscle cell plasma membrane. Moreover, the hydroxyl and keto groups on the β–end of the carotenoid molecule increase the strength of binding to the muscle (Henmi *et al*. 1989). This additional binding strength could explain the higher deposition of astaxanthin as compared to canthaxanthin, which lacks the hydroxyl function although it has a keto function. The weak, non-specific binding of astaxanthin with salmonid actomyosin requires the presence of binding sites. An absence or uneven distribution of binding sites within the muscle could result in poor or patchy flesh pigmentation and would also affect the efficiency of carotenoid retention.

Integrally linked to pigmentation is growth and muscle texture, which is greatly influenced by muscle fibre recruitment and hyperplasia. Pioneering work by Johnston *et al*. (2000) has begun to elucidate the relationship between muscle fibre density and pigment visualisation and meat texture. The visualisation of pigment is greater in fish with a higher muscle fibre density compared to those with a lower density, for a given pigment concentration. Earlier studies by Nickell and Bromage (1998a,b), suggested a relationship between the deposition of astaxanthin and muscle development in the rainbow trout. Muscle growth is an important factor influencing pigment deposition and subsequent colour visualisation. Since muscle growth does not proceed at a constant rate throughout the development of the fish (Johnston 1982; Totland *et al*. 1987; Kiessling *et al*. 1991) seasonal variations in pigment utilisation (Nickell 1997), deposition (Torrissen *et al*. 1995; Einen & Thomassen 1998) and subsequent colour visualisation could be expected. Ultimately, this could have implications for the implementation of optimal pigmentation regimes.

Maturation has an effect on flesh carotenoid concentrations and colour visualisation (Torrissen & Naevdal 1988; Bjerkeng *et al.* 1992; Hatlen *et al.* 1995). The effects of maturation on flesh pigment levels in Atlantic salmon can start five to six months prior to the event (Torrissen & Torrissen 1985) and in rainbow trout differences in flesh carotenoid loss and recovery have been reported between sexes (Choubert & Blanc 1993). It is therefore important to understand the stage of maturation of any one stock when applying specific pigmentation regimes.

Carotenoid deposition and pigmentation regimes

The complex nature and variety of possible limitations in the pigmentation process affect the overall efficiency of carotenoid deposition. The efficiency of utilisation of dietary astaxanthin for flesh pigmentation in Atlantic salmon and rainbow trout rarely exceeds 10–17% (Torrissen *et al.* 1989; Storebakken & No 1992; Nickell & Bromage 1998a) and would suggest that there are inherent limits in the process of salmonid flesh pigmentation. However, pigmentation retention efficiency is affected by carotenoid dose and feeding duration. Several studies have demonstrated that low doses of astaxanthin fed over a long period of time resulted in high retention efficiencies (Smith *et al.* 1992), although not necessarily attaining the target level of carotenoid in the muscle (Streiff 1984 unpublished observation; Torrissen 1995). Other work has demonstrated that 40–60 mg/kg astaxanthin is the most efficient dietary dose for optimal pigment retention in Atlantic salmon (Storebakken *et al.* 1987; Torrissen 1995; Torrissen *et al.* 1995), while in freshwater rainbow trout the level has been reported to range from 50–70 mg/kg astaxanthin (Bjerkeng *et al.* 1990, 1992; March & MacMillan 1996). Consequently, there is a balance between maximising pigmentation efficiency and obtaining maximum flesh carotenoid levels.

There are two different models that describe the rate of astaxanthin deposition in salmonid flesh (Torrissen *et al.* 1989). The linear model of retention predicts increasing retention efficiency with increasing fish size and has been described in short-term studies (Spinelli & Mahnken 1978; Torrissen 1985; Foss *et al.* 1987; Storebakken *et al.* 1987). Linear retention efficiency would enable adequate flesh pigmentation to be achieved later on in the production cycle by feeding high astaxanthin concentrations. In contrast, the curvilinear model suggests constant retention efficiency throughout the life of the fish, and predicts a rapid initial increase in flesh pigment concentration for small fish followed by a plateau level, as the fish become larger. Furthermore, this model would indicate that a low dietary astaxanthin dose should be fed over a long feeding period for most efficient flesh pigmentation (Storebakken *et al.* 1986; Torrissen *et al.* 1989, 1995).

In addition there are clear differences in carotenoid retention efficiencies among salmonid species; the retention of carotenoids by Pacific salmonids (including rainbow trout) is much greater than that of Atlantic salmon (March & MacMillan 1996). Furthermore, intra-specific differences in pigmentation have been reported in chinook salmon (McCallum *et al.* 1987) and as such these inter and intra-specific differences are probably genetically predetermined (Gjerde & Gjedrem 1984; Torrissen & Naevdal 1988; Iwamoto *et al.* 1990; Blanc & Choubert 1993). Consequently, recommending the optimal dietary astaxanthin concentration for maximum pigment retention is difficult due to these inter and intra-specific differences, and is further complicated by factors such as carotenoid feeding duration, deposition rate and the required target flesh colour.

The target flesh colour of salmonids can be determined visually using the Roche *Salmo-*Fan™, which is an internationally recognised standard colour assessment tool. It is important that the different salmon growing countries are able to match the standard colour requirements of the different markets worldwide. However, for different salmonid species to attain similar *Salmo*Fan™ scores, different levels of astaxanthin have to be deposited in the flesh (Table 6.1). To achieve these species-specific flesh astaxanthin levels requires the implementation and constant management of particular pigmentation regimes.

Table 6.1   Estimated minimum level of flesh astaxanthin to reach target colour level (unpublished Roche Chile, 1997; guide only)* Large sea-reared.

Species	Roche colour card	Roche *Salmo*Fan™	Flesh astaxanthin (mg/kg)
Coho salmon	16–17	31–33	16–18
Rainbow trout*	17–18	31–33	20–22
Atlantic salmon	15–16	29–31	8–10

At present, there are many pigmentation regimes employed worldwide to obtain optimal flesh colour while satisfying the requirements of the target market. In some countries there is a general trend to reduce dietary astaxanthin levels from 60–70 mg/kg to 30–70 mg/kg once the fish have achieved approximately 7.5 mg/kg astaxanthin in the flesh at a wet weight of approximately 2 kg. The general trend in salmon farming is for growth rates to increase and food conversion ratios (FCR) to reduce. This would indicate that over a shorter grow-out period the fish would receive less dietary pigment. Unless there are major gains in terms of the biological utilisation of carotenoids, the net effect is one of a downward trend in pigment intake. Moreover, studies have demonstrated the positive relationship between dietary astaxanthin dose and plasma astaxanthin concentrations (Choubert *et al.* 1994b) and the subsequent flesh carotenoid retention (Storebakken & Goswami 1996). Reducing dietary carotenoid concentrations to low levels during the major growth period of the fish would imply that flesh pigmentation would reduce or not increase accordingly. The dangers of maximising retention efficiency to such an extent while growth rates and FCR continue to improve could result in poorly pigmented fish and subsequent economic loss.

Pigmentation variation

The variation in flesh pigmentation within and among fish is an increasingly important issue due to the increase in processed products (e.g. fillets) and will have a major impact on pigmentation regimes. The chemically assessed variation in Atlantic salmon and rainbow trout has been reported to range from 20–30% irrespective of dietary dose and duration of feeding pigment (Torrissen, 1995; Nickell & Bromage 1998b) and appears common among species (March & MacMillan 1996). Nevertheless, increasing dietary dose will increase flesh pigment concentration (Torrissen 1995) and directly affect the visualisation of colour. At flesh carotenoid concentrations above 6–8 mg/kg, the human eye has less ability to distinguish differences between colour (Sinnott 1989), which suggests that visually perceived colour variation would be reduced.

The inherent variability of flesh pigmentation will always be present in any stock of fish. However, ensuring that the variability falls within a range which allows the poorest pigmented fish to attain the required colour specification following processing, requires the farmer to pigment all fish to a minimum level. However, high dietary carotenoid intake will not guarantee that all fish will be above the minimum pigment level acceptable to the market. Limits in the process of digestion, absorption, blood transport and/or deposition in any one fish may result in sub-standard pigmentation irrespective of dietary dose and feeding duration. The problems associated with variation in flesh pigmentation will most likely always be present, since the market demands an homogenous product which is counter to the characteristics of a normally distributed population.

Summary

⅃ Colour is an important sensory attribute. When purchasing food the appearance of the products is one of the main criteria on which consumers base their choice. From a marketing perspective it is therefore important that farmed salmonids have the same distinctive pink-red colour as their 'wild' counterparts.

Given the extremely competitive nature of the food industry it is critical that fish farmers and feed manufacturers optimise feed and pigment efficiency in order to attain maximum flesh colour, with minimal variation, via the use of the correct pigment type and form within controlled, managed pigmentation regimes. Pigmentation regimes are influenced by many biotic and abiotic factors. It is therefore most important that careful monitoring of such a significant flesh quality parameter is conducted throughout the growth of the stock and that adjustments to the regime are made when necessary. Most importantly, the industry must satisfy its customers' colour requirements as efficiently and economically as possible.

References

Aas, G.H. (1998) *Astaxanthin in salmonid nutrition, A focus on utilisation, blood transport and metabolism.* 127 pp. PhD thesis, Agricultural University of Norway.

Ando, S., Takeyama., T., Hatano, M. & Zama, K. (1985) Carotenoid-carrying lipoproteins in the serum of chum salmon (*Oncorhynchus keta*) associated with migration. *Agricultural Biological Chemistry*, **49**, 2185–7.

Ando, S., Takeyama, T. & Hatano, M. (1986) Isolation and characterisation of a carotenoid-carrying lipoprotein in the serum of chum salmon (*Oncorhynchus keta*) during spawning migration. *Agricultural Biological Chemistry*, **50**, 907–14.

Ando, S., Osada, K., Hatano, M. & Saneyoshi, M. (1989) Comparison of carotenoids in muscle and ovary from four genera of salmonid fishes. *Comparative Biochemistry and Physiology*, **93B**, 503–8.

Andrews, A.G. & Starr, M.P. (1976) (3R,3'R)-astaxanthin from the yeast *Phaffia rhodozyma*. *Phytochemistry*, **15**, 1009–12.

Babin, P.J. & Vernier, J-M. (1989) Plasma lipoprotein in fish. *Journal of Lipid Research*, **30**, 467–89.

Bird, J.N. & Savage, G.P. (1989) The absorption of astaxanthin by chinook salmon (*Oncorhynchus tshawytscha*). *Proceedings of the Nutrition Society of New Zealand*, **14**, 174–7.

Bjerkeng, B. (1997) Chromatographic analysis of synthesised astaxanthin – a handy tool for the ecologist and the forensic chemist? *Progressive Fish Culturist*, **59**, 129–40.

Bjerkeng, B., Storebakken, T. & Liaaen-Jensen, S. (1990) Response to carotenoids by rainbow trout in the sea, resorption and metabolism of dietary astaxanthin and canthaxanthin. *Aquaculture*, **91**, 153–62.

Bjerkeng, B., Storebakken, T. & Liaaen-Jensen, S. (1992) Pigmentation of rainbow trout from start feeding to sexual maturation. *Aquaculture*, **108**, 333–46.

Bjerkeng, B., Følling, M., Lagocki, S., Storebakken, T., Olli J.J. & Alsted, N. (1997) Bioavailability of all-E-astaxanthin and Z-isomers of astaxanthin in rainbow trout (*Onchorhynchus mykiss*). *Aquaculture*, **157**, 63–82.

Blanc, J-P. & Choubert, G. (1993) Genetic variation of flesh colour in pan-sized rainbow trout fed astaxanthin. *Journal of Applied Aquaculture*, **2**, 115–23.

Chichester, C.O. (1981) In: *Carotenoids as Colorants and Vitamin A Precursors*, (ed. J.C. Bauernfiend). Academic Press, London.

Choubert, G. & Blanc, J-M. (1993) Muscle pigmentation changes during and after spawning in male and female rainbow trout, *Oncorhynchus mykiss*, fed dietary carotenoids. *Aquatic Living Resources*, **6**, 163–8.

Choubert, G. & Heinrich, O. (1993) Carotenoid pigments of the green alga *Haematococcus pluvialis*: assay on rainbow trout, *Oncorhynchus mykiss*, pigmentation in comparison with synthetic astaxanthin and canthaxanthin. *Aquaculture*, **112**, 217–26.

Choubert, G. & Storebakken, T. (1989) Dose response to astaxanthin and canthaxanthin pigmentation of rainbow trout fed various dietary carotenoid concentrations. *Aquaculture*, **81**, 69–77.

Choubert, G. & Storebakken, T. (1996) Digestibility of astaxanthin and canthaxanthin in rainbow trout as affected by dietary concentration, feeding rate and water salinity. *Annals of Zootechnology*, **45**, 445–53.

Choubert, G., Guillou, A. & Fauconneau, B. (1987) Absorption and fate of labelled canthaxanthin 15, 15'-^{3}H$_2$ in rainbow trout (*Salmo gairdneri* Rich.). *Comparative Biochemistry and Physiology*, **87A**, 717–20.

Choubert, G., de la Nöue, J & Blanc, J-M. (1991) Apparent digestibility of canthaxanthin in rainbow trout, effect of dietary fat level, antibiotics and number of pyloric caeca. *Aquaculture*, **99**, 323–9.

Choubert, G., Milicua, J-C. G. & Gomez, R. (1994a) The transport of astaxanthin in immature rainbow trout *Oncorhynchus mykiss* serum. *Comparative Biochemistry and Physiology*, **108A**, 245–8.

Choubert, G., Gomez, R. & Milicua, J-C. G. (1994b) Response of serum carotenoid levels to dietary astaxanthin and canthaxanthin in immature rainbow trout *Oncorhynchus mykiss*. *Comparative Biochemistry and Physiology*, **109A**, 1001–6.

Choubert, G., Milicua, J-C. G., Gomez, R., Sance, S., Petit, H., Negre-Sadargues, G.,

Castillo, R. & Trilles, J-P. (1995) Utilisation of carotenoids from various sources by rainbow trout, muscle colour, carotenoid digestibility and retention. *Aquaculture International*, **3**, 205–16.

Davies, J.B., Jackman, L.M., Siddons, P.T. & Weedon, B.C.L. (1960) The structures of phytoene, phytofluene, ξ-carotene and neurosporene. *Proceedings of the Chemical Society*, 1961, pp. 261–3.

Einen, O. & Roem, A.J. (1997) Dietary protein/energy ratios for Atlantic salmon in relation to fish size, growth, feed utilisation and slaughter quality. *Aquaculture Nutrition*, **3**, 115–26.

Einen, O. & Skrede, G. (1998) Quality characteristics in raw and smoked fillets of Atlantic salmon, Salmo salar, fed high-energy diets. *Aquaculture Nutrition*, **4**, 99–108.

Einen, O. & Thomassen, M.S. (1998) Starvation prior to slaughter in Atlantic salmon (*Salmo salar*) II. White muscle composition and evaluation of freshness, texture and colour characteristics in raw and cooked fillets. *Aquaculture*, **169**, 37–53.

Foss, P., Storebakken, T., Schiedt, K., Liaaen-Jensen, S., Austreng, E. & Streiff, K. (1984) Carotenoids in diets for salmonids. I. Pigmentation of rainbow trout with individual optical isomers of astaxanthin in comparison with canthaxanthin. *Aquaculture*, **41**, 213–26.

Foss, P., Storebakken, T., Austreng, E. & Liaaen-Jensen, S. (1987) Carotenoids in diets for salmonids. V. Pigmentation of rainbow trout and sea trout with astaxanthin and astaxanthin dipalmitate in comparison with canthaxanthin. *Aquaculture*, **65**, 293–305.

Fremont, L. & Marion, D. (1982) A comparison of lipoprotein profiles in male trout *Salmo gairdneri* before maturity and during spermiation. *Comparative Biochemistry and Physiology*, **73B**, 849–55.

Gentles, A. & Haard, N.F. (1991) Pigmentation of rainbow trout with enzyme-treated and spray-dried *Phaffia rhodozyma*. *The Progressive Fish Culturist*, **53**, 1–6.

Gjerde, B. & Gjedrem, T. (1984) Estimates of phenotypic and genetic parameters for carcass traits in Atlantic salmon and rainbow trout. *Aquaculture*, **36**, 97–110.

Gobantes, I., Choubert, G., Laurentie, M., Milicua, J-C. G. & Gomez, R. (1997) Astaxanthin and canthaxanthin kinetics after ingestion of individual doses by immature rainbow trout *Oncorhynchus mykiss*. *Journal of Agricultural Food Chemistry*, **45**, 454–8.

Gouveia, L., Gomes, E. & Empis, J. (1996) Potential use of microalga (*Chlorella vulgaris*) in pigmentation of rainbow trout (*Oncorhynchus mykiss*) muscle. *Zeitschrift für Lebensmittel Untersuchung Forschung*, **202**, 75–9.

Guillou, A., Choubert, G. & de la Nöue, J. (1992) Absorption and blood clearance of labelled carotenoids ((14C)Astaxanthin, (3H)canthaxanthin and (3H)Zeaxanthin) in mature female rainbow trout (*Oncorhynchus mykiss*). *Comparative Biochemistry and Physiology*, **103A**, 301–6.

Hardy, R.W., Torrissen, O.J. & Scott, T.M. (1990) Absorption and distribution of 14C–labelled canthaxanthin in rainbow trout (*Oncorhynchus mykiss*). *Aquaculture* **87**, 331–40.

Hatlen, B., Aas, G.H., Jorgensen, E.H., Storebakken, T. & Goswami, U.C. (1995) Pigmentation of 1, 2 and 3 year old Arctic charr (*Salvelinus alpinus*) fed two different dietary astaxanthin concentrations. *Aquaculture*, **138**, 303–12.

Haxo F. (1950) Carotenoids of the mushroom *Cantharellus cinnabarinus*. *Botanical Gazette*, **112**, 228–32.

Henmi, H., Iwata, T., Hata, M. & Hata, M. (1987) Studies of the carotenoids in the muscle of salmons. I. Intracellular distribution of carotenoids in the muscle. *Tohoku Journal Agricultural Research*, **37**, 101–111.

Henmi, H., Hata, M. & Hata, M. (1989) Astaxanthin and/or canthaxanthin–actomyosin complex in salmon muscle. *Nippon Suisan Gakkaishi*, **55** (9), 1583–9.

Henmi, H., Hata, M. & Hata, M. (1990) Combination of astaxanthin and canthaxanthin with fish muscle actomyosins associated with their surface hydrophobicity. *Nippon Suisan Gakkaishi*, **56**, 1821–3.

Isler, O., Ruegg, R. & Schwieter, U. (1967) Carotenoids as food colourants. *Pure and Applied Chemistry*, **14**, 245–63.

Iwamoto, R.N., Myers, J.M & Hershberger, W.K. (1990) Heritability and genetic correlations for flesh coloration in pen-reared coho salmon. *Aquaculture*, **86**, 181–90.

Johnson, E.A. & An, G-H. (1991) Astaxanthin from microbial sources. *Critical Reviews in Biotechnology*, **11**, 297–326.

Johnson, E.A., Villa, T.G. & Lewis, M.J. (1980) *Phaffia rhodozyma* as an astaxanthin source in salmonid diets. *Aquaculture*, **20**, 123–34.

Johnston, I.A. (1982) Physiology of muscle in hatchery raised fish. *Comparative Biochemistry and Physiology*, **73B**, 105–24.

Johnston, I.A., Alderson, R., Sandham, C., Dingwall, A., Mitchell, D., Selkirk, C., Nickell, D.C., Baker R.T.M., Robertson W., Whyte, D. & Springate, J. (2000) Muscle fibre density in relation to the colour and texture of smoked Atlantic salmon (*Salmo salar* L.). *Aquaculture* (in press).

Kamata, T., Neamtu, G., Tanaka, Y., Sameshima, M. & Simpson, K.L. (1990) Utilisation of *Adonis aestivalis* as a dietary pigment source for rainbow trout *Salmo gairdneri. Nippon Suisan Gakkaishi*, **56** (5), 783–8.

Kiessling, A., Storebakken, T., Asgard, T. & Kiessling, K.-H. (1991) Changes in the structure and function of the epaxial muscle of rainbow trout (*Oncorhynchus mykiss*) in relation to ration and age. I. Growth dynamics. *Aquaculture*, **93**, 335–56.

Kiessling, A., Dosanjh, B., Higgs, D., Deacon, G. & Rowshandeli, N. (1995) Dorsal aorta cannulation: a method to monitor changes in blood levels of astaxanthin in voluntarily feeding Atlantic salmon, *Salmo salar* L. *Aquaculture Nutrition*, **1**, 1–8.

Kitahara, T. (1983) Behaviour of carotenoids in the chum salmon (*Oncorhynchus keta*) during anadromous migration. *Comparative Biochemistry and Physiology*, **76B**, 97–101.

Latscha, T. (1990) *Carotenoids in Animal Nutrition*. Roche Print, 2175, Basel, Switzerland.

Lie, Ø., Sandvin, A. and Waagbo, R. (1994). Transport of alpha-tocopherol in Atlantic salmon (*Salmo salar*) during vitellogenesis. *Fish Physiology and Biochemistry*, **13**, 241–7.

March, B.E. & MacMillan, C. (1996) Muscle pigmentation and plasma concentrations of astaxanthin in rainbow trout, chinook salmon, and Atlantic salmon in response to different dietary levels of astaxanthin. *Progressive Fish Culturist*, **58**, 178–86.

McCallum, I.M., Cheng, K.M. & March, B.E. (1987) Carotenoid pigmentation in two strains of chinook salmon (*Oncorhynchus tshawytscha*) and their crosses. *Aquaculture*, **67**, 291–300.

McCarthy, I.D., Carter, C.G. & Houlihan, D.F. (1992) The effect of feeding hierarchy on individual variability in daily feeding of rainbow trout, *Oncorhynchus mykiss* (Walbaum). *Journal of Fish Biology*, **41**, 257–63.

Nakamura, K., Hata, M. & Hata, M. (1985) A study on astaxanthin in salmon *Oncorhynchus keta* serum. *Bulletin of the Japanese Society Science Fisheries*, **51**, 979–83.

Nickell, D.C. (1997) *Pigmentation in the rainbow trout (Oncorhynchus mykiss), Lipids and other dietary influences*. 254pp. PhD thesis, University of Stirling.

Nickell, D.C. & Bromage, N.R. (1998a) The effect of dietary lipid level on the variation of flesh pigmentation in the rainbow trout (*Oncorhynchus mykiss*). *Aquaculture*, **161**, 237–51.

Nickell, D.C. & Bromage, N.R. (1998b) The effect of timing and duration of feeding astaxanthin on the development and variation of fillet colour and efficiency of pigmentation in rainbow trout (*Oncorhynchus mykiss*). *Aquaculture*, **169**, 233–46.

No, H.K. & Storebakken, T. (1991) Pigmentation of rainbow trout with astaxanthin at different water temperatures. *Aquaculture*, **97**, 203–16.

Østerlie, M., Bjerkeng, B. & Liaaen-Jensen, S. (1999) Accumulation of astaxanthin all-E, 9Z and 13Z geometrical isomers and 3 and 3' RS optical isomers in rainbow trout (*Oncorhynchus mykiss*) is selective. *Journal of Nutrition*, **129**, 391–8.

Ostrander, J., Martinsen, C., Liston, J. & McCullough, J. (1976) Sensory testing of pen-reared salmon and trout. *Journal of Food Science*, **41** (2), 386–90.

Renstrom, B., Berger, H. & Liaaen-Jensen, S. (1981) Esterified, optically pure (3S,3'S)-astaxanthin from flowers of *Adonis annua*. *Biochemical Systematic Ecology*, **9**, 249–250.

Schiedt, K. (1987) *Absorption, retention and metabolic transformations of carotenoids in chicken, salmonids and crustacea*. 291 pp. PhD thesis, University of Basel, Switzerland.

Schiedt, K. (1998) Absorption and metabolism of carotenoids in birds, fish and crustaceans. In: *Carotenoids. Volume 3, Biosynthesis and metabolism,* (ed. G. Britton, S. Liaaen-Jensen & H. Pfander) pp. 285-358. Birkhäuser Verlag, Basel, Boston and Berlin.

Schiedt, K. Leueunberger, L.J. & Vecchi, M. (1981) Natural occurrence of enantiomeric and meso-astaxanthin in wild salmon (*Salmo salar* and *Onchorhynchus*). *Helvetica Chimica Acta*, **64**, 449–57.

Schiedt, K., Leuenberger, F.J., Vecchi, M. & Glinz, E. (1985) Absorption, retention and metabolic transformations of carotenoids in rainbow trout, salmon, and chicken. *Pure and Applied Chemistry*, **57**, 685–92.

Schiedt, K., Vecchi, M. & Glinz, E. (1986) Astaxanthin and its metabolites in wild rainbow trout (*Salmo gairdneri* R.). *Comparative Biochemistry and Physiology*, **83B**, 9–12.

Shahidi, F., Meusalach, & Brown, J.A. (1998) Carotenoid pigments in seafoods and aquaculture. *Critical Reviews in Food Science*, **38**, 1–67.

Sinnott, R. (1989) Keep them in the pink to stay competitive. *Fish Farmer*, **12**, 23–6.

Smith, B.E., Hardy, R.W. & Torrissen, O.J. (1992) Synthetic astaxanthin deposition in pan-sized coho salmon (*Oncorhynchus kisutch*). *Aquaculture*, **104**, 105–19.

Sommer, T.R., Potts, W.T. & Morrissy, N.M. (1991) Utilisation of microalgal astaxanthin by rainbow trout (*Oncorhynchus mykiss*). *Aquaculture*, **94**, 79–88.

Spinelli, J. & Mahnken, C. (1978) Carotenoid deposition in pen-reared salmonids fed diets containing oil extracts of red crab (*Pleuroncodes planipes*). *Aquaculture*, **13**, 213–23.

Storebakken, T. & Goswami, U.C. (1996) Plasma carotenoid concentration indicates the availability of dietary astaxanthin for Atlantic salmon, *Salmo salar*. *Aquaculture*, **146**, 147–53.

Storebakken, T. & No, H.K. (1992) Pigmentation of rainbow trout. *Aquaculture*, **100**, 209–29.

Storebakken, T., Foss, P., Huse, I., Wandsvik, A. & Lea, T.B. (1986) Carotenoids in diets for salmonids. III. Utilisation of canthaxanthin from dry and wet diets by Atlantic salmon, rainbow trout and sea trout. *Aquaculture*, 51, 245–55.

Storebakken, T., Foss, P., Schiedt, K., Austreng, E., Liaaen-Jensen, S. & Mainz, U. (1987) Carotenoids in diets for salmonids. VI. Pigmentation of Atlantic salmon with astaxanthin, astaxanthin dipalmitate and canthaxanthin. *Aquaculture*, 65, 279–92.

Torrissen, O.J. (1985) Pigmentation of salmonids, factors affecting carotenoid deposition in rainbow trout (*Salmo gairdneri*). *Aquaculture*, 46, 133–42.

Torrissen, O.J. (1995) Strategies for salmonid pigmentation. *Journal of Applied Ichthyology*, 11, 276–81.

Torrissen, O.J. & Braekkan, O.R. (1979) The utilisation of astaxanthin forms by rainbow trout (*Salmo gairdneri*). In: *Finfish Nutrition and Fishfeed Technology*, (eds J.E. Halver & K. Tiews), Vol.II, pp. 377–82. Heenemann, Berlin.

Torrissen, K.R. & Torrissen, O.J. (1985) Protease activities and carotenoid levels during the sexual maturation of Atlantic salmon (*Salmo salar*). *Aquaculture*, 50, 113–22.

Torrissen, O.J. & Naevdal, G. (1988) Pigmentation of salmonids – variation in flesh carotenoids of Atlantic salmon. *Aquaculture*, 68, 305–10.

Torrissen, O.J. & Christiansen, R. (1991) Astaxanthin and canthaxanthin as pigment sources for salmonids. *Proceedings of the IV International Symposium on Fish Nutrition*, Biarritz, France, pp. 4–15.

Torrissen, O.J., Tidemann, E., Hansen, F. & Raa, J. (1981) Ensiling in acid – a method to stabilise astaxanthin in shrimp processing by-products and improve uptake of this pigment by rainbow trout (*Salmo gairdneri*). *Aquaculture*, 26, 77–83.

Torrissen, O.J., Hardy, R.W. & Shearer, K.D. (1989) Pigmentation of salmonids – carotenoid deposition and metabolism. *Review of Aquatic Sciences*, 1, 209–25.

Torrissen, O.J., Hardy, R.W., Shearer, K.D., Scott, T.M. & Stone, F.E. (1990) Effects of dietary lipid on apparent digestibility coefficients for canthaxanthin in rainbow trout (*Oncorhynchus mykiss*). *Aquaculture*, 88, 351–62.

Torrissen, O.J., Christiansen, R., Struksnaes, G. & Estermann, R. (1995) Astaxanthin deposition in the flesh of Atlantic salmon, *Salmo salar* L., in relation to dietary astaxanthin concentration and feeding period. *Aquaculture Nutrition*, 1, 77–84.

Totland, G.K., Kryvi, H., Jödestöl, K.A., Christiansen, E.N., Tangeras, A. & Slinde, E. (1987) Growth and composition of the swimming muscle of adult Atlantic salmon (*Salmo salar* L.) during long-term sustained swimming. *Aquaculture*, 66, 229–313.

Wallaert, C. & Babin, P.J. (1994) Age-related, sex-related, and seasonal changes of plasma lipoprotein concentrations in trout. *Journal of Lipid Research*, 35, 1619–33.

Chapter 7

Eating Quality of Farmed Rainbow Trout (*Oncorhynchus mykiss*)

L. Johansson

Department of Domestic Sciences, Uppsala University, Uppsala, S-75337, Sweden

Introduction

Eating quality as considered in this chapter comprises sensory properties, total lipid content and fatty acid composition of the fillet of farmed rainbow trout. The section of the fillet was chosen to represent what is eaten, that is the cooked fish freed from skin and bones and also from dorsal and ventral lipid depots (Fig. 7.1).

The main material originates from the thesis *Eating Quality of Rainbow Trout* (Johansson 1992) which aims to elucidate how farmed rainbow trout was affected regarding sensory characteristics, total lipid content and fatty acid composition when subjected to different treatments. These included types of heating (conventional versus microwave heating) to different final temperatures (55°, 65° and 75°C) (Johansson *et al.* 1992a), feeding versus starvation followed by icing and frozen storage (Johansson & Kiessling 1991), different feed ration levels (Johansson *et al.* 1995), altered feed ration levels (Johansson *et al.* 2000), and feed admixed with different levels of leaf nutrient concentrate (LNC) (Johansson *et al.* 1991).

Methodological considerations

Sampling

In order to reduce the effect of differences in the proportion of red versus white muscle (Kiessling *et al.* 1991a,b,c) and water and collagen contents along the fillet, only the middle part of the fish was used, from 20 mm behind the gill cover behind the pectoral fins back to the vent, this being the most uniform part in composition, weight and shape. This allowed uniform cooking of the fish and made possible the presentation of equivalent samples to the assessors, thus giving the latter the best possible opportunity of discerning even small differences between samples.

The samples from sites 1 and 6 (Fig. 7.1) were tested (Johansson & Kiessling 1991) regarding variation in lipid content and fatty acid composition due to site. There was no significant difference due to sample site in the total lipid content of the dorsal fillet ($p < 0.05$) or in the fatty acid composition ($p < 0.05$) of rainbow trout weighing approximately 1000 g.

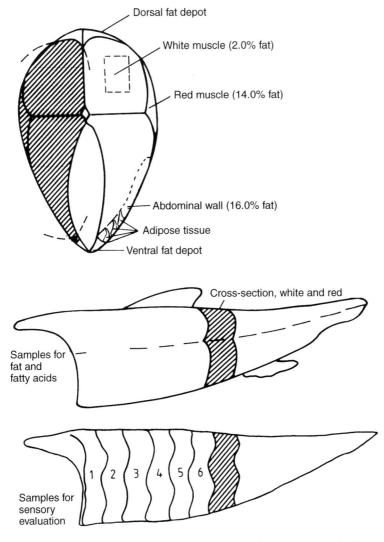

Dorsal fat depot

White muscle (2.0% fat)

Red muscle (14.0% fat)

Abdominal wall (16.0% fat)

Adipose tissue

Ventral fat depot

Cross-section, white and red

Samples for fat and fatty acids

Samples for sensory evaluation

1 2 3 4 5 6

Fig. 7.1 Cross-section of rainbow trout and sample sites for sensory evaluation.

Panel

The panel was selected and trained according to the International Standard 8586-1 (ISO 1993). Each panel member received her sample from the same part (Fig. 7.1) of the fish fillet (Johansson & Kiessling 1991). It was therefore impossible to have the panel statistically tested for concordance due to confounding of the effects of assessor and sample. The concordance within the panel was therefore tested separately, where each panel member was presented with samples from all sample sites to find out if differences in the assessments as regards the textural attributes, firmness, juiciness and tenderness were due to the panel members not being unanimous in their assessments or the samples along the fillet being different. The result showed that the assessors were in full agreement assessing each other's samples.

Optimum internal temperature for rainbow trout cooked in conventional and microwave ovens

Recommended internal temperatures for fish cookery have been scarce and those available vary widely, from 50° to 73°C. Furthermore those concerning fish in which sensory analyses were included were not comparable because of different fish species investigated, differences in sizes and cuts of fish within species, in choice of final temperatures (Gall et al. 1983; Brady et al. 1985; Madeira and Penfield 1985), in treatment of fish prior to cooking and in sampling procedure. Comparisons of sensory qualities of microwave and conventionally cooked fish have been carried out in the case of flounder and haddock cooked to an internal temperature of 70°C (Brady et al. 1985) and in the case of turbot cooked to internal temperatures within the range of 77° to 88°C (Madeira & Penfield 1985).

A comparative test (Johansson et al. 1992b) on rainbow trout baked in microwave (full power, 740 W) and conventional (175°C) ovens to final temperatures 55°, 65° and 75°C showed that the internal temperature had a greater effect on the eating quality than did the heating method (Table 7.1). With an increase in final internal temperature the sensory properties surface moistness, tenderness, juiciness and flavour declined significantly (Table 7.1).

Table 7.1 Least squares-means for sensory evaluation scores[1] of baked[2] rainbow trout.

Sensory attribute	Final temperature °C			Temp. effect	Oven type		Oven effect
	55	65	75		Micro	Conv.	
Odour	4.4	4.0	4.2	ns	4.3	4.1	ns
Surface appearance							
opacity	3.9	4.0	4.2	ns	4.3^a	3.8^b	*
moistness	4.2^a	3.5^b	3.5^b	*	3.6	3.8	ns
Texture							
cohesiveness	4.2	3.7	3.7	ns	3.9	3.8	ns
tenderness	4.2^a	4.0^a	3.3^b	*	4.1	3.6	ns
juiciness	3.5^a	3.0^{ab}	2.3^b	*	3.1	2.8	ns
Flavour	4.2^a	4.1^a	3.2^b	**	4.2^a	3.5^b	*

Levels of significance: $* = p \leq 0.05$, $** = p \leq 0.01$, $*** = p \leq 0.001$, ns = $p > 0.05$. Means with the same superscript letter are not significantly different, $p > 0.05$.
[1] A 5-point category scale was used (1 = low, 5 = high).
[2] Final temperatures were 55°, 65° and 75°C.

Post-cooking adjustments

Post-cooking temperature rise is dependent on type of heating (conventional versus microwave oven), conventional oven temperature or microwave power setting, and also weight of the fish. This must be adjusted to enable valid comparisons to be made.

Post-cooking temperature rise in whole rainbow trout (600–800 g) prepared in the conventional oven (175°C) was adjusted for by removing the fish at 3°, 4° and 5°C, respectively, below the desired final temperatures of 55°, 65° and 75°C. In parallel, post-cooking temperature rise in the same weight of microwaved fish was adjusted for by removing the fish at 12°C below the desired same final internal temperatures as above.

Keeping samples hot

In catering it is inevitable that food needs to be kept hot for some time and it is common practice in the case of conventionally prepared foods, including fish. Ample consideration needs to be taken also when scheduling the serving procedure of hot samples to the panel in order not to keep the samples hot too long. The effect on the sensory properties of whole rainbow trout baked and kept hot at 55°, 65° and 75°C for 0, 30 and 90 minutes (Johansson 1992) showed that total odour intensity, total flavour intensity and the texture attributes – cohesiveness, tenderness and juiciness – had not changed after 30 minutes, while they were all significantly lower for fish kept hot for 90 minutes, independent of holding temperatures (Table 7.2).

Table 7.2 Least squares-means for sensory evaluation scores of baked rainbow trout kept hot for 0, 30 and 90 minutes.

Sensory attribute	Keeping hot, minutes			Time effect
	0	30	90	
Odour	4.4^a	4.2^{ab}	3.9^b	*
Surface appearance				
opacity	4.6	4.1	4.2^b	ns
moistness	4.6^a	3.9^b	3.9^b	**
Texture				
cohesiveness	4.3^a	4.1^{ab}	3.8^b	*
tenderness	4.7^a	4.4	4.3^b	*
juiciness	4.0^a	3.6^{ab}	3.2^b	*
Flavour	4.5^a	4.1^{ab}	3.8^b	*

Levels of significance: * $= p \leq 0.05$, ** $= p \leq 0.01$, *** $= p \leq 0.001$, ns $= p > 0.05$. Means with the same superscript letter are not significantly different, $p > 0.05$.

In summary, the best eating quality as regards surface moistness, tenderness, juiciness and flavour will be achieved if the final internal temperature does not exceed 65°C, independent of heating method used. Fish should not be kept hot longer than 30 minutes.

Cooking losses

The type of heating (conventional versus microwave oven), the final temperature of the fish, the oven temperature/power setting and the weight all had an effect on the cooking losses of rainbow trout (Johansson *et al.* 1992) and thus affected the eating quality of the fish.

Type of heating

The type of heating affected the weight loss of the rainbow trout (Johansson 1992). The weight loss of the microwaved rainbow trout was twice the weight loss of the conventionally heated fish (12.9% versus 5.8%) and the lipid loss eight times higher for the microwaved trout (2.4% versus 0.3%, Table 7.3).

Table 7.3   Least squares-means for cooking losses of rainbow trout prepared in a conventional oven (150°C) and microwave oven (full power) to different final temperatures.

Variable	Oven		Oven effect	Final temperature, °C			Temp. effect
	Conventional	Microwave		53–57	58–67	68–83	
Weight loss							
g	37.9	83.5	***	58.1[a]	49.8[a]	74.1[b]	***
%	5.8	12.9	***	8.5[a]	7.9[a]	11.6[b]	***
Lipid loss[1]							
g	2.2	15.1	***	7.2[a]	5.3[a]	13.5[b]	**
%	0.3	2.4	***	—	—	—	
C[1]%				0.3	0.2	0.5	ns
M[1]%				1.8[a]	1.4[a]	3.9[b]	***
Liquid loss							
g	35.6	71.0	***	51.0[a]	44.0[a]	65.0[b]	**
%	5.5	10.8	***	7.5[a]	6.9[a]	10.0[b]	**

Levels of significance: $* = p \leq 0.05$, $** = p \leq 0.01$, $*** = p \leq 0.001$, $ns = p > 0.05$. Means with the same superscript letter are not significantly different, $p > 0.05$.
[1] Ovens (M = microwave oven; C = conventional) are presented separately because there is an interaction between oven and final temperature.

Final temperature

An increase in final temperature for rainbow trout affected most cooking loss variables (Johansson 1992), independent of type of heating. The weight loss increased significantly when the temperature exceeded 67°C: 8.5% weight loss for the temperature range 53–67°C as compared to 11.6% for the temperature range 68–83°C (Table 7.3).

Oven temperature

An increase in conventional oven temperature from 150° to 175°C caused an increase in weight loss (Johansson 1992) for the rainbow trout from 4.5% to 7.0% and in lipid loss from 6.7% to 18.9% (Table 7.4).

Table 7.4 Least squares-means for cooking losses of rainbow trout prepared at different conventional oven temperatures to final temperatures 55°C and 65°C.

Variable	Oven temperature, °C		Oven temp. effect	Final temperature, °C		Final temp. effect
	150	175		55	65	
Weight loss						
g	43.3	52.9	*	48.0	39.2	ns
%	4.5	7.0	**	6.0	5.5	ns
Lipid loss						
g	2.3	10.0	**	6.8	5.6	ns
%[1]	6.7	18.9	***	14.2	14.3	ns
Liquid loss						
g	32.0	42.9	ns	41.3	33.6	ns
%	5.8	4.9	*	5.2	4.7	ns

Levels of significance: $* = p \leq 0.05$, $** = p \leq 0.001$, $*** = p \leq 0.001$, ns $= p > 0.05$. Means with the same superscript letter are not significantly different, $p > 0.05$.
[1] Collected lipid phase/total exudate (weight of raw fish – weight of cooked fish)

Fish weight

An increase in fish weight affected the percentage of rainbow trout weight loss (Johansson 1992) independent of final temperature: fish with the lower weight had significantly higher ($p < 0.05$) percentage of weight loss (Table 7.5).

Fatty acid composition

Neither the type of heating nor the final temperature had any effect on the fatty acid composition of either fish flesh or collected lipid.

Table 7.5   Least squares-means for cooking losses of rainbow trout of different weight prepared in a conventional oven (150°C) to different final temperatures.

Variable	Fish weight, g		Weight effect	Final temperature, °C			Temp. effect
	425–625	626–825		53–57	58–67	68–83	
Weight loss							
g	33.9	36.1	*	26.7[a]	27.9[a]	50.5[b]	***
%	6.1	5.2	*	4.4[a]	4.6[a]	8.0[b]	***
Lipid loss							
g	1.8	2.2	ns	1.5[a]	1.2[a]	3.2[b]	*
%[1]	0.3	0.3	ns	0.2[a]	0.2[a]	0.5[b]	*
%[2]	5.0	5.3	ns	4.9	4.4	6.0	ns
Liquid loss							
g	32.1	34.0	ns	25.2[a]	26.7[a]	47.3[b]	***
%	5.8	4.9	*	4.1[a]	4.4[a]	7.6[b]	***

Levels of significance: $* = p \leq 0.05$, $** = p \leq 0.01$, $*** = p \leq 0.001$, ns $= p > 0.05$. Means with the same superscript letter are not significantly different, $p > 0.05$.
[1] Lipid loss as % of total fish weight.
[2] Lipid loss as % of total cooling loss.

Preslaughter fasting (starvation)

Periods of starvation are normal in the life of wild fish, and the effects on different organs (Love 1980; Weatherley & Gill 1987) and on growth, metabolism, histochemistry and fibre dimensions of white and red muscle of rainbow trout (Kiessling *et al.* 1991a–c) have been the subject of research. A few investigations on the effects of starvation on eating and storage qualities of rainbow trout have included sensory evaluations (Lea 1981; Ludovico-Pelayo *et al.* 1983; Aksnes *et al.* 1985). Descriptive analysis, however, was not performed.

Fasting of farmed fish prior to slaughter, lasting from one to two weeks in the warm seasons and for as much as three to four weeks in the cold seasons of the year, is a normal handling regime in fish farming. It is a hygenic precaution taken in order that the farmed fish will keep as long as possible (Maeland 1979; Lea, 1981; Fiskeridirektoratet 1988). The effect on sensory characteristics of rainbow trout of feeding versus fasting for one and two months, iced for 0, 1 and 2 weeks and stored frozen for 0, 3 and 6 months, respectively, was studied by Johansson and Kiessling (1991).

Effects of starvation

Starvation affected the eating quality of the farmed rainbow trout: total odour intensity and acidulous odour were reduced, total taste intensity was weakened, bitterness declined and the fish was less juicy (Fig. 7.2).

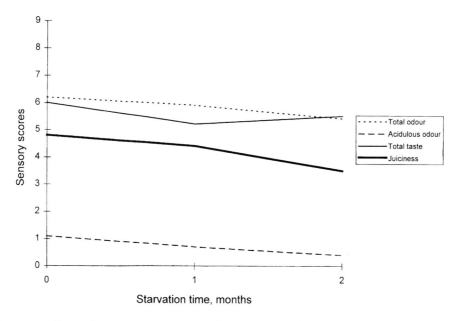

Fig. 7.2   Effects of starvation on sensory characteristics of rainbow trout.

Effects of icing

With increasing icing time the sensory properties of the starved rainbow trout declined: a decline in fresh odour was apparent after two weeks of icing, a less acidulous taste and a less fresh taste after one week of icing, and a continuous increase in intensity of putrescent taste with prolonged icing (Fig. 7.3).

Effects of frozen storage

Frozen storage of three and six months reduced total intensity of odour and taste, juiciness and firmness (Fig. 7.4).

Fatty acid composition

Non-feeding (starvation) caused a reduction in the relative level of MUFA (sum of monounsaturated fatty acids) after two months, whereas that of the relative level of PUFA (sum of the polyunsaturated fatty acids) increased after one month.

Feed ration levels

Feed composition and feed quantity affect both body composition and fatty acid profile of rainbow trout flesh (Papoutsoglou and Papaparaskeva-Papoutsoglou 1978; Alexis *et al.*

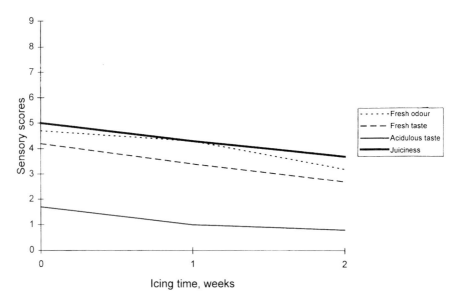

Fig. 7.3   Effects of icing time on sensory characteristics of rainbow trout.

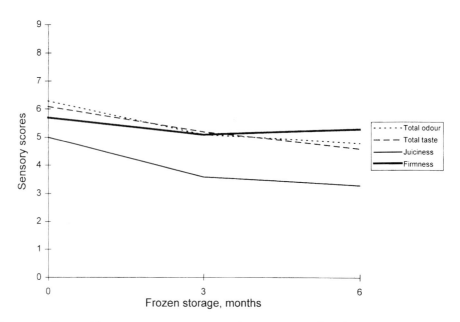

Fig. 7.4   Effects of frozen storage on sensory characteristics of rainbow trout.

1986; Johansson & Kiessling 1991; Johansson *et al.* 1991). The farming regimes, e.g. feed ration levels (RLs), had an effect on the eating quality of the rainbow trout (Grayton & Beamish 1977; Wartsbaugh & Davies 1977a,b; Huisman 1979; Reinitz 1983a,b). In a longitudinal investigation lasting from hatching to sexual maturity (2.4 years) Johansson *et al.* (1995) studied the effect of feed ration level (RL) on the eating quality of rainbow trout fillet (Fig. 7.5). Fish administered 75% (RL75), 100% (RL100) and 200% (RL200) of the necessary feed ration for adequate growth (100%) according to Storebakken and Austreng (1987) were subjected to sensory analysis at 2.0 years of age. Fish on RL75 (moderately restricted feed ration) did not differ from fish on RL100 and RL200 as regards total odour intensity and fresh odour but had a stronger total taste ($p < 0.01$) and tended to have a more fresh taste ($p = 0.0535$). They were firmer ($p < 0.01$) and had a less coarse flesh ($p < 0.001$).

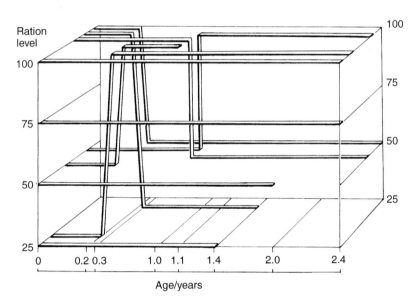

Fig. 7.5 Schematic of different feed ration levels to rainbow trout from hatching to 2.4 years of age.

Altered feed ration levels

Altering the feed ration level from RL50 to RL100 and *vice versa* after 0.2 or 1.0 years (Fig. 7.5) from hatching did not have any effect on the sensory properties of the fish.

Fatty acid composition

When there is a restriction of feed for rainbow trout (Johansson *et al.* 1995; Johansson *et al.* 1999) the depots of the flesh are used and the total lipid level of the muscle decreases. The triglyceride level (depot lipid) thus decreases and the phospholipid level (membrane lipid) is kept constant, leading to a decrease in the relative level of MUFA and an increase in the relative level of PUFA.

Leaf nutrient concentrate (LNC) in feed

Two studies were carried out with leaf nutrient concentrate (LNC) admixed in conventional rainbow trout feed (Johansson *et al.* 1991). In the first study fish were fed a control feed and two feeds with respectively 25% (LNC25) and 45% (LNC45) admixed. In the second study fish were fed the same control feed and two feeds with respectively 25% (LNC25$_1$) and 12.5% (LNC12.5) admixed.

The rainbow trout did not show satisfactory growth on either of the two LNC feeds in the first study. In the second study the LNC constituted a lactic acid fermented green leaf juice mixture. The strong grassy flavour of the LNC had an effect on the trout. Fish on LNC25$_1$ had a significantly higher atypical odour ($p < 0.05$) as compared to the control fish, whilst the fresh odour ($p < 0.05$) and the fresh taste ($p < 0.05$) decreased significantly. Fish on LNC12.5 did not differ from the control fish as regards fresh odour and fresh taste.

Fatty acid composition

Fish fed LNC12.5 and LNC25 had a significantly higher relative level of PUFA than fish on the control feed (28.6, 29.4 and 27.3 rel.%, respectively).

Acknowledgements

The chemical analyses were carried out at the Department of Domestic Sciences at Uppsala University in cooperation with the following: Department of Plant Physiology, University of Lund; Department of Animal Nutrition and Management, and Department of Aquaculture, Swedish University of Agricultural Sciences, Uppsala; Institute of Aquaculture, Research, Sunndalsöra, Norway.

References

Aksnes, A., Halvorsen, K. & Roald, S.O. (1985) Holdbarhet og kvalitets-bedømelse av iset oppdrettslaks. *Norsk Fiskeoppdrett*, **2**, 20–23.
Alexis, M.A., Theochari, V. & Papaparaskeva-Papoutsoglou, E.G. (1986) Effect of diet composition and protein level on growth, body composition, haematological characteristics and cost of production of rainbow trout (*Salmo gairdneri*). *Aquaculture*, **58**, 75–85.
Brady, P.L., Haughey, P.E. & Rothschild, M.F. (1985) Microwave and conventional heating effects on sensory quality and thiamin content of flounder and haddock fillets. *Home Economics Research Journal*, **14**, 236–40.
Fiskeridirektoratet (1988) *Melding fra fiskeridiretøren*, J-87-88. Bergen, Norway.
Gall, K.L., Otwell, W.S., Koburger, J.A. & Appeldorf, H. (1983) Effects of four cooking

methods on the proximate, mineral and fatty acid composition of fish fillets. *Journal of Food Science,* 48, 1068–74.

Grayton, B.D. & Beamish, F.W.H. (1977) Effects of feeding frequency on food intake, growth and body composition of rainbow trout (*Salmo gairdneri*). *Aquaculture,* 11, 159–72.

Huisman, E.A. (1979) Food conversion efficiencies at maintenance and production levels for carp, *Cyprinus carpio* L., and rainbow trout, (*Salmo gairdneri* R.). *Aquaculture,* 9, 259–76.

ISO (1993) *Sensory analysis – General guidance for selection, training and monitoring of assessors – Part I: Selectes assessors.* The International Organization for Standardization, Genève. ISO 8586-1, 1993 (E).

Johansson, L. (1992) *Eating quality of rainbow trout.* DPhil thesis, Uppsala University, Uppsala, Sweden.

Johansson, L. & Kiessling, A. (1991) Effects of starvation on rainbow trout. II: Eating and storage qualities of iced and frozen fish. *Acta Agriculturae Scandinavica,* 41, 207–16.

Johansson, L., Kiessling, A. & Carlsson, R. (1991) Eating quality and growth of rainbow trout (*Onchorhynchus mykiss*) on feed with different admixtures of leaf nutrient concentrate. *Journal of the Science of Food and Agriculture,* 57, 217–34.

Johansson, L., Beilby, R.I. & Rudérus, H. (1992a) Optimum internal temperature established by sensory evaluation for fish prepared in the conventional and microwave ovens. *Home Economics Research Journal,* 21, 192–205.

Johansson, L., Berglund, L. & Rudérus, H. (1992b) Cooking losses of rainbow trout – effects of final internal temperature, fish weight, heating method and oven temperature. In: *Eating quality of rainbow trout,* DPhil thesis, Uppsala University, Uppsala, Sweden.

Johansson, L., Kiessling, A., Åsgård, T. & Berglund, L. (1995) Eating quality of rainbow trout – Effects of ration level on sensory characteristics, lipid content and fatty acid composition. *Aquaculture Nutrition,* 1, 59–66.

Johansson, L., Kiessling, A., Kiessling, K-H. & Berglund, L. (2000) Effects of altered ration levels on sensory characteristics, lipid content and fatty acid composition of rainbow trout (*Onchorhynchus mykiss*). *Food Quality and Preference,* 11, 247–54.

Kiessling, A., Kiessling, K-H., Storebakken, T. & Åsgård, T. (1991a) Changes in the structure and function of the epaxial muscle of rainbow trout (*Onchorhynchus mykiss*) in relation to ration and age I. Growth dynamics. *Aquaculture,* 93, 335–6.

Kiessling, A., Kiessling, K-H., Storebakken, T. & Åsgård, T. (1991b) Changes in the structure and function of the epaxial muscle of rainbow trout (*Onchorhynchus mykiss*) in relation to ration and age II. Activity of key enzymes in the energy metabolism. *Aquaculture,* 93, 357–72.

Kiessling, A., Åsgård, T., Storebakken, T., Johansson, L. & Kiessling, K-H. (1991c) Changes in structure and function of the epaxial muscle of rainbow trout (*Onchorhynchus mykiss*)) in relation to ration and age III: Chemical composition. *Aquaculture,* 93, 373–87.

Lea, T.B. (1981) Virking av sult på vekst, kjemisk sammen setning, oksygenforbruk og indre kvalitet hos regnbueaure (*Salmo gairdneri*). *Hovedoppgave ved Inst. for fjärfe og pelsdyr.* Inst. for naturforvaltning, Norges Landbrukshøg-skole, Ås, Norge.

Love, R.M. (1980) Maturation. In: *The Chemical Biology of Fishes II,* pp. 36–46. Academic Press, London.

Ludovico-Pelayo, L., Hume, A. & Love, R.M. (1983) Seasonal variations in flavour change of cold-stored rainbow trout. *Proceedings of Cost 91, Seminar,* November 1983, Athens.

Madeira, K. & Penfield, M.P. (1985) Turbot fillet sections cooked by microwave and conventional heating methods: objective and sensory evaluation. *Journal of Food Science*, **50**, 172–7.

Maeland, K.T. (1979) Kvalitetsmål og faktorer som virker på dem. In: *Oppdrett av laks og aure*. Landbruksforlaget, Oslo.

Papoutsoglou, S.E. & Papaparaskeva-Papoutsoglou, E.G. (1978) Comparative studies on body composition of rainbow trout (*Salmo gairdneri* R.) in relation to type of diet and growth rate. *Aquaculture*, **13**, 235–43.

Reinitz, G. (1983a) Relative effect of age, diet and feeding rate on the body composition of young rainbow trout (*Salmo gairdneri*). *Aquaculture*, **35**, 19–27.

Reinitz, G. (1983b) Influence of diet and feeding rate on the performance and production costs of rainbow trout. *Trans American Fish Society*, **12**, 830–33.

Storebakken, T. & Austreng, E. (1987) Ration level for salmonids. II. Growth, feed intake, protein digestibility, body composition, and feed conversion in rainbow trout weighing 0.5–1.0 kg. *Aquaculture*, **61**, 207–21.

Wartsbaugh, W.A. & Davies, G.E. (1977a) Effects of temperature and ration level on the growth and food conversion efficiency of *Salmo gairdneri* Richardson. *Journal of Fish Biology*, **11**, 87–98.

Wartsbaugh, W.A. & Davies, G.E. (1977b) Effects of fish size and ration level on the growth and food conversion efficiency of rainbow trout, *Salmo gairdneri* Richardson. *Journal of Fish Biology*, **11**, 99–104.

Weatherley, A.H. & Gill, H.S. (1987). *The Biology of Fish Growth*, pp. 443. Academic Press, New York.

Chapter 8

Microbiological Problems in the Salmon Processing Industry

L.A. Laidler

Marine Harvest (Scotland) Ltd, Lochailort, Inverness-shire, PH38 4LZ, UK

Introduction

With an annual production in the region of 800 000 tonnes, salmon farming (predominantly Atlantic salmon, *Salmo salar*, but also Pacific, *Onchorhynchus* spp.) occurs in temperate zones of both hemispheres. Norway, Scotland and Chile are the principal producing nations. Salmon is a wholesome and healthy foodstuff and processors are required to exercise due diligence in providing a product which is safe for consumption. In the current era of heightened public awareness of food safety issues, failure to do so may have serious legal and commercial consequences.

Microbiological problems which concern salmon processors are especially associated with bacteria and fall under two headings: as a cause of spoilage and as potential human pathogens which pose a health risk to the consumer. Salmon is frequently eaten raw or after only light processing such as cold smoking (brining followed by exposure to wood smoke not exceeding 28°C) or gravad (raw salmon rubbed with a mixture of sugar, seasoning and dill and refrigerated prior to packing). These processes do not usually inactivate contaminating micro-organisms and such products, with a refrigerated shelf-life of up to three weeks, may be especially liable to microbial spoilage or pose a food poisoning risk.

This chapter discusses the role of micro-organisms in the spoilage of processed salmon, highlights potential food poisoning organisms and reviews control strategies.

Microbial contamination and spoilage

On removal from the marine environment at harvest, the internal and external bacterial flora of salmon will largely reflect that of the surrounding seawater (Cahill 1990). Although the microflora of salmon is not well known, it is likely to be similar to that of other temperate water marine fish. Dominant genera are Gram-negative psychrotrophic bacteria which will include representatives of *Aeromonas*, *Flavobacterium*, *Pseudomonas*, *Shewanella* and *Vibrio* (Gram & Huss 1996). Gram-positive organisms such as *Bacillus*, *Micrococcus*, *Clostridium*, *Lactobacillus* and *Corynebacterium* can be found in varying proportions (Gram & Huss 1996). Total viable counts (TVC) on skin and gut contents of salmon taken from net pens showed respective mean figures of 10^4 and 10^5 colony forming units (cfu) per g (Ben Embarek *et al.* 1997).

During harvesting and subsequent evisceration, the level of surface contamination of the fish is minimised by chilling carcasses to limit bacterial multiplication and by thorough attention to the hygienic status of contact surfaces, equipment and personnel. The aim is to avoid loss of product quality through spoilage, which in fresh fish is due to the action of autolysis and microbial growth and activity (Gram & Huss 1996). Being poikilothermic, the body temperature of salmon at harvest will be the same as that of the surrounding seawater. Chilling immediately post-slaughter constitutes the first step in reducing spoilage, especially during the summer and autumn months when ambient seawater temperatures are raised.

Spoilage impacts directly on the processing industry through reduction in shelf-life. Lightly preserved salmon products such as cold-smoked and gravad pose particular problems of microbial spoilage. They are consumed as ready-to-eat products without prior heat treatment which would inactivate contaminating bacteria or autolytic enzymes. Storage of vacuum-packed cold-smoked salmon at chill temperatures may result in development of a large microflora reaching levels of 10^7–10^8 cfu per g. However, this does not necessarily coincide with the onset of spoilage. No relation has so far been found between total number of microorganisms, sensory quality and shelf life (Truelstrup Hansen *et al.*1996). Initial quality deterioration is largely caused by autolytic changes and is unrelated to microbial action (Gram & Huss 1996). The variation in flora is wide and is probably related to the different production processes involved (Truelstrup Hansen *et al.* 1996; Leroi *et al.* 1998). Gram negative organisms isolated from vacuum packed salmon include *Vibrio* spp, *Enterobacteriaceae*, *Serratia liquefaciens*, *Rahnella aquatilis*, *Shewanella putrefaciens* and *Photobacterium phosphoreum* (Truelstrup Hansen *et al.* 1996; Leroi *et al.* 1998). Gram-positive organisms are dominated by lactic acid bacteria, especially *Carnobacterium piscicola* (Leroi *et al.* 1998; Paludan-Muller *et al.* 1998). Yeasts and moulds were reported to be rare on smoked salmon stored at 8°C (Leroi *et al.* 1998).

Processed salmon and pathogenic micro-organisms

Huss *et al.* (1995) divided bacterial pathogens which may occur on cold smoked salmon into two groups: indigenous bacteria, commonly found in the aquatic or general environment – members of this group include *Clostridium botulinum*, *Listeria monocytogenes*, *Aeromonas hydrophila* and *Vibrio* spp.; and non-indigenous bacteria, those belonging to the animal–human reservoir and not naturally present on fish and fish products – members include *Salmonella* spp., *Escherichia coli* and *Staphyloccus aureus*. Presence of the latter group of bacteria is routinely monitored on processed salmon and the lack of published data on their occurrence suggests that their presence is probably infrequent.

C. botulinum type E has been shown to be the most prevalent toxin type in fish products and has a worldwide distribution in temperate geographical areas (Hyttia *et al.* 1998). A survey of fish and fish products in Finland showed that contamination was widespread in various fish species and positive product samples included hot smoked salmon (Hyttia *et al.* 1998). According to Huss *et al.* (1995), growth of *C. botulinum* type E should be prevented in cold smoked salmon which is maintained at refrigeration temperature and contains a minimum of 5% sodium chloride. However, any deviation from these conditions could lead to growth of the organism, with potentially serious consequences for consumers.

One member of the indigenous group which has attracted a great deal of industry and academic attention is *Listeria monocytogenes*. Although this is the only significant pathogen of the genus, the presence of other *Listeria* species on product or in the processing environment is often considered by the industry to be significant: harmless *Listeria* species may occur concurrently with *L. monocytogenes* (Eklund *et al*. 1995). However, evidence to the contrary was presented by Rorvik *et al*. (1997), who surveyed 40 smoked salmon smoking plants and showed that detection of other *Listeria* spp. was not positively correlated with detection of *L. monocytogenes*.

Although listeriosis is a rare disease, it may have a high mortality rate among certain susceptible groups. Predisposing conditions include pregnancy, extremes of age and immunosuppression (Farber & Peterkin 1991). *L. monocytogenes* is easily and frequently isolated from a variety of seafoods (Fuchs & Reilly 1992; Gibson 1992) but association between consumption of seafood and human listeriosis has been very rare and no clear epidemiological link has been established. The ability of *L. monocytogenes* to survive the brining and smoking process and multiply at refrigeration temperatures has been well documented (Guyer & Jemmi 1991; Hudson & Mott 1993; Cortesi *et al*.1997). Lightly-preserved products such as cold-smoked and gravad are considered to pose a particular risk because they are ready-to-eat items with a long shelf-life during refrigeration (Loncarevic *et al*. 1996).

It should be emphasised that no case of human listeriosis has to date been positively linked with the consumption of processed salmon products. However, a possible association between human cases of listeriosis and consumption of gravad salmon was suggested by a study in Sweden by Loncarevic *et al*. (1998). Using pulsed-field gel electrophoresis (PFGE), a typing method with good discriminatory power for strain characterisation, they showed that two salmon isolates of *L. monocytogenes* shared the same electrophoretic type (ET) as five human isolates. The incidence of *L. monocytogenes* on cold-smoked and gravad salmon is variable although a number of studies have shown a contamination rate of about 10% (Jemmi 1993; Rorvik *et al*. 1995). To control contamination by *L. monocytogenes*, processors must identify the source: are salmon naturally contaminated in the marine environment or through processing itself?

Listeria monocytogenes in the fish farm environment

There is no evidence that *L. monocytogenes* is a natural inhabitant of the marine environment. However, it is widespread in the terrestrial environment, on plants and excreted by animals (Fenlon *et al*. 1996). Entry to water may be by run-off from agricultural land, direct faecal contamination by animals and release of sewage effluent (Fuchs & Reilly 1992). The incidence in seawater is variable. Soontharanont and Garland (1995), surveying the temperate inshore waters of a bay in Tasmania, found that only 6.6% of water samples were positive for *L. monocytogenes*, even though it received the organism from rivers, creeks and effluents. They suggested that the low occurrence may be due to factors such as dilution, UV radiation and competition by marine bacteria.

Seawater taken from deep water and analysed prior to use in a salmon slaughterhouse was negative for *L. monocytogenes* but 9% of seawater samples taken outside the same plant were positive for the organism (Rorvik *et al.* 1995). Ben Embarek *et al.* (1997) did not detect *L. monocytogenes* in water samples from a salmon farm or from the skin, gills and intestinal contents of salmon removed from net pens. Seagulls are regularly present around marine fish farms and may represent a source of contamination, since their faeces have been shown to harbour *L. monocytogenes* (Fenlon 1985; Laidler unpublished data). The possibility that salmon could become contaminated through ingestion of infected feed was addressed by Bremer and Cooke (1993). A survey of New Zealand salmon farms showed that properly stored feed was negative for *Listeria* spp., but the organisms could be recovered from feed which had been spilled or stored in damaged bags. In a laboratory experiment, salmon were fed with pellets which had been surface inoculated with *Listeria innocua*. The organisms were subsequently isolated from the water column but not from the surface of the fish or from the gut 96 hours after exposure. Contamination of processed product with *Listeria* via salmon faeces is therefore unlikely because it is routine industry practice for fish to undergo seven days of fasting prior to harvest.

Listeria monocytogenes in the processing environment

There are only a limited number of reports on the isolation of *L. monocytogenes* from salmon during primary processing. In a longitudinal study, Rorvik *et al.* (1995) traced the level of contamination through a process plant containing a slaughterhouse and smokehouse. They failed to detect the organism in 50 pools of surface-swabs taken from slaughtered fish in the slaughterhouse but 6/83 (7%) swabs from the slaughterhouse environment and 4/24 (17%) of unfilleted fish prior to entry to the smokehouse were positive for *L. monocytogenes*. Ben Embarek *et al.* (1997) did not detect the organism on the skin and gills of three salmon from a slaughterhouse; similarly small numbers of water and ice samples were also negative.

A study of a Scottish salmon processing plant and one of its supplying harvest sites (D.R. Fenlon, pers. comm.) showed *L. monocytogenes* to be present on fish during processing and widespread in the associated environment. Analysis of 31 isolates by PFGE showed that those from fish and the environment were dominated by two ETs, 031 and 032, with incidences of 39% and 58% respectively (Table 8.1). Furthermore, three isolates from the harvest site had different ETs to those in the process plant.

Table 8.1   Location of electrophoretic types (ETs) of *L. monocytogenes* isolates from a marine salmon farm and primary processing plant.

Source	No. of isolates	ET	No. of isolates
Harvest equipment	2	017	1
		033	1
Seawater	1	033	1
Salmon during processing	2	032	2
Processing plant	29	018	1
		031	12
		032	16

High incidences of contamination with *L. monocytogenes* have been recorded when salmon undergo further processing by filleting and subsequent smoking. In the processing plant study by Rorvik *et al.* (1995) referred to above, the incidence of *L. monocytogenes* rose to 26% and 29% of fillets and smokehouse environment samples respectively. No fillets were positive immediately post-smoking and before further handling, suggesting that the cold smoking process had an effect on the bacteria. The number of samples was low and the authors stated that further investigations were needed before a conclusion could be drawn. Samples of the vacuum-packed final product showed an 11% incidence of the organism. Typing of isolates using multilocus enzyme electrophoresis showed that one (ET-6) was predominant in the smokehouse and was the only ET found on packed final product. Isolates in the process chain before smoking were almost all different ETs from those in the smokehouse and final product.

Eklund *et al.* (1995) showed high incidences of *L. monocytogenes* on smokehouse samples but attributed this to contamination from raw product. Analysis of suppliers' frozen gutted salmon showed that 30/46 (65%) skin samples were positive for *L. monocytogenes*, suggesting contamination during harvesting or evisceration.

L. monocytogenes therefore appears to have the ability to colonise salmon processing plants (Rorvik *et al.* 1995). Loncarevic *et al.* (1998) demonstrated that, from the same processing plant, identical ETs were shared by two isolates of *L. monocytogenes* obtained from vacuum-packed smoked rainbow trout in 1989 and gravad salmon in 1993. These findings indicate that a specific ET of *L. monocytogenes* can survive in a process plant for several years and continually contaminate the products.

Control of microbiological hazards in the process environment

In recent years, the hazards and risks associated with food processing have been influenced by use of the HACCP (Hazard Analysis and Critical Control Point) concept. The basic principle is the identification of potential hazards within a process and their control at critical points. The development and use of the concept in fish processing was described by Huss (1992) and advocated for fish smoking plants by Jemmi and Keusch (1994).

In salmon processing, most of the published information on identification and control of risk factors concerns cold smoked salmon. Rorvik *et al.* (1997) identified job rotation amongst processing personnel as the strongest risk factor for isolation of *L. monocytogenes* from smoked salmon. Well maintained facilities and the use of vats for salting fillets showed a preventative effect.

Garland (1995) provided a detailed identification of the specific hazards causing bacterial contamination and growth in production of cold smoked salmon. Recommendations to control these included the use of chlorinated process water, control of carcass temperature, cleaning and disinfection of contact surfaces, the physical separation of butchery and high-risk dry areas, strict attention to hygiene by staff and restricted entry to the process plant.

Contamination of carcasses at the evisceration stage carries a serious risk that further processed product will be similarly contaminated. Eklund *et al.* (1995) identified elimination or reduction of *L. monocytogenes* on the external surfaces of fish prior to filleting as the first step in controlling the organism in cold smoked fishery products. Removal of contaminated slime layers and inactivating *L. monocytogenes* on the fish surfaces were being studied but details were unpublished. It was proposed that unspecified treatments could be applied to skin and belly cavity linings (discarded during processing) which could not be applied after filleting without damaging product quality.

The aim of controlling *L. monocytogenes* in salmon processing may be to produce ready-to-eat products entirely free from the organism to comply with 'zero tolerance' regulations imposed by some countries (Maple 1995). In Australia, the instigation of zero tolerance in the processed seafood industry (including salmon processors) resulted in some companies going out of business while others rose to the challenge and product contamination levels fell significantly. This was not achieved without substantial expenditure in plant modifications, increased use of hygiene chemicals, staff training and additional sample analyses (Maple 1995).

An alternative strategy was proposed by Gilbert (1995). He questioned whether zero tolerance in foods was either necessary or realistic and argued that because the organism was ubiquitous and very difficult to eliminate from processing plants, the application of zero tolerance would lead to the rejection of many foods. However, attractive though this argument may be to processors whose products are contaminated with *L. monocytogenes*, ultimately they have to comply with conditions imposed by customers or regulatory authorities. For reasons of due diligence, there is a significant bias towards zero tolerance.

Although there is clear evidence for the persistent contamination of salmon processing plants by *L. monocytogenes*, the mechanism of contamination has not been elucidated. However, comparison with other food processing businesses suggests that the presence of biofilms on plant contact surfaces may be a significant factor and would explain why control is so difficult to achieve. Biofilms consist of micro-organisms and their extra-cellular products adhering to a surface (Carpentier & Cerf 1993) and cause problems in many branches of industry (Mattila-Sandholm & Wirtanen 1992). The problems of biofilm contamination in the food industry arise through direct contact of food products with the bacterial population or detachment of fragments of biofilm.

Biofilms may contain a range of organisms which could constitute spoilage or pathogen hazards in processed salmon products including lactic acid bacteria, enterobacteria, *Pseudomonas* spp. and *L. monocytogenes* (Mattila-Sandholm & Wirtanen 1992). In a clear analogy with salmon processing plants, dairy and meat processing facilities are highly susceptible to contamination with *Listeria* spp. and *L. monocytogenes* was shown to be able to grow in multispecies biofilms at 10°C (Dong Kwan Jeong & Frank 1994).

Control of biofilms is usually attempted with a combination of detergents, biocides and mechanical cleaning (Mattila-Sandholm & Wirtanen 1992); the use of a single disinfectant or cleaning agent for eradication of food spoilage and pathogenic bacteria is unlikely to be effective (Wirtanen 1995; Arizcun *et al.* 1998). Bacteria are much more resistant to antimicrobial agents when contained in a biofilm than when dispersed in liquid medium (Carpentier & Cerf 1993) and the resistance increases with the age of the biofilm.

From an industrial standpoint, choice of contamination control methods may be empirical or advice may be sought from hygiene chemical suppliers or external consultants. Process contact surfaces should be microbiologically sampled and monitored (preferably by an accredited laboratory) on a regular basis to assess the effectiveness of hygiene chemicals and procedures. Research into the underlying causes of persistent plant contamination by *L.*

monocytogenes should be undertaken by the industry. If the causes can be identified, more effective control procedures should lead to an improvement in the microbiological status of processed products.

Conclusions

Microbiological contamination may originate in the marine environment and be augmented by processing practices. However, control measures including strict attention to operative and plant hygiene and maintenance of product chilling at all stages of the process chain, should ensure that problems are minimised. *Listeria monocytogenes* remains the greatest perceived threat to ready-to-eat products such as cold smoked salmon. Although no cases of listeriosis have so far been directly linked to such products, the industry must protect its commercial interests by making every effort to safeguard the health of the consumer. In practice, this means continual vigilance in the detection and effective control of the organism and sources of contamination. The fact that *L. monocytogenes* has been widely reported on salmon products is evidence that processors have some way to go in achieving this goal.

References

Arizcun, C., Vasseur, C. & Labadie, J. C. (1998) Effect of several decontamination procedures on *Listeria monocytogenes* growing in biofilms. *Journal of Food Protection,* 61, 731–4.

Ben Embarek, P.K., Hansen. L.T., Enger, O. & Huss, H.H. (1997) Occurrence of *Listeria* spp. in farmed salmon and during subsequent slaughter: comparison of the Listertest™ Lift and the USDA method. *Food Microbiology,* 14, 39–46.

Bremer, P.J. & Cooke, M.D. (1993) A laboratory based study on the accumulation and excretion of *Listeria* spp. in King salmon (*Onchorhynchus tshawytscha*). *Journal of Aquatic Food Product Technology,* 2, 67–78.

Cahill, M.M. (1990) Bacterial flora of fishes: a review. *Microbial Ecology,* 19, 21–41.

Carpentier, B. & Cerf, O. (1993) Biofilms and their consequences, with particular reference to hygiene in the food industry. *Journal of Applied Bacteriology,* 75, 499–511.

Cortesi, M.L., Sarli, T., Santoro, A., Murru, N. & Pepe, T. (1997) Distribution and behaviour of *Listeria monocytogenes* in three lots of naturally-contaminated vacuum-packed smoked salmon stored at 2° and 10°C. *International Journal of Food Microbiology,* 37, 209–14.

Dong Kwan Jeong & Frank, J.F. (1994) Growth of *Listeria monocytogenes* at 21°C in biofilms with micro-organisms isolated from meat and dairy processing environments. *Lebensmittel-Wissenschaft und Technologie,* 27, 415–24.

Eklund, M.W., Poysky, F.T., Paranjpye, R.N., Lashbrook, L.C., Peterson, M.E. & Pelroy, G.A. (1995) Incidence and sources of *Listeria monocytogenes* in cold-smoked fishery products and processing plants. *Journal of Food Protection,* 58, 502–508.

Farber, J.M. & Peterkin, P.I. (1991) *Listeria monocytogenes*, a food-borne pathogen. *Microbiological Reviews,* 55, 476–511.

Fenlon, D.R. (1985) Wild birds as reservoirs of *Listeria* in the agricultural environment. *Journal of Applied Bacteriology,* 59, 537–43.

Fenlon, D.R., Wilson, J. & Donachie, W. (1996) The incidence and level of *Listeria mono-cytogenes* contamination of food sources at primary production and initial processing. *Journal of Applied Bacteriology*, **81**, 641–50.

Fuchs, R.S. & Reilly, P.J.A. (1992) The incidence and significance of *Listeria monocytogenes* in seafoods. In: *Quality Assurance in the Fish Industry* (ed. H.H. Huss *et al.*) pp. 217–29, Elsevier Science Publishers, London.

Garland, C.D. (1995) Microbiological quality of aquaculture products with special reference to *Listeria monocytogenes* in Atlantic salmon. *Food Australia*, **47**, 559–63.

Gibson, D.M. (1992) Pathogenic microorganisms of importance in seafood. In: *Quality Assurance in the Fish Industry* (ed. H.H. Huss *et al.*) pp.197–209. Elsevier Science Publishers, London.

Gilbert, R.J. (1995) Zero tolerance for *Listeria monocytogenes* in foods – is it necessary or realistic? In: *Proceedings of the XII International Symposium on Problems of Listeriosis*, pp. 351–6. Perth, Western Australia.

Gram, L. & Huss, H.H. (1996) Microbiological spoilage of fish and fish products. *International Journal of Food Microbiology*, **33**, 121–37.

Guyer, S. & Jemmi, T. (1991) Behavior of *Listeria monocytogenes* during fabrication and storage of experimentally contaminated smoked salmon. *Applied and Environmental Microbiology*, **57**, 1523–7.

Hudson, J.A. & Mott, S.J. (1993) Growth of *Listeria monocytogenes, Aeromonas hydrophila*, and *Yersinia enterocolitica* on cold-smoked salmon under refrigeration and mild temperature abuse. *Food Microbiology*, **10**, 61–8.

Huss, H.H. (1992) Development and use of the HACCP concept in fish processing. In: *Quality Assurance in the Fish Industry* (ed. H.H. Huss *et al.*) pp. 489–500. Elsevier Science Publishers, London.

Huss, H.H., Ben Embarek, P.K. & Jeppesen, V.F. (1995) Control of biological hazards in cold smoked salmon production. *Food Control*, **6**, 335–340.

Hyttia, E., Hielm, S. & Korkeala, H. (1998) Prevalence of *Clostridium botulinum* type E in Finnish fish and fishery products. *Epidemiology and Infection*, **120**, 245–50.

Jemmi, T. (1993) *Listeria monocytogenes* in smoked fish: an overview. *Archiv für Lebensmittel-hygiene*, **44**, 1–24.

Jemmi, T. & Keusch, A. (1994) Occurrence of *Listeria monocytogenes* in freshwater fish farms and fish-smoking plants. *Food Microbiology*, **11**, 309–16.

Leroi, F., Joffraud, J-J., Chevalier, F. & Cardinal, M. (1998) Study of the microbial ecology of cold-smoked salmon during storage at 8°C. *International Journal of Food Microbiology*, **39**, 111–21.

Loncarevic, S., Tham, W. & Danielsson-Tham, M.-L. (1996) Prevalence of *Listeria mono-cytogenes* and other *Listeria* spp. in smoked and 'gravad' fish. *Acta Veterinaria Scandinavica*, **37**, 13–18.

Loncarevic, S., Danielsson-Tham, M.-L., Gerner-Smidt, P., Sahlstrom, L. & Tham, W. (1998) Potential sources of human listeriosis in Sweden. *Food Microbiology*, **15**, 65–9.

Maple, P.C. (1995) *Listeria monocytogenes* in fish and fish product exports. In: *Proceedings of the XII International Symposium on Problems of Listeriosis*. pp.279–83. Perth, Western Australia.

Mattila-Sandholm, T. & Wirtanen, G. (1992) Biofilm formation in the industry: a review. *Food Reviews International*, **8**, 573–603.

Paludan-Muller, C., Dalgaard, P., Huss, H.H. & Gram, L. (1998) Evaluation of the role of *Carnobacterium piscicola* in spoilage of vacuum- and modified-atmosphere-packed cold-smoked salmon stored at 5°C. *International Journal of Food Microbiology*, **39**, 155–66.

Rorvik, L.M., Caugant, D.A. & Yndestad, M. (1995) Contamination pattern of *Listeria monocytogenes* and other *Listeria* spp. in a salmon slaughterhouse and smoked salmon processing plant. *International Journal of Food Microbiology*, **25**, 19–27.

Rorvik, L.M., Skjerve, E., Knudsen, B.R. & Yndestad, M. (1997) Risk factors for contamination of smoked salmon with *Listeria monocytogenes* during processing. *International Journal of Food Microbiology*, **37**, 215–19.

Soontharanont, S. & Garland, C.D. (1995) The occurrence of *Listeria* in temperate aquatic habitats. In: *Proceedings of the XII International Symposium on Problems of Listeriosis*, pp.145–46. Perth, Western Australia.

Truelstrup Hansen, L., Gill, T., Rontved, S.D. & Huss, H.H. (1996) Importance of autolysis and microbiological activity on quality of cold-smoked salmon. *Food Research International*, **29**, 181–8.

Wirtanen, G. (1995) Biofilm formation and its elimination from food processing equipment. *VTT Publications*, No. 251, 165pp.

Chapter 9

Medicine and Vaccine Residues in Aquaculture

E. Branson[1] and P. Southgate[2]

[1]Red House Farm, Llanvihangel, Monmouth, Gwent, NP25 4HL, UK
[2]The Fish Veterinary Group, Rowandale, Dunscore, Dumfries, DG2 0UE, UK

Introduction

The almost inevitable consequence of the use of veterinary medicines and vaccines in farmed food animals is the appearance of residues in their flesh. In this context, residues are the remains of a medicine and/or its metabolites, persisting in the flesh during treatment and after treatment has ceased. Residues in general terms can also include materials from other sources, such as pollutants, but this is outside the scope of this chapter. As treated animals and/or their products provide food for human consumption, it is vital that any potential residues are assessed for their effect on human health, and that appropriate checks are put in place to ensure that animal products containing unacceptable levels of residues are not placed on the market.

Over the past few years there has been an increase in awareness and concern about the effect on consumer safety of medicine residues arising from legitimate and illegitimate use of medicines in agriculture and aquaculture. As a consequence, not only are effective methods of control on medicine use and surveillance of residues important, but it is vital that any such systems are seen to be effective by the general public.

In the UK, partly in response to European Union (EU) Directives, limits have been put on the type and allowable levels of residues in the flesh of farmed animals; and the presence of residues has been monitored by a mixture of industry and government, in the form of Ministry of Agriculture, Fisheries and Food (MAFF) departments. However, in the wake of several human health issues, this process has been deemed to be insufficient, and a Food Standards Agency (FSA) has been established as an independent body to oversee the general state of safety of food products.

This chapter will consider the types of medicines and vaccines currently licensed worldwide for use in fish farmed for human consumption. The means by which residues of these treatments are assessed for their effects on human safety will also be discussed by comparison with the system in existence within Europe. Withdrawal periods and the monitoring programmes, if any, which ensure that safe residue levels are not exceeded, will also be considered.

Treatments used in aquaculture

Treatments used in aquaculture are usually given by bath or the oral route but in certain circumstances may be given by injection, for example in the case of valuable broodstock. Medicines currently in use are antibiotics (baths, oral, occasionally injection), other anti-bacterials (bath), anti-fungals (bath), parasiticides (baths and oral), anaesthetics (bath) and vaccines (bath, oral and injection).

When fish are treated with a medicine, other fish might also come into contact with the medication. For example, fish may be held in the effluent water from the treated population, so they may come into direct contact with the medicine in the case of bath treatments; or they might consume uneaten medicated feed in the case of oral treatments. In addition fish, or other water users, in the same water course but downstream from a treated population, could be exposed to the effluent from the treated farm. This exposure is significant because, although the in-contact fish will be exposed to sub-therapeutic levels of medicine, there is a risk of residues appearing in their flesh. Therefore residue monitoring needs to take into account any fish which may have been in contact with a treatment, as well as the treated fish.

If fish are kept close to other industries, such as intensive agriculture or industrial plant, there is a potential risk for a range of compounds to appear in the flesh of fish exposed to effluent from these processes, for example polychlorinated biphenyls (pcbs). These factors too must be taken into account when residue testing takes place.

The availability of medicines varies widely between countries, ranging from some countries which have no medicines licensed for use in fish at all, for example Spain, to others in which many are licensed, for example Japan (see Table 9.1). Within the EU legislation has been introduced to standardise the licensing procedures for medicines and, ultimately, to enable a range of medicines to be available to all member states. However, as can be seen from Table 9.1, this legislation has not yet had time to take effect, with the result that medicine availability varies greatly between EU countries as well as between the EU and the rest of the world.

Where no medicines are licensed for use in fish, there will usually be some method by which medicines licensed for use in other food producing species, or unlicensed medicines, can be used legally; but the absence of licensed medicines may encourage the illegal use of medicines.

In the EU, in clearly defined circumstances, veterinary surgeons are able to prescribe medicines for use in fish which are not actually licensed for that use via the 'prescribing cascade'. This is laid out in EU Directive 90/676 and is incorporated into UK law in The Medicines (Restrictions on the Administration of Veterinary Medicinal Products) Regulations 1994 (SI 1994/2987). These state that:

> 'Where no authorised veterinary medicinal product exists for a condition in a particular species, and in order to avoid causing unacceptable suffering, veterinary surgeons exercising their clinical judgement may prescribe for one or a small number of animals under their care in accordance with the following sequence:
> (1) a veterinary medicine authorised for use in another species, or for a different use in the same species;
> (2) a medicine authorised in the UK for human use;

Table 9.1 Medicines licensed for use in aquaculture.

	Country	No. licensed	Medicines licensed for use in fish
	UK	11	Oxolinic acid, oxytetracycline, amoxycillin, potentiated sulphonamide, sarafloxacin, dichlorvos, azamethiphos, hydrogen peroxide, cypermethrin, teflubenzuron, MS222
	Ireland	4	Oxolinic acid, oxytetracycline, potentiated sulphonamide, dichlorvos
EU countries	Greece	3	Potentiated sulphonamide, oxolinic acid, oxytetracycline
	Spain	0	
	France	4	Oxytetracycline, potentiated sulphonamide, flumequine, oxolinic acid
	Italy	4	Potentiated sulphonamide, flumequine, oxytetracycline, amoxycillin
	Norway	7	Oxolinic acid, oxytetracycline, flumequine, potentiated sulphonamide, praziquantel, azamethiphos, diflubenzuron
		4	General exemption – hydrogen peroxide, cypermethrin, deltamethrin, teflubenzuron
	Canada	5	Tricaine methanesulfonate, oxytetracycline, potentiated sulphonamide, formalin, florfenicol
		3	Emergency drug release – Azamethiphos, hydrogen peroxide, teflubenzuron
	US	5	Tricaine methanesulfonate, oxytetracycline, potentiated sulphonamide, formalin, sulfamerazine
Non-EU countries	Japan	28	Ampicillin, erythromycin, oxytetracycline, oxolinic acid, kitasamycin, spiramysin, sulphamonomethoxine, sulphamonomethoxine with ormetoprim, sulphadimethoxine, thiamphenicol, florfenicol, amoxycillin, colistin sulphate, sulphisoxazole, tetracycline, doxycycline, lincomycin, josamycin, nalidixic acid, sodium nifurstyenate, sodium novobiocin, promidic acid, flumequine, sodium polystyrene sulphonate, miloxacin, bicozamycin, phosphomycin, oleandomycin
	China	14	Acriflavine, chloramphenicol, chloramine-T, erythromycin, furazolidone, norfloxacin, oxytetracycline, rifampin, sodium chloride, sodium dichloroisocyanurate, potentiated sulphonamide, tetracycline, trichloroisocynuric acid, trichlorphon
	Chile	9	Iodine, flumequine, oxolinic acid, benzocaine, erythromycin, oxytetracycline, amoxycillin, potentiated sulphonamide, chloramine-T

(3)   a medicine to be made up at the time on a one-off basis by a veterinary surgeon or a properly authorised person.'

In addition, veterinarians treating food-producing animals should only use medicines whose pharmacologically active ingredients are contained in a product already licensed for use in UK in food-producing animals. This is to ensure that residue implications have been evaluated.

Because the licensing of medicines for use in aquaculture is a relatively recent event, many unlicensed chemicals have been used for many years, apparently safely, and the industry has come to rely on such treatments. Examples of these are formalin and malachite green. In the case of malachite green, this is seen to pose a significant risk to human health and, as such, has been effectively banned in most countries. However, substances used in the treatment of other species (such as formalin for the treatment of foot rot in sheep), which are thought not to leave significant residues, have been accepted by many countries for use outside the normal legal framework. These are seen as being necessary for the welfare of fish in farms, not likely to be licensed by any company and harmless to fish and consumer. It is likely that this type of use will disappear as more medicines are introduced.

Control systems

In common with the large variability of available medicines from country to country, there is also considerable variation in the level and type of controls on the licensing of medicines, and on the monitoring of their use. Generally speaking all countries involved with aquaculture have some form of medicine licensing system and, in principle, only licensed treatments should be used for the treatment of fish. However, the quality of the licensing system used, and the level of control over the use of non-licensed medicines, varies from country to country. Where no products are licensed for use in aquaculture, such use of non-licensed products is inevitable.

Licensing of medicines

Within the EU, medicines can be licensed by national bodies for individual countries, or can go through a centralised procedure resulting in a Marketing Authorisation (MA) which is recognised by all countries within the EU. The national body in the UK is the Veterinary Medicines Directorate (VMD), a part of MAFF. Licensing takes into account the quality, safety and efficacy of a compound. The safety element includes safety of the target animal, the operator, the environment and, most important for the purposes of this chapter, the consumer.

In the EU the assessment for consumer safety is carried out according to EU Directive 86/469/EC and Regulation 2377/90 (the MRL regulation). The Committee for Veterinary Medicinal Products (CVMP) considers the safety of the medicine with respect to the consumer, and sets a Maximum Residue Level (MRL) for the medicine. This MRL refers to the pharmacologically active ingredient of the medicine, or an appropriate marker metabolite, and applies throughout the EU once set.

The MRL is set by initially determining the No Observed Effect Level (NOEL) by using a series of appropriate toxicological tests. The most sensitive and appropriate NOEL is then divided by a safety factor to give an Average Daily Intake (ADI). The ADI is the amount of a medicine which can be consumed safely each day by the 'average 65 kg consumer'. Once the ADI has been set, this amount is divided up between all the constituents of an 'average shopping basket' which might contain the medicine. The amount of medicine allowed for each constituent in the basket is the MRL for that product.

The 'average shopping basket' assumes that 300 g of fish is consumed per day by the 'average 65 kg consumer'. This assumption raises several issues:

(1) The size of the average consumer will vary from country to country;
(2) The amount of fish consumed by the average consumer will vary from country to country because some countries have more of a fish eating culture than others. For example, the average per capita consumption of seafood in Germany in 1997 was 14.5 kg, compared to 36.5 kg in Spain for the same period (information from the Scottish Salmon Growers Association);
(3) The proportion of farmed fish, the product most likely to contain medicinal residues, to total amount of fish eaten, will vary from country to country. For example, the amount of salmon eaten, as a percentage of the per capita consumption of seafood, in 1997 was 7% in France and 2% in Italy (information from the Scottish Salmon Growers Association).

The consequence of this is that the MRL system must assume the worst case, that is the smallest consumer eating the maximum amount of farmed fish every day, and as this consumer is unlikely to exist, this adds an extra safety factor to the MRL equation.

One obvious problem arises with the MRL system where countries have no medicines licensed for use in fish, so that medicines licensed for use in other species and even unlicensed medicines may be used. In these circumstances, if a medicine is used which contains active ingredients for which an MRL for fish has been set, there should be no problem. However, if a product is used which has no MRL for fish, or even for any species, potential problems might arise.

If an MRL has been set for species other than fish, or even exclusively for one fish species, this MRL may not be appropriate for another fish species: the metabolic pathway in the treated fish species might be different from that in the species for which the MRL has been set, so the marker residue may be different. In these cases, and with the use of unlicensed compounds, a potential problem can arise if analysis is carried out for the wrong residues.

Once a medicine has been assessed for an MRL, it is normally assigned to one of four annexes according to its status. Annex I contains medicines for which an MRL has been set. Annex II contains medicines for which no MRL is deemed necessary, for example common salt. Some products are granted a provisional MRL pending completion of the assessment, usually due to some information being inadequate, and these are put into Annex III. In some cases medicines are considered to be too dangerous to be used for the treatment of fish at any time, and in this case they are placed in Annex IV. Chloramphenicol is one of these.

Once an MRL has been set it is taken into account by the national licensing authorities when an application for an MA is being considered. The granting of the MA will include the setting of a withdrawal period, namely the time taken, after the cessation of treatment, for residues within the treated animal to fall to the MRL or less. Usually this withdrawal period will be stated in °C days (withdrawal period (°C days) = temperature (°C) × time (days)), although in some countries, such as Norway, a withdrawal period in days, independent of temperature, may be given. The concept of a temperature related withdrawal period is usually applied because, as fish are poikilotherms, their physiological processes are dependent on ambient temperatures and, as such, it is likely that medicine elimination will also be temperature dependent. As stated above, at present each European country sets its own withdrawal period, but eventually this will change as licensing is harmonised within the entire EU.

Organic fish are grown in some countries, and avoidance of residues is a major issue with all the schemes under which these are produced. Where treatments are necessary, most schemes insist on the withdrawal period being increased beyond the legal requirement, often doubling it, to ensure the absence of residues.

Where medicines are used under the 'prescribing cascade' system, i.e. a medicine is used to treat a condition and/or an animal for which it is not licensed, a standard withdrawal period must be applied under EU law. In the case of fish this is a minimum of 500°C days, which can be extended, at the discretion of the veterinary surgeon, if there is any concern about the persistence of residues. This withdrawal period is longer than the withdrawal period for most licensed medicines, and its length will have implications on when the fish can be slaughtered for consumption. In these circumstances a welfare problem may arise. Treatment of animals may be withheld if the withdrawal period of the medicine to be used is likely to interfere with the sale of the fish. This conflict may not occur if the withdrawal period was appropriate for the medicine in use rather than an arbitrary 'catch all' figure, and the problem will only disappear once adequate licensed medicines are available.

In the USA veterinary surgeons can use information provided by the Food Animal Residue Avoidance Database (FARAD), which makes available data for the use of medicines in minor species, including fish. This allows a wide range of medicines to be legally prescribed and their withdrawal periods to be varied according to the medicine used, whilst still safeguarding against excessive residues appearing in the flesh. Although this system is not legal within the EU, it is still based on an MRL type approach, and most countries in the world involved in aquaculture use a system similar to this EU MRL system in order to ensure consumer safety (see Table 9.2).

Table 9.2 Medicines licensed for use in aquaculture – consumer safety.

Country	Consumer safety considered	Residue monitoring
EU	MRL system	Statutory
Norway	MRL system but zero residue policy	Statutory
Canada	MRL system	Statutory
US	Tolerances – MRL equivalent	Statutory
Japan	Zero residue policy	Statutory
China	Safety assessed for licensing	Export only
Chile	MRL equivalent	Statutory

Residue monitoring

Once the acceptable residue level for a pharmacologically active ingredient within a product has been set, and a licence granted for its use, the next control phase is the monitoring of residues to ensure that acceptable levels are not exceeded.

Within the UK a statutory residue surveillance scheme has operated for some years in other farmed species according to EU Directive 86/469/EC but, until 1 January 1999, monitoring of fish was only by means of a non-statutory scheme. This situation has now changed. EU Directive 96/23/EC extended the surveillance of other species to include, amongst other things, fish throughout the EU. There is now a statutory National Surveillance Scheme for Residues in Farmed Fish in operation within the UK. This scheme randomly tests farmed salmon and trout for a range of chemicals, looking for violations of MRL levels, but also for prohibited medicines and possible contaminants from the environment, such as pcbs. Certain groups of substances and some specific materials for which testing must be carried out are detailed in an annex of the Directive, but mostly the actual substances tested are set by the member state.

In addition, producer organisations (POs) and most major retailers will have in place random testing programmes to check these same issues. Some POs require a positive release scheme to ensure the absence of unacceptable residue levels. In this system any stock having received a medicine will automatically be tested for the presence of that medicine before being released to the market. The statutory testing scheme within the UK only covers salmon and trout farmed within the UK. Similar schemes within other EU countries will apply to their main farmed species in order to comply with the Directive. A non-statutory scheme also still exists within the UK, and this randomly tests tiger prawns and tilapia, but non-salmonid species farmed within the UK and other aquaculture imports from outside the UK are significantly absent from this testing. This programme is purely a UK scheme, and is designed to support and complement the statutory testing programme, but is not an EU requirement.

When third countries wish to export aquaculture goods to the EU, there is an EU requirement (Directive 97/78/EC, the 'veterinary checks' Directive) that whatever checks are necessary within the EU should also apply to the third country. Officials visit the third country, including individual establishments within the countries, to ensure that controls are in place and are being applied to the correct standard. Directive 97/78/EC also requires that border controls are applied at the place of importation, designated border inspection posts (BIPs), to ensure compliance. This Directive requires that a minimum proportion of consignments are subjected to physical checks, the actual number dependent on the product. Physical checks must be carried out on a minimum of 20% of consignments of fresh, frozen, dry, salted or hermetically sealed fish/fishery products, or 50% of consignments of other fish/fishery products. A minimum of 1% of consignments which are subjected to these physical checks must also be tested for residues. The actual numbers and types of tests carried out are, however, left entirely to the discretion of the attending official veterinary surgeon (OVS).

One obvious problem which arises with this situation is over the use of a medicine in a third country for which no MRL has been granted within the EU, either because it has not been assessed or has been placed in Annex IV as a banned substance. Such medicines should not be present within the flesh of a product on the market within the EU, even if the residues are not detectable, because the medicine in question has not been assessed for consumer safety. Normal testing procedures approved by the EU and carried out at BIPs may not detect such residues, and almost certainly would not be able to detect historical use. Consequently it is possible for products to be imported into the EU having been treated with medicines whose use would be illegal within the EU, but whose use is impossible to detect. For example, chloramphenicol is an EU Annex IV designated medicine, meaning that it is considered to be too dangerous for residues to be present in the flesh of food producing animals at any stage of their life cycle. If this medicine is used, but residues are undetectable at the point of sale of the fish, there will be no way of knowing whether or not the medicine has been used, and a potentially dangerous product could be available for sale.

Problems such as these would, of course, disappear if illegal medicines were unavailable but, unfortunately, some use of such medicines is probably always going to occur. There can never be any certainty of a medicine having not been used unless testing is sufficiently rigorous to detect *any* previous use of the medicine in question.

Residue testing schemes operated by most major retailers will give an added level of security when considering imports from third countries, and they may well also monitor their suppliers in third countries. However, this will be constrained by the limits on the range of residues for which analyses are carried out.

In addition to all the above controls there are also various committees considering the safety of medicine residues within food producing species. The Department of Health has a committee on toxicity, mutagenicity and carcinogenicity, and examines medicines under consideration for use in food producing animals and highlights problems to the appropriate authorities. An FAO/WHO Joint Expert Committee on Feed Additives (JECFA), amongst other things, recommends MRLs for consideration in setting internationally recognised MRLs.

A Food Standards Agency has been proposed for the UK. The exact responsibility of the FSA is still under debate, but it will carry out monitoring of farming processes which impinge on human health. The VMD will retain responsibility for the monitoring of residues, but there will be cooperation with the FSA. The FSA will have an input into the surveillance programme and on the way results are dealt with. The FSA will also retain the right to carry out separate monitoring if deemed necessary. This same arrangement will apply to use of pesticides. Any direct implications of residues on health will require collaboration between MAFF, FSA and the Department of Health to ensure an effective method of control.

An additional safeguard on the use of fish medicines in the UK is the surveillance programme carried out by the Scottish Environmental Protection Agency (SEPA) in Scotland and the Environment Agency (EA) in England and Wales. This monitoring may detect any illegal use of chemicals and, consequently, helps to constrain the illegal discharge of chemicals into the environment.

Most of the discussions above have centred on finfish, and the controls in this area of aquaculture are generally quite good. However, there is a whole section of aquaculture, that concerned with the production of shellfish and crustacea, where few if any controls are in place.

Vaccines

Current fish vaccines, by their nature, tend to be a lesser issue than medicines where residues are considered. In addition, the range of vaccines used worldwide is much smaller than the range of medicines (see Table 9.3). Generally they tend to consist of an antigen, either in an aqueous form or in conjunction with an adjuvant. According to EU Directive 81/852/EEC, the only parts of vaccines of relevance to consumer safety are adjuvants and preservatives. Antigens associated with the usual fish diseases are of no concern to consumer safety. However this may change if DNA vaccines are introduced, but so far these are not in use in fish or in general.

Table 9.3   Vaccines licensed for use in aquaculture.

	Country	Vaccines licensed for use in fish
EU countries	UK	ERM, Vibriosis, Furunculosis
	Ireland	ERM, Vibriosis, Furunculosis, Multi-valent oil adjuvanted vaccines (Furunculosis, Virbriosis, Hitra)
	Greece	Vibriosis and Pasteurellosis
	Spain	ERM
	France	ERM
	Italy	ERM and Vibriosis
	Norway	Hitra, Vibrio anguillarum, Vibrio viscosus, Furunculosis, IPN, all with oil adjuvants
Non-EU countries	Canada	Multi-valent oil adjuvanted vaccine (Vibriosis, Hitra, Furunculosis), Autogenous ISA vaccine
	Japan	Enterococcus seriolicida, Vibrio anguillarum, Vibrio ordalii
	China	None
	Chile	ERM, Piscirickettsia

Obviously the presence of preservatives could be an issue, but they are generally not used in fish vaccines. Formalin may be present, but this is usually only a trace, the result of the inactivation process, not as a preservative, and so is considered to be harmless from the point of view of residues. Adjuvants in general use for vaccines have also been assessed for residue implications and are considered harmless; they are included in Annex II of the MRL Regulations.

Within the EU all vaccines should be licensed before use, and the licensing procedure is similar to that for medicines. Use of vaccines in other countries, as with medicine use, is not necessarily restricted as in the EU. However, consumer safety is not such an issue with these products, for the reasons given above.

Conclusions

The principles behind the licensing of medicines and testing for residues are sound, and there would be few, if any, residue problems if all medicines used for treatment were licensed, and withdrawal periods were all strictly adhered to. However, in practice there are many treatments used which are not licensed, or are used outside the terms of their licence, and withdrawal periods may be inappropriate. In fact control and monitoring methods used in the various centres of aquaculture vary, from almost non-existent to very strict.

It has been shown above that, within the EU, there are laws which require strong controls on the use of medicines in aquaculture and the monitoring of such use. However, even with these controls problems can arise. Medicines used via the 'prescribing cascade' may not have been rigorously tested in the target species of fish, and it is possible that the pharmacokinetics of the medicine in the treated fish could give rise to a chemical whose safety implications to the consumer are not known. The result of this is that there could be a problem with residue testing if the wrong compound is looked for. It is also theoretically possible that some of these by-products could be toxic.

As monitoring, even under the EU Directive 96/23/EC within the EU is, to a certain extent, under the control of the member state, individual monitoring programmes may not test for all relevant residues. Third countries, outside the EU, may use a whole range of medicines not licensed, or even used, within the EU. Consequently, although these countries may have control systems, their systems may be less effective or rigorous than the EU systems. As testing at BIPs into the EU is minimal, residues may also not be detected at this stage. This will be especially true in cases where medicines which are not recognised within the EU are used in the third country. Importing bodies such as supermarkets introduce an extra level of checks to ensure that their imports are 'wholesome', but these may not be sufficiently rigorous or may be too narrow in scope.

There is also the issue of the testing methods used. Methods are continually improved and are generally increasing in sensitivity. For example, the replacement of the microbiological method for determining antibiotic residues by the HPLC method has greatly reduced the limits of detection and quantification. This increase in sensitivity may detect the illegal use of medicines which had previously been undetectable, and consequently increases the safety factor of the monitoring programme, as long as the correct residues are looked for. All the above comments refer to finfish, but there is a whole section of aquaculture – concerned with the production of shellfish and crustacea – where few if any controls are in place.

How are these problems going to be resolved? The only way that these issues will be dealt with is by standardisation on a worldwide basis. The EU has made a start on this process, but even within Europe it will take some considerable time to achieve the necessary level of standardisation. Beyond Europe it is necessary to standardise practices on a worldwide basis – standardisation of medicine licensing, use and surveillance – in order to ensure that all the food we eat, wherever it is grown, is of sufficiently high quality.

Acknowledgements

The authors would like to acknowledge the help of many people, too numerous to mention, from many different countries, for their assistance with information for this chapter.

Chapter 10

Ethical Considerations in the Handling and Slaughter of Farmed Fish

A.J. Wall

Fish Vet Group, 22 Carsegate Road, Inverness, IV3 8EX, UK

Introduction

This chapter will discuss preslaughter management practices and slaughter methods, and how these might influence both the welfare of the fish and the quality of the final product. Generally speaking there is no conflict between the welfare of the fish and quality. Good welfare and good quality go hand in hand.

Preslaughter management

Feeding strategies

It was quite common 10 years ago to starve fish prior to slaughter for up to two weeks. This was in the mistaken belief that flesh quality could be significantly changed during this period, hopefully by reducing the oil level in the muscle. It is now realised that fish must be starved for a much longer period to achieve this oil reduction. Feeding a reduced amount can cause over competition in a cage of fish with a resulting bimodal distribution of live weight in the population as the less dominant fish are out-competed. A more satisfactory way of achieving a reduction in fat levels would be the introduction of a pre-harvest diet which would be fed at the usual daily rate (usually to appetite). This diet would have a much lower oil level than is fed throughout the rest of the production cycle. If, however, fish were monitored throughout the production cycle for pigment, oils etc., the use of such emergency diets might be unnecessary as fine tuning would have taken place beforehand.

In practice now, most fish are starved for one to three days, depending on water temperature, i.e., just long enough to ensure complete emptying of the gut. The removal of faeces is important for two reasons: first to reduce the occurrence of food spoilage organisms and second, to ensure that during the preslaughter crowding period, oxygen levels remain high. These can be jeopardised by the presence of organic material.

Stocking density

There is a tendency prior to slaughter to allow stocking densities to creep up. It is well recognised that high densities can lead to stress which may affect the quality of the fish. External abrasion of flanks, tails and snouts is commonly seen in this situation. Any sub-clinical or carrier disease may become an overt clinical condition which will not only affect quality but can be difficult to treat due to the imposition of withdrawal times of medicines and chemicals. Finally, fish held at high stocking densities seem to be more susceptible to predation, particularly seals in sea cages (Fig. 10.1).

Fig. 10.1   Seal attack on salmon ready for slaughter. This attack took place in a cage with a high biomass.

Transport

In some systems fish will need to be moved from the growing site to the slaughter area. This is more common in salmon production systems where cages may be towed, or fish are moved by well-boat. Towing cages can lead to unexpected problems often caused by winds and tides. Planning and patience are all important. Towing too quickly, especially against a tide, will cause fish to become damaged as they become crowded at the back of the cage.

Well-boat transfer is becoming an industry norm for transporting fish prior to harvest. The usual consequences such as damage by pumps, transfer of disease and the need to maintain good water quality are self evident. Most importantly in any transport operation, it will be necessary to allow the fish to be rested prior to slaughter. Failure to implement this resting period can affect the quality of the flesh.

Crowding prior to slaughter

In order to deliver the fish to the killing area it is necessary to crowd the fish to pump, braille or air-lift them (Fig. 10.2). This procedure is often responsible for damaging and killing fish. The common factor involved in these episodes is often lack of oxygen in the water. If the fish

Fig. 10.2   Fish crowded prior to slaughter. Note clean nets are present and the pump-pipe is wide enough to allow even large fish to enter comfortably. Crowding should be carefully monitored, especially the dissolved oxygen levels.

are too crowded oxygen can fall causing an increase in excitability with an even faster decrease in the oxygen. If the nets are dirty, this organic material can contribute significantly to the low oxygen. This increase in excitability will cause burrowing, with abrasions leading to scaling, skin ulceration, eye and snout damage and bruising (Figs 10.3 and Fig. 10.4). Fish should not be too tightly crowded and it is preferable to draw more fish into the crowding area using a sweep net every few hours than allow all fish to be crowded for too long. The regular monitoring of the oxygen levels at this time in the crowding area is important. If necessary, in times of high water temperatures, adding oxygen by diffusers may be necessary. Experience shows that crowding damage is one of the prime causes of both poor welfare and poor flesh quality.

Fig. 10.3   Scaling and abrasions as a result of prolonged crowding with low oxygen levels.

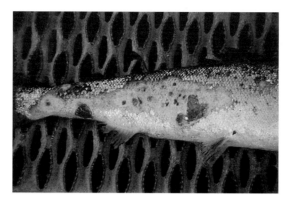

Fig. 10.4   Scaling and abrasions as a result of prolonged crowding with low oxygen levels.

Moving fish to the killing area

This movement can be carried out by pumping, air-lift or brailling. Both pumping and air-lift should have inlet funnels and pipes wide enough to accommodate the fish easily. Brailling should always be carried out with modified nets that allow the fish to be transferred in water. With any method, fish should not be allowed to drop onto hard surfaces, there should be no projections likely to wound the fish and crane operators should be experienced when brailling. Let the novice crane operator practice on inanimate objects, not on live animals! In summary, the fish that are presented for killing should be rested, fasted only long enough to ensure gut emptying, and have no external injuries or muscle bruising.

Slaughter

Killing fish requires special considerations. Large numbers of fish are often killed in a short time. Over 10 000 trout may be killed at one time and practices should be in place to ensure each fish is humanely killed. Fish are resistant to hypoxia, especially at lower temperatures. Thus to deprive fish of oxygen as a method of slaughter is neither efficient nor humane. Equally, decapitation of fish without destruction of the brain does not eliminate normal brain function and is therefore unlikely to be humane. Small fish are difficult to handle individually. The corollary of this is that systems should be in place to effectively and humanely slaughter small fish without the need for individual handling. There are considerable species differences which make universal methods difficult to adopt. Eels are very difficult to handle and are notoriously resistant to hypoxia, as are carp. The shape of the flat-fish make percussive stunning difficult to achieve, and as well as damaging eyes, it can also cause a downgrading in quality.

Slaughter objectives

(1)   *Increase efficiency.* Can more fish be killed per unit time without compromising welfare or quality? For instance, can we double the number of percussive stunners and so halve the

time of crowding? Changing/rotating personnel is important in this work especially as staff need to be mentally alert (monitoring the crowding area), manually dextrous (percussive stunning) and patient (during carbon dioxide narcosis).

(2) *Decrease fear*. Fish must be delivered to the killing area in a calm, controlled manner to reduce the levels of fear. The adjunct of this is that any handling should be done quickly and fish should be out of water for as short a period as possible.

(3) *Decrease pain*. It is reasonable, until evidence is shown to the contrary, that fish feel pain in a similar manner to the higher animals. It is for this reason that death or insensibility should intervene quickly and that no damage occurs to the fish during handling and crowding.

(4) *Improve quality*. It is fortuitous that the above three points, absence of pain and fear and increased efficiency, will make for a better final product. Pain can be caused by bruising and skin ulceration which will lead to down-grading. High levels of stress resulting from fear can lead to pH changes as increased muscular activity will affect flesh quality. It is well recognised that an efficiently killed fish will lead to a high quality product.

Slaughter methods

Whatever method is used it is necessary to do the job properly. The Farm Animal Welfare Council document on Welfare of Fish noted:

> 'Staff employed in the killing of fish must have the knowledge and skill to perform the task humanely and efficiently regardless of the method employed as required by the Welfare of Animals (Slaughter) Regulations 1995.'

It may be necessary in the future to train personnel involved in the slaughter of fish just as mammal and poultry slaughterers are trained and accredited.

Percussive stunning

Percussive stunning is mainly used on salmon and other large fish. The fish is stunned one or more times with a hard dense club, usually plastic (Fig. 10.5). The blow is often applied anterior to the brain, but the shearing forces produced cause a massive brain disruption and the fish is instantly insensible. Haemorrhage will usually occur later. It is possible in theory for the fish to recover and it is important that this method is followed up by exsanguination. However, the main reason for exsanguination is to improve flesh quality. This method is fast and humane. But the operator can become easily tired and should be rotated. Inaccurate blows can lead to eye damage, bruising and some down-grading (Fig. 10.6). Modifications of this method have been devised and include some automatic aspects. At the moment they are in a trial stage.

Carbon dioxide narcosis

Fish are placed in a bath of water containing dissolved carbon dioxide. After about three minutes the fish become insensible whereupon they are removed from the water and the gills are cut to effect exsanguination and death. The carbon dioxide water is aversive to the fish and they show great excitability in the first few moments. Immobility usually precedes

Fig. 10.5   Percussive stunning.

Fig. 10.6   Eye damage due to incorrect percussive stunning.

insensibility so it is important that the fish are left in the bath for a full three minutes to ensure exsanguination does not take place in conscious fish. All four gill arches should be cut on one side (Fig. 10.7).

It is easy to criticise this method on the grounds that the bath water is very aversive to the fish. However it must be remembered that this is a very efficient, quick method of narcotising large numbers of fish of all sizes and is effective in many species. This could be the basis of new, more humane killing methods in the future, possibly using a controlled carbon dioxide delivery system or even using inert gases to promote anoxia.

Exsanguination alone
Fish are removed from the water, the gills are cut and the fish are either returned to the water or put into ice slurry. The fish show signs of aversion for up to 10 minutes with violent head shaking and exaggerated opercular movements. This method has, quite correctly, been banned by some producer groups.

Fig. 10.7   Exsanguination should be achieved by cutting all four gill arches.

Dying in air

Fish, usually table-sized trout at about 400 g, are removed from the water and are allowed to die in air. The period before insensibility ensues is variable and will largely depend on temperature (the lower the ambient air temperature, the longer it takes to die) and the species of fish involved (carp and eels are more resistant to hypoxia than Salmonids). At normal summer temperatures in Scotland insensibility will occur within four minutes in rainbow trout. As well as the welfare concerns associated with this prolonged slaughter method the physical activity of the fish in the early stages may have an adverse affect on flesh quality.

Dying in ice slurry

This method, mainly employed on rainbow trout, is commonly used in the UK. Fish are introduced directly from the water into tubs of ice and water. This has the effect of rapidly reducing the core body temperature of the fish with a similar reduction in the metabolic rate. The oxygen requirements of such fish are minimal and, coupled with the ability of the fish to tolerate hypoxia, these fish can remain alive for long periods. On the face of it, this method, although it produces good quality fish, is not acceptable because of the prolonged period before death supervenes. Indeed, the Farm Animal Welfare Council in its recent report, was very critical of this method and said:

> 'Trout taken out of water and cooled on ice take much longer to die but remain sensible to stimuli and we recommend that this method should be prohibited.'

However, the advantages are significant (quick, easy and good flesh quality) and it may be that research could be directed at improving this method. The use of carbon dioxide or inert gases in the ice slurry tubs could hasten death.

Electrical stunning

Electrical stunning has been used from time to time to kill fish. Undoubtedly it is effective but has the unfortunate side effects of causing muscle bruising (haemorrhages) and some spinal fractures. Achieving the correct voltages can mitigate these effects to some extent. Operator safety can be a problem in some of the older equipment.

In summary, research into methods for fish slaughter is urgently needed.

Conclusions

It is important that we should step back from accepting existing fish slaughter methods and take a long cool look. Perhaps we should be developing a number of pre-harvest strategies which would enable the fish to be killed more easily and would be associated with less stress. As well as the other measures mentioned in this chapter, it may be worth looking at two things. First, pre-slaughter cooling of fish would render the fish less active and more easily managed prior to slaughter. This would not be a slaughter method itself but would have the added benefit of improving flesh quality. Second, the use of sedative agents such as clove oil would to some extent have the same effect as pre-slaughter cooling. It remains to be seen whether this method would be acceptable to the consumer.

In killing large numbers of fish it may not be possible to achieve the mammalian ideal of immediate insensibility. But, if this entails high levels of stress prior to slaughter, it may not be desirable. A low stress system prior to the point of slaughter may be just as important as the actual method used.

Chapter 11

Ethical Considerations for the Production of Farmed Fish – the Retailer's Viewpoint

M. Cooke

Tesco House, P.O. Box 18, Delamare Road, Cheshunt, Hertfordshire, EN8 9SL, UK

Introduction

It may appear anomalous in a chapter about ethics in the context of fish quality, but I am focusing here on trade. Ethics, even in this context, is not just a synonym for animal welfare; it is the cornerstone of exchange. It can be argued that the production and marketing of farmed fish is the subject of four ethical domains:

(1)   Employment
(2)   Trading
(3)   Animal husbandry
(4)   Environment

Employment

I believe that, in the UK at least, the employment environment for people working in the fish farming industry is ethically acceptable. By that I mean that adequate standards of safety for staff are achieved and that workers' rights are protected. Nevertheless, it behoves all of us in the farmed fish production and marketing chain to ensure that ethical employment conditions are maintained. This is particularly true when farmed fish are obtained from other parts of the world with less robust health and safety and employment legislation than our own. Farmed fish is a global commodity. It should be seen as an ethical advantage that salmon farming contributes significantly to the rural economy of an otherwise relatively impoverished region in the UK.

Trading

To retailers, fresh animal protein products, such as meat, poultry and fish, are key to our business. Customers are attracted into our stores by the quality, variety and high standards of our own brand fresh food products. It is the benchmark by which customers judge the quality of their shopping experience.

We have an ethical responsibility to our customers to provide farmed fish of consistent high quality, safe to eat and at the least cost. I believe that we have achieved this, particularly with salmon. The MAFF fish and shellfish cultivation review concludes that salmon farming in the UK has been a success for farmers, retailers and consumers alike. Volumes of production have increased, prices have declined and UK consumption has increased dramatically.

However, customer attitudes and preferences are continually evolving. In such a fiercely competitive market place, like Lewis Carroll's Red Queen, we have to keep running hard just to maintain our position. The demand for consistent quality and low price almost goes without saying. As Sir John Harvey Jones says, we are always looking for more of better out of less.

Fish has a rather privileged status in many customers' minds, compared to many other protein products. It has an extremely clean, healthy, untarnished image. Fish is seen as wholesome and nutritious; it is brain food. It is part of our ethical responsibility towards our customers to act to protect that image. This was the philosophy behind the decision of many retailers to exclude growing sites affected by infectious salmon anaemia from our supply. Another example is the reduction of the use of canthaxanthin in salmonid diets, despite the fact that one would need to eat 20 kg of salmon a day to match the intake of carotenoid in the suntan pills that sparked the issue. By acting to reassure our customers we also defended our market and therefore the industry's market. Anthony Burgmans put it rather well, albeit in a slightly different context: if we fail to act we put our reputation at risk; we put the bond of trust with our customers at risk.

Welfare

Customers are becoming increasingly articulate in the process of choice. It is no longer only a question of quality and price for many consumers. They are concerned about the way in which animals are reared for food and also about the impacts of the farming process on the environment. It should be remembered that there is a small but significant body of opinion that rejects completely the farming of animals for food.

As mentioned earlier, ethics is not a synonym for animal welfare; however, satisfactory welfare standards are the cornerstone of an ethically acceptable farming system. Tony Wall addresses certain specific welfare issues in Chapter 10. I would like to focus on some broader aspects.

At Tesco we have taken as the benchmark for our animal welfare policy the five freedoms paradigm, proposed by the Farm Animal Welfare Council. These are:

- ❏ Freedom from hunger and thirst: by ready access to a diet to maintain their full health and vigour
- ❏ Freedom from discomfort: by providing an appropriate environment
- ❏ Freedom from pain, injury or disease: by prevention or rapid diagnosis and treatment
- ❏ Freedom to express normal behaviour: by providing sufficient space, proper facilities and company of the animal's own kind
- ❏ Freedom from fear and distress: by ensuring conditions and treatment which avoid mental suffering

I appreciate that these principles were devised with terrestrial animals in mind, but there is no reason for them not to be adapted to farmed fish as well. The key to this, and the challenge to our industry, is to understand the needs of the animals we keep.

There is a tendency in all intensive farming systems to adopt a utilitarian approach to animal welfare. That doctrine holds that the right action consists in the greatest good for the greatest number without regard to the distribution of benefits and burdens. It aims to maximise the total benefit and usually measures this in terms of some growth parameter, such as food conversion or daily liveweight gain.

However, animal welfare is an individual animal issue. There is a story which illustrates this. There has been a storm and all along the beach starfish are washed up in their thousands. A man is walking along and, to prevent them from dying, picking them up and throwing them back into the sea. A passer-by asks him, 'Why are you doing this? There are thousands of starfish along miles of beach and you can only throw back a few. What difference does it make?' The man replies, 'It makes a difference to this one and to this one and to this one...' This is easy to forget when each farming unit may contain 50 000 individuals.

Animal welfare issues are not always resolved by collective solutions. Television scare stories are made about the individual animals we have failed to protect, never the mass that we have looked after well.

Customer expectations of animal welfare are high. They do not want to think about ethics; they want to feel comfortable and not worried about the food they eat. Their loyalty to a brand or a product depends on our ability to allay their concerns. The key to our industry's success lies in being beyond reproach, to be the customers' champion and the animals' champion.

Environment

Customers are becoming increasingly concerned about the impact of farming systems on the environment. Coastal and riparian habitats are not pristine, but they are probably the closest thing we have to it. We therefore have an ethical obligation to ensure that fish farming activities do not harm these fragile ecosystems. The starting point is formal benthic monitoring under and around farm sites as well as water quality monitoring in freshwater environments. However, this is not necessarily enough. We need to understand more about the interaction between farmed and wild fish. And we need to understand more about the effects of fish farming activities on other organisms that share the environment. For example, marine conservation groups are becoming particularly concerned about the disturbance effects of acoustic deterrent devices, seal scrammers, on threatened harbour porpoise populations.

Another aspect of our ethical responsibility to the environment is the issue of sustainability. This relates particularly to the use of fishmeal in most farmed fish diets. It is possible to argue that farmed fish eat no more fish, in fact probably less, than they would do if they were wild and that fish farming is therefore more sustainable than wild harvest. This argument ignores the local effects both on the target species and on other non-target organisms that harvesting of fish for fishmeal may have.

Conclusions

To finish on a positive note, it must be ethically robust that modern finfish farming techniques are among the most efficient forms of animal protein biomass production yet devised. A 100 m Polar Cirkel cage, containing 250 tonnes of fish at harvest, equates to the production of 6250 lambs on over 500 ha. The challenge to the fish farming industry is to demonstrate that this amazing feat is conducted in an ethical and humane way. It is not enough just to be doing it. To trade ethically to the satisfaction of our customers' expectations we must be able to prove that we are doing it.

Chapter 12

Problems of Sea Bass and Sea Bream Quality in the Mediterranean

G. Smart

Culmarex SA, Muelle del Hornillo, 30880 Aguilas, Murcia, Spain

Introduction

Sea bream have long been considered high value fish in Mediterranean regions, yet the species is not well known in northern Europe or indeed even in the non-Mediterranean regions of Spain or France. On the other hand, sea bass are well known in Northern Europe and in the Mediterranean and – except in Greece – are considered a high quality product (see Pickett & Pawson 1994).

Despite the interest in bass and bream farming, current production is under 85 000 tonnes per year for the entire Mediterranean region (including Portugal, the Canary Islands and Atlantic coast areas of southern Spain). Table 12.1 shows the evolution of bass and bream farming during the 1990s and shows that Greece, by a clear margin, is the major producer with 30 000 tonnes estimated for 1998 production. Nearly all this production is carried out in cages moored in the well-protected, deep-water bays so characteristic of Greece. Although Greece is a major producer, it is not a major consumer; the most important markets are Italy and Spain. The majority of the fish produced are sold whole, without bleeding, packed in ice in polystyrene boxes. Initially, most of the trade was with fish wholesalers in the major markets of Milan, Rome, Paris and Madrid. In turn, they sell on to restaurants, fish shops and other assorted fish dealers. Typically, the fish packed in their boxes would be in circulation for 10–15 days before being consumed. In recent years, supermarkets have become increasingly important customers and some of the big producers are now selling gilled and gutted products in controlled atmosphere packs.

This chapter will describe some of the husbandry, harvesting and marketing factors which have influenced the quantity and quality of the market for bass and bream, and will also highlight some areas where further research is required to further enhance the quality of the products.

The production process

Hatchery

Work began on developing farming techniques for these species in the 1970s, notably in France and Italy. The process of industrialising the production has centred on the ability to

Table 12.1   Evolution of bass and bream farming in the Mediterranean region during the 1990s.

	1990 Bream	1990 Bass	1994 Bream	1994 Bass	1997 Bream	1997 Bass	1998 (est.) Bream	1998 (est.) Bass
Croatia	50		240	960	1000	1000	1000	1000
France	30	300	360	1440	1016	1650	1250	2300
Greece	530	1200	6700	6800	14000	12000	18000	12000
Italy	600	1200	2500	3500	3500	4300	4000	5000
Malta			1100	400	2000	500	2000	2000
Spain	565	31	2720	610	5530	829	6900	1300
Turkey	1031		6070	2229	8000	7000	9000	8500
Others	274		1250	300	4150	2552	4600	2725
Total	3080	2731	20940	16239	39196	29831	46750	34825
	5811		37179		69027		81575	

Derived from *Globefish Research Programme* (1994) Vol. 31. Markets in the European Union for turbot, sea bream and sea bass. FAO Rome. *Fish Farming International*, July 1998 and *Production en El Mediterraneo* supplied by Trouw Espana (1999).

produce large quantities of quality juveniles in a highly specialised hatchery. Figure 12.1 summarises the major stages in production of sea bass and sea bream and shows that the hatchery process is relatively complex. It takes 100–120 days to grow the just-hatched larvae up to 1 g, the size at which many are then shipped out to the on-growing site. Mostly in the Mediterranean, fish are on-grown in cages. However, it has only been in the last five years that survival rates have improved significantly and that consistent supplies of high quality juveniles have become available. The quality hatcheries now produce fish free of spinal and opercular deformities and with properly formed swim-bladders, and in consistent numbers throughout the year.

In the early days, manipulation of broodstock photoperiods and water temperature regimens allowed fertilised eggs to be produced all year round. However, survival rates were highly variable from one batch to another. Many juveniles developed without swim-bladders, had severe spinal deformities and, although they could be brought to market size of 300–350 g in about two years, were not always attractive marketable products. The on-growers adopted a policy of only buying juveniles from hatcheries guaranteeing a high percentage of fish with swim-bladders, backed up with X-ray evidence from random samples. Two major advances in the latter half of the 1980s revolutionised the production of juveniles.

First, there was a better understanding of the nutritional requirements of juvenile fish, particularly with regard to highly unsaturated fatty acids (HUFAs). Live feeds such as rotifers and artemia were subsequently enriched with HUFAs and other essential nutrients before being fed to the larvae. The larvae were subsequently stronger, resistant to disease with much improved survival rates (Watanabe *et al.* 1983; Sorgeloos *et al.* 1988).

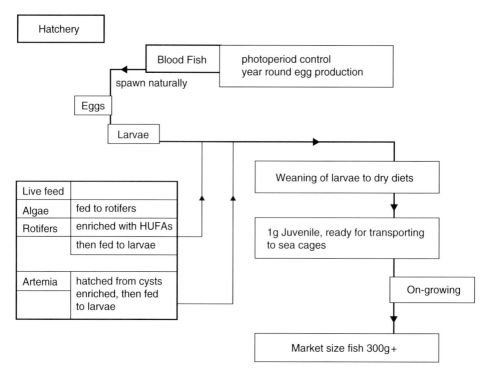

Fig. 12.1   Major stages in production of sea bass/sea bream.

Second, surface cleaners were introduced to free the water surface of oil. Previously, the tiny larvae had been unable to break through the surface tension of debris floating on the water surface. Before they are 10 days old, the larvae of both bass and bream need to break the water surface in order to gulp air to inflate their swim-bladder. The swim bladders can then develop normally and, as a consequence, relatively few spinal deformities result. By the beginning of the 1990s a major bottleneck to production of bass and bream had been overcome. There was then a consistent supply of juveniles.

On-growing

The next stage in the chain is the on-growing of the fish to market-size. Mostly, on-growers introduce their juveniles into their cages in spring/early summer just as the temperatures start to rise. The farmers are then able to take advantage of the high summer temperatures, as indicated in Fig. 12.2, which shows the growth curves of bass and bream for fish introduced into the cages at 5 g in the beginning of May. Growth is rapid until November/ December and is then very slow through the winter months. This slow-down is not only due to the colder temperatures but also to sexual maturation of the fish.

Bream are sexually mature from November until about March in their first year in the sea. They are protandrous hermaphrodites – initially they all are male, with some fish subsequently changing to female in their second or third year when they are around 1 kg.

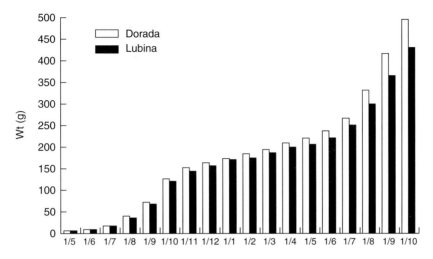

Fig. 12.2 Typical monthly growth profile – sea bass (*Lubina*)/sea bream (*Dorada*).

Bass mature later in the winter as daylength starts to increase. In south-east Spain for example this is normally in the period February to April. Hatchery produced bass have been shown to be at least 65% male which mature in their first winter at sea (Blazquez *et al*. 1998). Females mature in their second winter and thus their growth rates are considerably better than male fish. Although the fish eat during their period of maturation, much of the energy from the feed is channelled into gonad production. Growth resumes after maturation when the temperatures start to rise again. On average it takes 15–16 months for bream to reach 400 g and about 18 months for bass. About 80% of the growth occurs in the period June–October.

The consequences of these temperature and maturation effects are that large proportions of the annual production become available during the last four months of each year. This is also the period when there is significant wild catch available in the major markets. Table 12.2 illustrates the 1998 sales of sea bass and sea bream in the Merca Madrid, the biggest fish market in Spain and one of the largest in Europe. The late autumn bulge is perfectly illustrated with 44% of the annual sales compressed into the last 16 weeks of the year. By contrast, in the 16-week period April to July, only 23% of the total sales were represented.

This has considerable implications for the quality of the products that the farmer is putting into the market. A large proportion of the annual production is available during the last 16 weeks of the year. These fish are usually relatively large. The sales price of these fish can drop sharply during this period, not only from competition between farmers, but also due to the availability of wild-caught fish. Having stimulated a demand for cheap, large fish during the last quarter of the year (and helped by the Christmas festivities), the farmer soon runs out of market-size fish in the early part of the next year. Because little or no growth of his fish is being achieved, the average weight of his market-size fish falls as he is forced to delve into fish stocks of smaller size in order to meet his customers' demands. An alternative is to hold large quantities of market-size fish over winter, until reasonable growth resumes.

Table 12.2   Sales (tonnes) sea bream and sea bass – Merca Madrid 1998.

Week no.	2–4	5–8	9–12	13–16	17–20	21–24	25–28	29–32	33–36	37–40	41–44	45–48	49–53	Total
Sea bream	61	79	74	47	45	68	49	56	54	70	74	86	187	950
Sea bass	44	49	57	54	55	46	35	43	42	50	52	60	180	767
Total	105	128	131	101	100	114	84	99	96	120	126	146	367	1717

In practice, this strategy is not practical given the restricted concession space available in the Mediterranean and the increased risk of holding market-size fish over winter months.

Thus, a major step forward in levelling out the peaks and troughs of production would be for the hatcheries to produce sterile fish, or at least fish that mature only in their second winter at sea. Such a breakthrough could come from temperature-shocking the eggs to produce triploid fish or by genetic selection of late spawning stocks. Some trials have begun with these techniques, but as yet they are not available on an industrial scale. Another interesting hatchery project would be to determine if environmental factors rather than genetic factors influence the sexual differentiation of sea bass.

Husbandry procedures

A major difference between a 500 tonnes per year cage farm producing salmon and bass/bream is the number of fish held on the farm. Salmon farmers produce market-size fish of 3–4 kg and will therefore buy 150 000–200 000 juveniles per year. The bass/bream farmer sells fish of around 400 g and therefore needs to buy around 1.5 million juveniles each year. Furthermore, the salmon farmer transfers juvenile fish of 100 g to cages, whereas the bass/bream farmer inputs fish usually smaller than 5 g. Although there is a need to grade the bass and bream, it is often difficult to carry out this procedure in the summer months when the temperatures are very high. Farmers are loath to forego precious growth or risk a stressful husbandry procedure when the oxygen concentrations are low and there is a higher risk of disease outbreaks.

In reality, grading bream can usually be carried out without too many problems since they are hardy fish, resistant to crowding and handling. Bass, on the other hand, are very nervous, not at all resistant to crowding and can easily be put off feeding by noise or even by currents deflecting the cage net. Recent research carried out at the University of Murcia in Spain (Sanchez-Vazquez *et al.* 1998) has indicated that bass even prefer to feed at night during winter time, though revert to dusk and dawn feeding at higher temperatures. But perhaps one of the most difficult problems to overcome is the strongly developed hierarchical social behaviour. Laboratory tests with juvenile sea bass have clearly demonstrated pronounced social interactions between large dominant fish and smaller subordinate fish (Ajuzie 1998). In a sea cage containing 100 000 fish, this translates to large variations in size, cannibalism (often resulting in significant shortfalls in the number of fish, often 10%), poor food conversion and therefore a more costly fish to produce.

Grading bass is possible but has to be carried out with considerable care and is very time consuming. Normally, bass must be anaesthetised, mainly to prevent the sharp dorsal fin from rising when the fish are stressed by crowding. This raised fin punctures the belly of the fish above and the fish become susceptible to secondary infections. This problem with the dorsal fin also arises when the fish are being harvested and is discussed in a later section.

Nutrition and the mechanics of feeding play an important role in the quality of the end product. For many years bass and bream farmers used cheap, pelleted diets with relatively low energy content, low quality protein, low fat levels (around 12%) and relatively high carbohydrate contents. This resulted in overall conversion rates on a production scale of 2.2–2.5 : 1 for bream (see also Aksnes *et al.* (1997) for laboratory scale data) and as poor as 3 : 1

with bass. In winter time there was little or no growth and the fishes' food intake was reduced to maintenance level. However, in the last couple of years, significantly improved extruded diets have become available with higher energy levels due mainly to increased lipid content (around 22%) and higher grade protein. This has proved particularly beneficial during the over-wintering period – fish maintain their weight or even grow, without significant increases in visceral fat. Evaluation of various feeding strategies to maximise feed intake is an important research subject, particularly with sea bass, to ensure that not only dominant but also subordinate fish get access to feed, and that feeding is modulated with natural feeding rhythms (Azzaydi *et al.* 1999).

Harvesting

It is vital that the techniques used in harvesting the fish do not result in any decrease in the quality of the product. It has taken around 18 months to grow the fish to market-size and it would be absurd to treat the fish poorly during the last few days of such a long cycle. We have found that a number of actions can enhance the appearance and colour of the fish, and also extend the shelf-life in the market place:

(1) *Separation of fish for harvest.* The fish are swum from the main cage into smaller holding units. They are not netted at this stage. At high summer temperatures, the fish are highly active and these procedures must be carried out calmly and with the minimum of stress to the fish.

(2) *Stocking density.* The fish are maintained at typically 5–10 kg/m^3 until ready for cropping. This also helps to prevent the fish from becoming too dark.

(3) *Fasting.* Starvation is essential. The period of starvation depends on temperature. As a general guide, at temperatures below 20°C, starvation will be for two days, but above 20°C may only be for one day. The important point is to ensure that gut contents are evacuated. At higher temperatures, weight loss due to starvation can be significant – up to 1% per day. If sea bream are starved for too long they tend to chew the nets and can escape. Some farmers are inclined to crop directly from their production cages because in the summer months they do not wish to starve their fish during the peak growth periods. However, because fish are sold with guts in, their products can reach their customers with faeces trailing from the anus and this does not give a good impression.

(4) *Method of killing.* Most farmers use thermal shock for killing the fish. Fish are plunged directly into iced water at close to 0°C. The fish are then left in the ice/water mixture until all have reached a core temperature close to 0°C. This procedure must be monitored precisely and staff must be properly trained to ensure they carry out the procedures properly. Often, the water is taken directly from the sea and the ice added. This may be adequate in winter time but in summer can lead to a gross miscalculation of the true temperature of the water in the slaughter tank. We have found, for example, that water temperatures in the upper levels of the tank may be near 0°C but that at 1 m depth – below the floating but rapidly melting ice – the temperature may be as high as 8°C. Here, fish are not dying from thermal shock but from asphyxia. Their appearance, colour and muscle texture will consequently all be adversely affected.

Both species have a remarkable capacity to tolerate low oxygen concentrations and can live for a considerable time at oxygen concentrations around 2 mg per litre. Thus fish can arrive at the packaging plant having finally died from asphyxia, after violent struggling, and may still not be cooled to 0°C.

(5) *Special considerations for sea bass.* Sea bass are particularly vulnerable at the harvesting. Anaesthetics cannot be used at this stage to prevent the lifting of the dorsal fin. It is vital that crowding prior to cropping is reduced to the minimum and that fish are swiftly killed. Otherwise considerable damage and bleeding around the belly occurs. There can also be significant de-scaling and other skin lesions. These skin blemishes alone can result in a 20% drop in market price because of the poor appearance.

(6) *Packaging.* The next stage in the process is packaging. In Spain, bream are normally sold in 6 kg boxes. It is vital to ensure that fish do not warm up in their passage from the slaughter tank to the polystyrene box. Grading of sea bass is particularly important at this stage to ensure consistency of size in the boxes, a very important criterion for many customers. Sufficient ice must be put in the boxes to maintain the temperature close to 0°C; the customer definitely does not want to open a box of water at the fish market.

The care with which the above procedures are carried out is subsequently reflected in the appearance, freshness and the shelf-life of the product in the market place. These products are sold with guts in and may be in circulation for 10–15 days before being consumed. Fish traders quickly learn which producers can supply them with products with a good shelf-life and the shelf-life is directly related to the quality of treatment the fish have received in the harvesting and packaging stages.

Marketing

Finally, it is important to mention briefly the marketing of the products. It is vital that the customers appreciate the positive aspects of bass and bream produced from aquaculture. One of the most important is the guarantee of freshness of the product in the Mediterranean market place. At Culmarex, we have to assure a 24-hour delivery service to customers all over Spain. They can telephone us up to 24 hours before the planned delivery time and we can adjust our harvesting programme to take the fish out of water at the last possible moment. A reliable and comprehensive delivery system is essential, and in Spain there are thousands of lorries travelling the length and breadth of the country dedicated to servicing the country's enormous demand for fish. As the level of national production and imports from Greece, Turkey, France, Portugal and Morocco increase, the Spanish Association of Marine Fish producers (APROMAR) is mounting a national promotion campaign to encourage Spaniards to eat more farmed bass, bream and turbot. We need to ensure that we can supply high class fish all year round and not over-produce in autumn and under-supply in spring. We need to also ensure our products have consistency of size, colour, appearance and flesh quality.

Perhaps the most worrying aspect of current production trends is the risk of over-production of bream. Spain's bream production in 1998 was more than five times its bass production and yet, as mentioned in the introduction, bass has much greater pan-European appeal than bream.

References

Aksnes, A., Izquierdo, M.S., Robaina, L., Vergara, J.M. & Montero, D. (1997) Influence of fish meal quality and feed pellet on growth, feed efficiency and muscle composition in gilthead sea bream (*Sparus aurata*). *Aquaculture*, **153**, 251–61.

Ajuzie, C.C. (1998) Aspects of behaviour in European sea bass juveniles. *Aquaculture Magazine*, **24**, March/April, 37–44.

Azzaydi, M., Martinez, F.J., Zamora, S., Sanchez-Vazquez, F.J. & Madrid, J.A. (1999) Effect of meal size modulation on growth performance and feeding rhythms in European sea bass (*Dicentrarchus labrax*, L.). *Aquaculture*, **170**, 253–66.

Blazquez, M., Zanuy, S., Carillo, M., & Piferrer, F. (1998) Effects of rearing temperature on sex differentiation in the European sea bass (*Dicentrarchus labrax*, L.). *Journal of Experimental Zoology*, **281**, 207–16.

Pickett, G.D. & Pawson, M.G. (1994). *Sea Bass. Biology, exploitation and conservation.* Chapman and Hall, London.

Sanchez-Vazquez, F.J., Azzaydi, M., Martinez, F.J., Zamora, S., & Madrid, J.A. (1998). Annual rhythms of demand-feeding activity in sea bass: evidence of a seasonal phase inversion of the diel feeding pattern. *Chronobiology International*, **15**, 607–22.

Sorgeloos, P., Leger, P. & Lavens, P. (1988). Improved larval rearing of European and Asian seabass, seabream, mahi-mahi, siganid and milkfish using enrichment diets for *Brachionus* and *Artemia. World Aquaculture*, **19**, 78–9.

Watanabe, T., Tamiya, T., Oka, A., Hirata, M., Kitajima, C. & Fujita, S. (1983) Improvement of dietary value of live foods for fish larvae by feeding them on omega3 highly unsaturated fatty acids and fat-soluble vitamins. *Bulletin of the Japanese Society of Scientific Fisheries*, **49**, 471–9.

Chapter 13

Causes of Downgrading in the Salmon Farming Industry

I. Michie

Marine Harvest McConnell, Craigcrook Castle, Edinburgh, EH4 3TU, UK

Introduction

Downgrading represents a significant monetary loss to the salmon farming industry. I would estimate that downgrades may have cost the Scottish industry upwards of £3 million in 1998. In any year, up to 12% of harvested volume (gutted weight equivalent) may be downgraded during the progress of the product along the value chain to its final destination. In this chapter I firstly describe the sources I have used in collating my information. The causes of downgrading are not always accurately recorded and it is not easy sometimes to be specific or accurate in quantifying the information, particularly in the area of further processing. I also define what I mean by the terms 'downgrading', and 'primary' and 'secondary' processing, as these can mean different things to different people.

There are a number of reasons for downgrading, from the appearance of the fish to the intrinsic properties of the meat, and there are key segments in the value chain where downgrading occurs. In the main body of this chapter I describe the three most significant causes for downgrading both in primary and further processing as experienced by Marine Harvest McConnell, and illustrate the seasonal nature of their occurrence. I do not cover in any depth what is believed to be the underlying reason, or reasons, for these problems becoming manifest, as in many cases they are still not fully understood. Finally, I discuss the areas where efforts should be concentrated for improving on the current burden of loss sustained by the industry through downgrading of the product.

Definitions

Sources of information

Most of my data are taken from the quality assurance department's records of the company in which I am employed as a technical manager. In 1998, Marine Harvest McConnell produced 30 thousand tonnes of farmed Atlantic salmon in Scotland. This represents approximately 30% of the Scottish farmed salmon production in that year. The data are from company sources in Scotland; I do not use any of our Chilean figures. The information therefore is representative of only a portion – albeit a significant one – of the UK volume. The information on downgrading in the further processing and smoking sectors, in which Marine Harvest McConnell is not directly active, is derived from the feedback we receive from our customers. This can come systematically in the form of quality data, or can be more anecdotal in nature through comments and observations, sometimes solicited in the form of customer surveys. As such I feel this might present an understated picture, as a number of customers probably make what they can of the product without referring back to the primary supplier while incurring some form of downgrading loss. The information is for the calendar year 1998, and so on can be looked on as a snapshot sample of quality issues in salmon farming.

Terms

I define 'downgrading' as any physical aspect which causes loss of value for the product at any point in the process, or value chain, which brings it to the consumer's plate. Loss of value can be incurred where the product does not meet any agreed customer specification – referred to as a 'non-conformance', say for smoking. The characteristic which causes this loss of value for a smoked product, e.g. pale flesh, might not incur such loss of value where different characteristics have been specified, for instance for a salmon *en croûte* recipe dish where a deep colour is not of such significance.

There are two distinct stages through which a fish progresses once it has been harvested. Some form of product preparation is now normal practice, to the extent that few consumers now buy a whole salmon, or even a single fish gutted and cleaned. 'Primary processing' I define as evisceration, cleaning and weighing, prior to distribution to the next stage in the chain. A definitive break occurs here when the fish are boxed for transit. Downgrading at primary processing is based simply on the surface appearance of the fish (including the surface of the abdominal cavity). The grading is done by trained staff who make judgements against 22 separate criteria – from scale loss to knife cuts in the belly cavity.

'Secondary processing' I define as removal of the meat from the skeletal frame and pre-paration of the product for final presentation to the consumer, transformed into either the ready to eat or ready to cook form. Approximately 33% of Marine Harvest McConnell production went for smoking last year, which remains the most significant transformational process for the product. Downgrading during this stage is usually either rejection or realignment of material to a lower specification product, by the operatives on the production line. There are frequently several quality grades of smoked product.

It is obvious that a fish downgraded for having a defect of appearance externally at primary processing may well still possess excellent flesh quality characteristics at filleting and smoking.

Primary processing

Typically downgrading at primary processing can affect up to 10% of volume produced in a given year. The most significant causes in primary processing are heavily influenced by the growth profile of the fish, the stage in the seasonal cycle and directly by the effects of husbandry techniques such as feeding regimes, sea lice controls, harvest methods, etc. At this stage the specification criteria for the majority of the Scottish industry are principally the quality descriptors which have been developed by quality assurance schemes such as 'Scottish quality salmon'. Fig. 13.1 shows the relative causes of downgrading in primary processing.

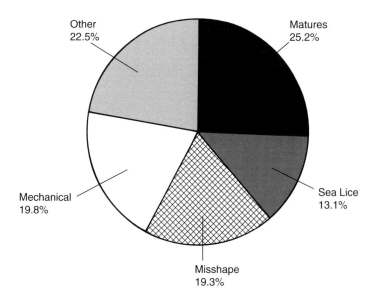

Fig. 13.1   Causes of downgrading in primary processing, 1998.

Maturation

Maturation accounts for 50% of downgrading in primary processing and is the single most significant cause. Secondary sexual characteristics begin to become apparent in the spring, in June, leading to downgrading from August onwards. The characteristics become progressively more pronounced from August through to November/December in preparation for spawning. The apparent secondary sexual characteristics are a darkening of the skin of the fish extending eventually round the whole characteristic silver belly of the salmon, reddish spots and finer scale definition as the scales sink and the skin takes on a thicker protective mucous layer. There is an elongation of the head and the development of the 'kype' in the lower jaw of the cock fish. As they develop, these characteristics further widen the gap between the consumer's perception of what the classic silver salmon should be like and the reality at that point.

Sexual characteristics are difficult to judge at grading on the farm in the early part of the season. Similarly, reversion after the early onset of maturity can also make the decision difficult and cause the human grader on the farm to reserve the fish for ongrowing if in doubt. The point then at which a certain judgement can be made is close to the point at which the fish has lost superior quality status. In addition, high maturing stocks can give rise to precocious male development during the first summer at sea, and although these fish revert in the spring they retain some of the secondary sexual characteristics, particularly the shape of the head and scale formation.

A further complication for graders in the processing factory is the natural darkening of the fish in response to a dark background. This can occur after prolonged rainfall has made the water murky with peat and/or when short winter days give little light, especially if the site is in the shadow of a hill and does not receive direct sunlight. This is unrelated to maturation and is a feature of some Scottish inshore sites.

Conversely, immature salmon in the season following the main mature harvests can be long and thin and can be downgraded due to their condition factor being less than 1.0 (ratio of weight to length). This partially reflects the dominance of the maturing fish in a population when competing for feed.

Spinal deformity

The next most important cause of downgrading results from spinal deformity which gives the fish a misshapen appearance. Essentially, it would seem that during growth the development of the spine is inadequate for the load placed on it and is susceptible to traumatic injury, the vertebrae collapsing and fusing to form a site of compaction. The ability of the fish to feed, swim and grow is not compromised and, again, grading for minor deformities on a grading table at sea is very difficult for farm staff. After evisceration and kidney removal the compaction of the vertebrae is more obvious, and the deleterious effect this would have on fillet quality causes the fish to be downgraded even when the misshape is not significantly pronounced in external appearance.

Research is progressing into the cause of this problem, which appears to have beleaguered the industry both in Scotland and Norway for many years. Although faster growing stocks seem to have a slightly higher incidence the underlying factor is more likely to have either a genetic, nutritional or developmental origin.

Disease

There are further interrelated health issues which can have a serious impact. These are principally to do with the effects of sea lice infestation, which accounts for 15% of the total downgraded at primary processing. At chronic levels lice infestation can result in patterns of small unsightly lesions along the belly. Heavier infestation can result in wider haemorrhagic lesions, and in severe cases there is an open wound across the back of the head. The only licensed treatments in Scotland at the time of writing (early 1999) are immersion treatments. Crowding of the pen to allow enclosure of the fish within a tarpaulin to create a bath can cause some eye damage and result in abrasion of the fish and scale loss through forcible rubbing against the netting. These effects are not usually attributed to sea lice as such by the graders at processing, and are included in the 'other' category (Fig. 13.2). Similarly, repeated stress of treatment and consequent innappetance, particularly in the late summer, can bring some immature fish below a condition factor of 1.0. The pattern of downgrades for sea lice damage across the year (Fig. 13.2) reflects the fact that fish take longer to heal skin wounds at lower temperatures, and the difficulty of treatment due to adverse weather.

Although furunculosis is not a problem of any significance now, a feature of disease

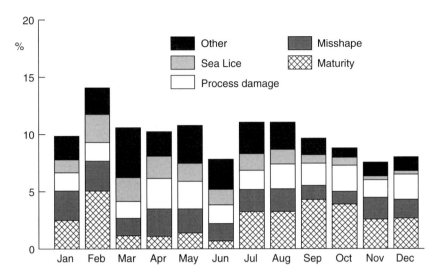

Fig. 13.2 Downgrades by cause by month in primary processing, 1998.

control is that of melanisation of small areas of the peritoneum ('staining') as a result of vaccination by injection. These areas look like bruises and are unsightly and so the fish are downgraded. This effect is becoming less significant year on year.

The remainder of the downgrade portion at primary processing consists of a number of lesser and more occasional issues which crop up, such as damage resulting from equipment defects or operator error both at farm and processing factory. The seasonal nature of downgrading by cause, most obviously for maturity, is illustrated with the monthly breakdown across the year (Fig. 13.2).

Secondary processing

I estimate that the loss of value at secondary processing can account for around 2% by weight as calculated from farm production (gutted weight equivalent). Downgrading stems from the characteristics exhibited by the muscle of the filleted and then further processed fish. The causes are likely to be related to stock characteristics, responses to season, maturation and growth history on the farm; or they can be caused during critical phases of handling the fish from harvest and in the processing and supply chain. At this stage the criteria used for determining the quality of the product are those set by the product raw material specifications, which differ both by customer and product type. Figure 13.3 shows the relative causes of downgrading in secondary processing.

Colour

At nearly 40%, the biggest single cause is poorly visualised colour, or paleness. This is measured by the Roche colour card or *Salmo*Fan™ and, generally, fillets are not acceptable

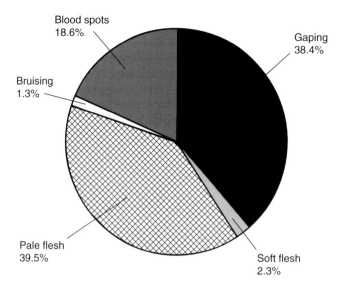

Fig. 13.3   Causes of downgrading in secondary processing, 1998.

when reading below 15 or 26 respectively. In addition, a condition can arise called 'striping' which looks even worse than a uniformly pale fillet. 'Striping' is a condition where there is greater than a three point difference in *Salmo*Fan™ scores between the lateral and dorsal areas of the fillet. The problem seems to be seasonal and linked to maturation. In preparation for spawning, the fish increase fat deposition as an energy store in the winter and spring of the year prior to spawning. The fat is deposited between the muscle blocks and at high levels has a tendency to mask the pigment in the muscle. As maturation progresses in late summer, mobilisation of chemical pigment out of the musculature and into the skin and eggs occurs, resulting in pale and patchy colour.

In the smoked product, colour loss during smoking is occasionally seen probably due to pigment oxidation. This is more likely in fish harvested at a young age which have not had a chance to accumulate much pigment in the flesh, past the point at which visual colour saturation occurs (approximately 7–8 ppm). If the loss of pigment during smoking reduces the level below the visual saturation threshold then the fillet will appear paler. Sometimes the oxidative effect can be one of discolouration following smoking, when the fish takes on a dark brown/yellow appearance.

Gaping

At 39%, 'gaping' is almost as important as paleness as a cause of downgrading. This occurs when the connective tissue holding the muscle blocks together has been compromised in some way and is ruptured as a consequence of exerted tension. When handled, gaps appear both along the length of the fillet, principally below the epaxial muscle group, and between the myotomes. Traditionally in wild caught fish this often calls into question the freshness of the fish, but is not the case with the condition in farmed salmon. Similarly 'gaping' is not always related to soft texture in the fish, which can be the result of stress at harvest, inadequate temperature control through processing and distribution, or simply decay through age (very uncommon). It is probably more correct to view gaping as a consequence of muscle tension in *rigor mortis* overcoming the resilience of the binding tissue between the blocks. Gaping is again seasonally influenced (Fig. 13.4) and appears to be associated with phases in the growth cycle. A similar effect can be seen at times after smoking when the slices break down to resemble fragments of 'lace curtain'. An overall weakness of the connective tissue both between the muscle blocks and along the length of the fibre bundles seems to be exacerbated by rapid hot smoking.

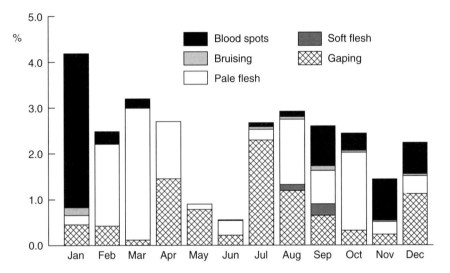

Fig. 13.4 Downgrades by cause by month in secondary processing, 1998.

Bloodspotting

A proportion of all harvested salmon contains some unsightly residual blood presence. The main blood vessels feeding the musculature from the dorsal aorta are ruptured at evisceration and filleting and inadequate drainage becomes apparent in the form of small stains or points of blood running the length of the fillet in the midline. Residual blood in the small vessels and capillaries networking the white muscle appears as sometimes indistinct areas of discolouration. All of this is of particular significance to smokers, where the smoking process and final presentation of the product make the blood residues much more visible by darkening and fixing them in thin slices of muscle. Bloodspotting can necessitate downgrading to a lower value smoked product, or accrue cost as a result of the laborious operation of bloodspot removal, although this is impossible with the lighter flecks of blood which remain, particularly in the belly wall area of the fillet. Although it is a chronic problem, the prevalence reflects the seasonal smoking demand, and there appear to be acute episodes from individual slaughter dates, suggesting either stress or environment as exacerbating factors.

There is also a suggestion that bloodspotting is actually most prevalent in the late summer months when water temperatures are high.

The melanisation associated with vaccination can penetrate through the peritoneum into the muscle and cause wider areas of staining, resulting in a similar look to that of bad bloodspotting or mechanical bruising.

Discussion

Traceback and the accurate allocation of figures to the appropriate causes of downgrading is important for identifying and prioritising the areas of greatest economic concern. Based on the current information available, I would like to suggest that to mitigate downgrading at primary processing, advances are necessary in the following areas:

(1) Prevent or control maturation in a predictable manner which does not detract from the image of the product
(2) Resolve the spinal deformity problem
(3) Develop stock husbandry techniques and therapeutants which eliminate the need for crowding and handling of the fish, particularly for sea lice treatments

At secondary processing the priorities are:

(1) Achieve greater consistency of visualised flesh colour, both across a population and between individuals in the population
(2) Develop pre-harvest husbandry, harvest and post harvest handling methods which minimise gaping
(3) Achieve a consistent, economic method of exsanguination

Even if progress is made on all of these fronts, it is inconceivable that inherent variability within populations and within individuals could be eliminated. However, progress with selective breeding programmes will in the long term go some way to reduce fish variability. Given that a downgrade to one man is a good fish to another, a desirable adjunct would be the development of objective in-line techniques at primary processing for the assessment of essential flesh attributes.

Acknowledgements

I would like to thank all the staff in the quality assurance department of Marine Harvest McConnell for their assistance, particularly Nicola Wright for helping me to interpret the available information. I would also like to thank the health services department and the development department, in particular David Mitchell, for invaluable technical advice. All the data and observations are from company files and internal research documents, not in the public domain.

Chapter 14

The Potential Impact of Genetic Change on Harvest and Eating Quality of Atlantic Salmon

R. Alderson

Alderson Aquaculture, 3 Lumsdaine Drive, Dalgety Bay, Fife, KY11 9YU, UK

Introduction

Genetic selection in plants and animals used for food production has a long history. It is estimated that domestication of wheat and barley may have begun 10 000 years BP and sheep, goats and pigs, 8000 years BP (Diamond 1998). In the last 100 years selection programmes for these species have become increasingly sophisticated. Salmon cultivation is a considerably younger activity, but programmes for the genetic improvement of the fish being farmed, using between and within family selection, were established in Norway in 1971, shortly after farming began (Gjedrem *et al.* 1991). In common with breeding programmes for other food animals, those developed for salmon concentrate on reducing the cost of production by increasing the rate of growth and thus shortening the time to harvest. Gjedrem (1997) estimates that the time taken to reach a harvest weight of 3.5 kg has been shortened by 1 month per generation as a result of the Norwegian breeding programme. This, he estimates, has led to a reduction in the production cost to the Norwegian salmon farming industry of between $28 million and $60 million.

This accent on increasing growth rate in food animal production is understandable. It is relatively easy to demonstrate, satisfies the accountants of the farming businesses, and is essential for business survival in the competitive market produced by ever decreasing prices. Price is however only one factor determining the demand for repeat purchase in the marketplace. Product appearance has long been regarded as an important factor, and today's discerning purchasers are demonstrating that they are also capable of discriminating on the basis of eating quality.

From the early beginnings in Norway, salmon genetic selection programmes are now being pursued in all the major salmon farming regions of the world. While most of the accent will remain on the increase in growth rate, this chapter suggests that attention needs also to be directed at all aspects of product quality. There are lessons to be learned from the effects on product quality of selection programmes in other areas of food production.

Quality aspects of salmon genetics

Salmon at the point of sale is now predominantly presented in the form of fillets or steaks. The customer therefore has the appearance of the raw flesh as one of the factors to be considered in making the buying decision. The red colour of salmon is of considerable importance, as also is fat level. At high levels the latter becomes clearly visible as banding in the flesh. Information from the Norwegian breeding programme indicates that both these factors are, at least in part, under genetic control. Rye and Gjerde (1996) give heritabilities of 0.12 to 0.18 and 0.30 to 0.38 respectively for these two factors. What these figures mean in practice may be illustrated by reference to the results presented in Figs 14.1 and 14.2. These data have been obtained as part of a large scale trial looking at muscle fibre recruitment and flesh quality in Atlantic salmon using individually tagged fish of different strains, and different families within the strains, from the Marine Harvest McConnell salmon breeding programme. Details of the trial protocols are given in Johnston *et al.* (2000a,b). The fish sampled to provide the data for Figs 14.1 and 14.2 were held in the same cage environment and fed on the same feed with the same level of astaxanthin pigment (Carophyl Pink[R]) and dietary oil.

Figure 14.1 shows the very large difference in flesh pigment level that can be found as a result solely of genetics. All the fish had reached similar weights after an identical time in the sea, but as shown in Fig. 14.1, fish of the two genetically different strains being compared had very different levels of pigment deposition in the flesh. Results from the same pro-

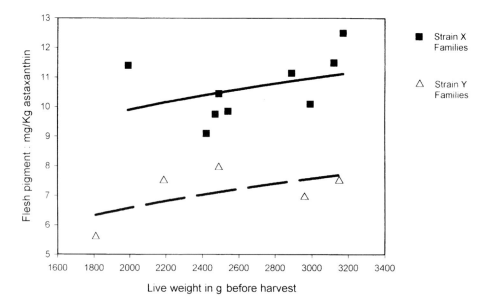

Fig. 14.1   The level of astaxanthin pigment in sections of fillet flesh taken from a vertical sample of 5 cm width taken immediately behind the dorsal fin. The results here are for fish of two strains, reared in the same 5 m × 5 m cage and fed on the same diet from smolt transfer in April 97 to sampling on 22 June 1998.

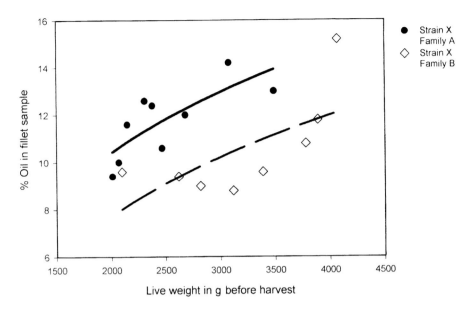

Fig. 14.2 The level of oil in sections of fillet flesh taken from a vertical sample of 5 cm width taken immediately behind the dorsal fin. The results here are for fish of two families from strain X, reared in the same 5 m × 5 m cage and fed on the same diet from smolt transfer in April 97 to sampling on 4 May 1998.

gramme shown in Fig. 14.2 also demonstrate that fat levels in two families within one of the strains in the trial could also show a marked difference. These results indicate that with the heritability figures quoted, there is a clear potential for change and improvement in product by appropriate genetic selection. They also demonstrate that some of the variability in the product currently being harvested is due to the variability that is present in and between the strains presently being farmed. In addition, they should also warn of the consequences of selection of strain, or individual within strain, for broodstock use without knowledge of the flesh quality characteristics.

Pigment level alone is not, however, the only determinant of visual colour in salmon. Johnston *et al.* (2000a) demonstrated that when visual colour was measured with the Roche *Salmo*Fan™, muscle fibre density explained between 27% and 42% of the variation in colour independent of pigment level. The same work also demonstrated that the genetic strain was a significant determinant of differences in muscle fibre density at a given weight. Selection for improved flesh colour can therefore have two components: improving the efficiency of deposition, and improving the visualisation of the pigment deposited.

As part of the same investigation, taste panel assessments were made on harvested fish that had been smoked by a commercial smoker (Pinneys of Scotland Ltd). As shown in Johnston *et al.* (2000a), significant positive correlations were seen between muscle fibre density and four measures of texture assessed by the taste panel. Since muscle fibre density was seen to be different between the two genetic strains of salmon, texture, at least of the smoked product, could also become the subject of a selection programme to improve flesh firmness.

These examples all serve to demonstrate that some of the key quality parameters for salmon are, at least in part, under genetic control. If incorporated into selection programmes, as some have been in Norway (Anon 1995) and in Scotland (Johnston *et al.* 2000a), appropriate selection can lead to an improvement in the quality of the salmon being produced. The converse is also true, that where selection programmes concentrate on growth rate alone, ignoring the potential for change in these other parameters could lead to unwanted changes or an increase in product variability. Experience in other areas of food production has also shown that product quality should not be ignored when operating breeding programmes.

Experience from pig meat selection programmes

At a personal level, I have believed that over my lifetime the eating quality of the pork that is generally available has deteriorated, while the product has got ever cheaper, a sentiment also echoed by Blanchard *et al.* (1999). The pig meat industry, like the salmon industry of today has had to respond to the competition to reduce production costs. The key selection targets for the industry in recent years have therefore been: faster growth, increased meat yield from the carcass, increased lean meat production, and improved feed utilisation. Glodek (1982) reviewed the improvements that had been seen by way of increasing average daily weight gain and reducing feed per kg of gain in the Danish industry over the period from 1908 to 1964. Over this period daily weight gain increased from 560 g/day to 680 g/day. Today, weight gain can be as high as 914 g/day (Blanchard *et al.* 1999). This increase involves improvements in husbandry and nutrition as well as genetic selection.

Barton-Gade (1990) gave more specific details of the Danish pig breeding experience and how this has affected some aspects of eating quality. Over the period she describes, 1983 to 1988, daily weight gain as a result of selection increased by up to 93 g/day and meat content was also improved by up to 1–2%. At the same time some measurements were made of factors that were known to affect eating quality. For some breeds, these showed that significant declines in eating quality had occurred over the same period. The level of intramuscular fat (% IMF) is one factor associated with good eating quality, and in her example, for Landrace pigs, this declined from an average of 1.57% to 1.27% and in Large White pigs, from 1.75% to 1.20%. At the same time, the shear force measured in newtons in the *Longissimus dorsi* muscle, an indication of toughness, increased from 80 to 105.5 in Landrace pigs and from 87.4 to 102.5 in Large White pigs. She indicated that the eating quality of pork becomes poor when IMF falls below 1% and shear force exceeds 125 newtons.

Figure 14.3 shows that the proportion of Landrace pigs showing each of these characteristics increased over this period. Barton-Gade quoted heritabilities for % IMF from a number of sources ranging from 0.50 to 0.81. The decrease in % IMF therefore occurred despite the fact that it would have been relatively easy to have incorporated into the programme selection to at least maintain the original low level of pigs with meat which had IMF% > 1. She concluded that it might be necessary to include this factor in breeding goals in the future. The danger of losing eating quality while applying selection solely to improving the economics of production was also highlighted by Lundström (1990). She stated that, 'Due to the negative correlations with some production traits, selection without including meat quality might lead to negative consumer responses in the future.'

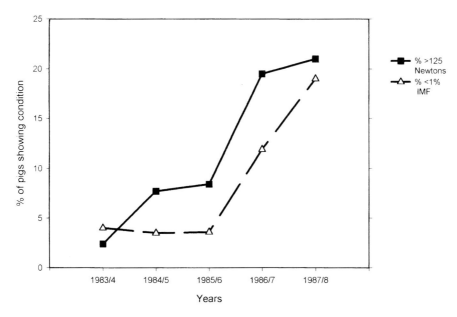

Fig. 14.3   Graph plotted from data given by Barton-Gade (1990) showing the change in % of sampled pigs of the Landrace race which had flesh quality measurements judged to be indicative of poor eating quality. Toughness, as measured by a shear force of >125 newtons required to cut the *Longissimus dorsi* muscle, and lacking in succulence as indicated by the % IMF < 1%.

These examples indicate that selection in the pig industry has concentrated on reducing production costs but also, by increasing the proportion of lean meat, has improved appearance and acceptability to 'health conscious' consumers. This has however been achieved, in at least some cases, at the expense of ignoring changes in a factor such as % IMF that is known to correlate with eating quality.

Warriss *et al.* (1996) looked at pork quality from a number of different pig breeds, and texture, % IMF and flavour were measured in taste panels. When the results from their work are ranked on flavour score (Table 14.1) it is clear that high scores for good eating flavour are not dependent solely on % IMF. Duroc and Berkshire breeds have similar % IMF, but have flavour scores that are at both ends of the range measured. Other more complex flavour attributes are therefore involved in determining consumer appeal and appear also to have a genetic component. Measuring 'flavour' is however much more difficult and expensive than measuring % IMF and it is this that led Lundström (1990) to comment that 'sensory evaluation of meat quality is not used in any breeding programme due to the high costs'. Nevertheless she pointed out that despite this obvious difficulty and the generally low heritabilities for sensoric traits, 'acceptance by the consumer is the most important quality trait'.

Returning to my personal observations over the years, the genetics programmes for the production of most of the pork that is available appear to have been eminently successful in

Table 14.1   Data from Warriss *et al.* (1996) rearranged and ranked on flavour score. Flavour and texture scores were obtained from taste panels assessing cooked loin joints. The range of score was from 1 to 7 with higher scores indicating better quality. The data show that % IMF is not the only factor of importance in determining whether a sample of pork from a particular breed will score highly for flavour.

Breed	Flavour score	% IMF	Texture score
Duroc	3.46	2.19	1.26
Tamworth	3.39	1.46	1.73
Hampshire	3.08	1.30	1.69
Pietrain	2.96	1.44	0.50
Saddleback	2.96	1.37	0.96
Large Black	2.86	1.58	0.64
Large White	2.68	0.89	1.18
Berkshire	2.50	2.14	1.24
Landrace	2.48	1.39	0.91
Gloucester Old Spots	2.48	1.32	1.05

increasing growth and reducing production costs. They have however done this at the expense of eating quality and have produced a product that may be very cheap, but which I for one avoid buying.

The flavour of tomatoes

Again, a personal recollection with regard to the flavour of tomatoes is that in the 1960s and 1970s the attention of producers and breeders of commercial fruit was directed towards reducing production costs, improving uniformity of size and appearance, and improving disease resistance. In this they were very successful, but again, as with the pig industry, some aspects of eating quality were lost in the process. It was not until the 1980s that serious attention began to be directed at eating quality, principally in the smaller, cherry tomatoes. The original older varieties such as 'Gardeners Delight' lacked resistance to common tomato pathogens and so breeding programmes were established to introduce disease resistance while trying to maintain the attractive eating characteristics of the original varieties (Hobson & Bedford 1989). These authors reviewed the early developments and described the results of taste panel assessments of the varieties then being brought into cultivation. The two key factors in determining consumer acceptance were judged to be sugar and acidity and it was varieties with balanced levels of these that were the ones preferred. While taste panel assessment was required to establish the nature of the factors involved in determining consumer preference, simple measurements of fruit chemistry could then be used to aid in evaluating fruit in breeding programmes.

The commercial development of these varieties with good flavour characteristics has continued to expand and today most supermarkets provide a range of tomatoes, with the highly flavoured types identified by the name of the variety. There is no doubt that these smaller fruited tomatoes are more expensive to produce, being lower yielding and more expensive to harvest. Hobson and Bedford (1989) indicate that yield will be half that of standard, large, round fruited varieties. Nevertheless it is clear that consumers are prepared to pay a high price where purchase of a named variety will guarantee them a tomato of good eating quality. Table 14.2 gives the results of three snapshot surveys of retail tomato prices in one supermarket chain in Edinburgh. Production costs may be higher but the most expensive variety, Santa, was retailing at between 4.4 and 5.5 times the price of the standard, flavourless, tomatoes.

Table 14.2   The result of three surveys of tomato prices at separate dates in the Edinburgh branch of a large multiple retail store chain. Not all varieties were on sale on each of the occasions that the prices were checked. Prices are given in £/kg.

Tomato variety	January 1997 (£)	August 1997 (£)	June 1998 (£)
Un-named standard variety	1.08	1.30	1.52
Cherry	4.23	3.06	
Baby cherry	5.96		5.96
Melrow		5.83	4.26
Santa	5.96	6.63	6.63

The lessons for salmon

Salmon is a new species in cultivation and initially many companies had their own in-house salmon broodstock derived from different wild sources. Today, production is increasingly being dominated by a few strains, and organised breeding programmes are being established in all of the major salmon producing regions. The initial objectives of these programmes, in the face of continuing price competition, will undoubtedly be to select for improved growth rate to reduce production costs. As in Norway however (Anon 1995), and as indicated here in at least one Scottish company, attention will then shift onto the aspects of flesh quality that determine appearance and presentation. Important as these issues are, final eating quality, as in other foods, is assessed by the consumer by the texture and flavour of the product as it is eaten. The genetic control of one aspect of texture is now being studied in depth (Johnston *et al.* 2000a,b). However, who has even begun to look at the genetics controlling the distinctive salmon flavour? It is not going to be as easy to measure as the sugar and acidity of tomato fruits, and that will make it expensive to incorporate into any breeding programme. Nevertheless, from the example of tomatoes, some consumers are prepared to pay a high price for a product that they have a reasonable guarantee will taste good. If on the other hand, as in the pig industry, flavour is sacrificed in the production of an ever cheaper product, then it will become difficult to maintain into the future the quality eating image of this excellent food.

Acknowledgements

Part of the information presented here was derived from work supported by a grant under the Natural Environment Research Council of the UK under the LINK Aquaculture Programme. It involved the active and enthusiastic collaboration of Ian Johnston (Gatty Marine Laboratory, St Andrews), Billy Robertson, Craig Selkirk, David Mitchell and Remi Baker (Marine Harvest McConnell Ltd), Alistair Dingwall (Pinneys of Scotland Ltd), David Nickell and John Springate (Roche Products Ltd), David Whyte (Aquasmart UK Ltd) and Patrick Campbell (BioMar Ltd).

References

Anon (1995) Avl pa farge oker lonnsomheten [Breeding in colour improves profitability]. *Norsk Fiskeoppdrett,* 4/95, 41.

Barton-Gade, P.A. (1990) Danish experience in meat quality improvement. In: *Proceedings of the 4th World Congress on Genetics applied to Livestock Production, Edinburgh,* pp. 511–20.

Blanchard, P.J., Ellis, M., Warkup, C.C., Hardy, B., Chadwick, J.P. & Deans, G.A. (1999) The influence of rate of lean and fat tissue development on pork eating quality. *Animal Science,* 68, 477–85.

Diamond, J. (1998) *Guns, Germs and Steel: A Short History of Everybody for the Last 13 000 Years.* Vintage, London.

Gjedrem, T. (1997) Selective breeding to improve aquaculture production. *World Aquaculture,* 28, 33–45.

Gjedrem, T., Gjoen, H.M. & Gjerde, B. (1991) Genetic origins of Norwegian farmed Atlantic salmon. *Aquaculture,* 98, 41–50.

Glodek, P. (1982) Selection responses in pigs: results and implications. In: *Proceedings of the 2nd World Congress on Genetics applied to Livestock Production, Madrid,* pp. 568–77.

Hobson, G.E. & Bedford, L. (1989) The composition of cherry tomatoes and its relation to consumer acceptability. *Journal of Horticultural Science,* 64, 321–9.

Johnston, I.A., Alderson, R., Sandham, C., Dingwall, A., Mitchell, D., Selkirk, C., Nickell, D., Baker, R., Robertson, B., Whyte, D. & Springate, J. (2000a) Muscle fibre cellularity in relation to flesh quality in fresh and smoked Atlantic salmon (*Salmo salar* L.). *Aquaculture.* (in press).

Johnston, I.A., Alderson, R., Sandham, C., Mitchell, D., Selkirk, C., Dingwall, A., Nickell, D., Baker, R., Robertson, B., Whyte, D. & Springate, J. (2000b) Muscle fibre recruitment in early and late maturing strains of Atlantic salmon (*Salmo salar* L.) fillets. *Aquaculture,* (in press).

Lundström, K. (1990) Genetics of meat quality. In: *Proceedings of the 4th World Congress on Genetics applied to Livestock Production, Edinburgh,* pp. 507–10.

Rye, M. & Gjerde, B. (1996) Phenotypic and genetic parameters of body composition traits and flesh colour in Atlantic salmon, *Salmo salar* L. *Aquaculture Research,* 27, 121–33.

Warriss, P.D., Kestin, S.C., Brown, S.N. & Nute, G.R. (1996) The quality of pork from traditional pig breeds. *Meat Focus International,* 5, 179–82.

Chapter 15

Effects of Husbandry Stress on Flesh Quality Indicators in Fish

T. G. Pottinger

NERC Institute of Freshwater Ecology, Windermere Laboratory, The Ferry House, Far Sawrey, Ambleside, Cumbria, LA22 0LP, UK

Introduction

The aim of this chapter is to examine how husbandry practices in the aquaculture environment may affect the quality of flesh of farmed fish. It will focus in particular on effects on muscle composition of fish which can be caused by exposure to stressful conditions and by the imposition of forced exertion. Clearly, it should be emphasised that the physiological changes associated with stress are only part of a range of factors which must be considered when evaluating potential modifiers of flesh quality in fish. The topics covered by other contributors to this volume reveal the full diversity of the factors which influence flesh quality in fish.

Indicators and determinants of flesh quality

The characteristics which define flesh quality are discussed elsewhere in this book. The following factors are among those listed by Gjedrem (1997) as important indicators of flesh quality:

(1) Fat content, which must be appropriate for the species and market
(2) Distribution of fat within the body – the deposition of excess fat in depots may reduce the value of the fish
(3) Flesh colour – pigmentation is an aesthetic consideration but can be important to the consumer
(4) Flesh texture, which can be described in terms of firmness, cohesiveness and elasticity

What practical factors, under control of the grower, are important in determining flesh quality? Obviously nutrition, discussed in detail by other contributors to this book, plays a critical role and is arguably the single most critical element in rearing fish for the table. Genetics, in terms of the pedigree of the fish, is also an important consideration. There can be substantial differences in the performance characteristics of different strains of salmonid fish and flesh quality traits may be improved in selective breeding programmes (Gjedrem 1997).

The husbandry practices to which farmed fish are exposed are also important. The term husbandry refers to a range of activities in which disturbance to the fishes' environment is a common factor. Husbandry is inclusive of stocking densities, frequency of handling and disturbance, water quality, transport, slaughter methods, prophylactic and therapeutic treatment, the occurrence of parasites and pathogens, and the overall well-being or welfare of the fish. Many husbandry procedures are unavoidably associated with a degree of stress and may also be accompanied by the imposition of physical exertion on the fish. This may range from rapid swimming to attempt to evade capture or escape the source of a disturbance, to a prolonged period of struggling against restraint, or in air, during procedures involving netting or transport.

Even the most minor of disturbances can have profound effects on the physiological status of fish and these elements may contribute to undesirable changes in the characteristics of the flesh of fish post-slaughter. To understand the mechanisms by which stress and exertion can affect flesh quality in fish, it is necessary to appreciate some of the changes in the physiology and metabolism of fish which stress and/or exercise can cause.

Physiological changes in fish following exposure to stressors

Stress and homeostasis

Fish, in common with all animals, must maintain a stable internal environment in order to develop, grow and reproduce normally. All the biochemical and physiological processes associated with maintaining normal functions operate most successfully within a specific range of pH, temperature and solute concentration and the maintenance of this stable environment is termed homeostasis. If homeostasis is threatened, for example by an abrupt change in environmental conditions, then a neuroendocrine response, which triggers a highly effective array of physiological changes, is invoked to help the animal overcome or avoid the challenge to stability. This protective mechanism is found in all vertebrates from fish through to man and is colloquially termed the stress response. Its retention in all vertebrate groups demonstrates that the response is of significant adaptive value.

The major features of the neuroendocrine stress response are well-characterised in fish (Wendelaar Bonga 1997). The response comprises two main elements: a rapid neurally-mediated release of catecholamines (adrenaline, noradrenaline) into the blood stream from the chromaffin cells (homologous to the adrenal medulla in higher vertebrates), which is accompanied by a slightly slower endocrine-mediated release of the steroid hormone cortisol into the blood from the interrenal tissue (homologous with the adrenal cortex in higher vertebrates) (Sumpter 1997). As a direct consequence of these events, often referred to as the primary stress response, a wide range of secondary changes are evoked in behaviour, metabolism, respiration, the immune system, and in other elements of the endocrine system (Barton 1997).

Adaptive value of catecholamine and cortisol elevation during stress
The beneficial effects of elevating blood catecholamines to a fish which is facing a severe challenge can be easily identified. Most of the effects of catecholamines in fish during stress

are associated with maintaining cardiorespiratory performance under challenging conditions, together with an important role in the mobilisation of energy (see Randall & Perry 1992 for a full account).

It is more difficult to identify the advantage gained by the fish of elevating cortisol levels. Cortisol in fish does not appear to have the clearly defined energy mobilising role which glucocorticoids perform in other vertebrates (van der Boon *et al.* 1991), although it may have a role to play in gluconeogenesis and in mobilising lipids (Suarez & Mommsen 1987; Pickering and Pottinger 1995; Wendelaar Bonga 1997).

Acute stress – an adaptive response

To characterise the stress response as an essentially adaptive mechanism is at odds with a general perception of stress as a harmful condition. However, a distinction must be made between the effects of stressors, or stressful stimuli, of different duration. In the natural environment most stressors of natural origin are assumed to be short in duration – the fish either overcomes or avoids the challenge, or succumbs. We presume that this is the type of challenge with which the stress response evolved to cope. This short-term response is referred to as an acute stress response; following the perception of a threat, there is a rapid neuroendocrine response, the threat is removed or avoided and the response subsides within hours. There are numerous published studies which describe this response profile in fish (see Barton 1997 for references).

Chronic stress – a maladaptive response

Primarily as a result of man's activities, fish are also exposed to stressors which are chronic in duration or intermittent in frequency and for which the stress response fails to provide an adequate response. A good example is the imposition of adverse conditions upon fish reared within the aquaculture environment. When the threat or stressor is prolonged the fish has no way of avoiding or escaping the challenge and therefore the physiological response is also prolonged. It is in fish exposed to a chronic stressor, or to repeated intermittent acute stressors, that the adverse effects associated with stress such as cessation or reduction of growth, reproductive dysfunction and increased disease susceptibility become apparent (Pickering & Pottinger 1989; Pankhurst & Van Der Kraak 1997). Paradoxically, many of these adverse effects of chronic stress are associated directly with elevated cortisol levels (Pickering & Pottinger 1995).

As will be discussed below, because the net effect of activating the stress response can broadly be described as a switch from an anabolic state to a catabolic state, a major effect of stress on flesh quality is related to the redistribution of stored energy resources. These effects are more pronounced in fish which are exposed to chronic or intermittent stressors than in fish which are exposed to isolated acute stressors.

Physiological changes during severe exercise

As already noted, the stress response anticipates the likelihood that a significant increase in the activity of the fish will be required by mobilising metabolic fuel (respiratory substrates), thereby 'priming' the fish. It is not clear whether enforced exercise is itself inherently stressful or whether under aquaculture conditions the severity of imposed exercise will approach 'exhaustion' levels. However, imposed levels of exercise may be severe and may also result in a cognitive perception of threat by the fish. Both catecholamines and cortisol have been reported to be elevated during and after exhaustive exercise (Milligan 1996) and it is probably impossible to distinguish physiological changes due solely to stress from those due solely to exercise. Although the adaptive value of elevated catecholamine levels to an exercising fish are clear, the role of elevated cortisol levels following exercise is unresolved. High post-exercise cortisol levels appear to retard the recovery of exercised fish in terms of muscle metabolite and acid-base status (Pagnotta *et al.* 1994; Eros & Milligan 1996).

Whether perceived as stressful by the fish or not, important features of exercise are the mobilisation of energy sources and accumulation of waste products. Thus, many biochemical and metabolic events are common to both exercised fish and fish exposed to stressors. With regard to possible effects on flesh quality, it should be noted that disturbances caused to metabolic processes by exercise in fish are proportionately greater than those observed in mammals. Fish skeletal muscle consists predominantly of white (anaerobic) fibres and constitutes 40–60% of body mass (Bone 1978). In high intensity burst exercise, in which white muscle is recruited, most of the fishes' body mass is affected in contrast to mammals in which only a relatively small proportion of body mass is recruited during burst exercise (Milligan 1996).

Because the metabolic changes taking place in muscle tissue during exercise appear to be directly responsible for alterations in flesh quality, it is appropriate to summarise briefly their major features. Driedzic and Hochachka (1978) and Hochachka (1985) provide a detailed account of metabolism in fish muscle, and an overview of the major biochemical alterations which occur in muscle following exercise in fish is presented by Milligan (1996). Hydrolysis of adenosine triphosphate (ATP) during metabolic activity (e.g. muscle contraction) liberates energy contained within high energy phosphate bonds and generates adenosine diphosphate (ADP) together with free phosphate. ATP is regenerated from ADP by replenishment from a pool of phosphagens which in muscle is primarily phosphocreatine. However, this supply is limited and may become depleted.

In 'red' muscle, which has a predominantly aerobic metabolism, ATP is generated from the complete oxidation of carbohydrates, fats and amino acids in the tricarboxylic acid cycle. However, the majority of muscle mass in fish is adapted to anaerobic metabolism. In this tissue ATP is generated by the catabolism of glucose to lactic acid via the glycolytic pathway. Glycogen stored within the muscle itself provides a primary source of glucose to fuel the anaerobic pathway in muscle. Lactic acid accumulates within the muscle and during recovery is utilised to replenish the depleted muscle glycogen. Thus the net effect of exercise within the bulk of fish skeletal muscle is a depletion of phosphagen and glycogen reserves and an accumulation of lactic acid with a resulting decline in pH.

Energy mobilisation in fish during stress and exercise

The changes in blood and/or tissue levels of glucose, glycogen, free amino acids, free fatty acids, lactate, phosphocreatine and ATP which occur during stress and exercise are well-documented. Elevation of blood glucose and depletion of tissue (liver and muscle) glycogen levels are features common to both stressed (Pottinger & Machin, unpublished data) and severely exercised fish (Wang *et al.* 1994). The utilisation of free fatty acids (FFA) during exercise-induced anaerobic activity and subsequent recovery is not fully characterised in fish (Milligan & Girard 1993; Wang *et al*. 1994). Blood FFA are elevated during stress in Atlantic salmon (*Salmo salar*; Waring *et al*. 1992) but depleted in exhaustively exercised rainbow trout (*Oncorhynchus mykiss*; Dobson & Hochachka 1987). However, lipids appear to be a more important source of energy during aerobic swimming than during anaerobic burst activity (Lauff & Wood 1997; Kieffer *et al*. 1998). Blood amino acid levels in rainbow trout are known to increase following both exhaustive exercise (Milligan 1997) and acute stress (Pottinger & Machin, unpubl.; Morales *et al*. 1990) and it has been suggested that amino acid oxidation following exercise may fuel ATP and glycogen replenishment (Milligan 1997). Certainly, protein oxidation is a significant element in supporting aerobic activity in fish (Weber & Haman 1996; Kieffer *et al*. 1998). Changes in the muscle adenylate pool of rainbow trout following exercise have been fully characterised (Wang *et al*. 1994). Figures 15.1 and 15.2 provide a simplified picture of some of the physiological disturbances caused by stress and exercise.

The effects of stress and exercise on flesh quality

Having outlined the extent and duration of physiological changes which arise as a result of exposure to stressful stimuli and/or exercise, the question arises: how exactly do these changes impact on flesh quality? There has been a considerable amount of research into alterations in meat quality arising from stress and imposed exercise in cattle and pigs, and to a lesser extent in poultry, but relatively little research on the problem in fish. It is appropriate therefore to consider some of the principles which have been established from studies on mammals and poultry before looking at fish in more detail.

Effects of stress and exercise on flesh quality in cattle, pigs and poultry

The effects of a range of husbandry factors on meat quality have been widely studied in commercially reared meat producing animals. Factors which have been examined include handling of cattle (Warriss 1990; Lahucky *et al*. 1998) and pigs (D'Souza *et al*. 1998a,b), transportation (Becker *et al*. 1989) and pre-slaughter treatment of pigs (Brown *et al*. 1998) and transport and pre-slaughter disturbance in poultry (Froning *et al*. 1978; Ngoka *et al*., 1982; Kannan *et al*. 1997a). Although some studies have failed to demonstrate a direct link between husbandry stress/disturbance and meat quality (Kannan *et al*. 1997b; Brown *et al*. 1998) there is a general consensus that stress and exertion are linked to reduced flesh quality and that this is associated with disturbances prior to slaughter (Warriss, 1990; D'Souza *et al*. 1998a).

Two main flesh quality conditions have been identified in meat producing animals – pale soft exudative meat (PSE) and dark firm dry meat (DFD). The DFD state is related to excessive depletion of muscle glycogen. After death, glycogen continues to be converted to lactic acid in the muscle via the glycolytic pathway. Because stress and exercise deplete muscle glycogen, as discussed in the preceding sections, post-mortem lactate levels are reduced in the muscle of animals which have been exercised or exposed to stressful stimuli, compared to rested animals. This results in an abnormally high muscle pH which in turn

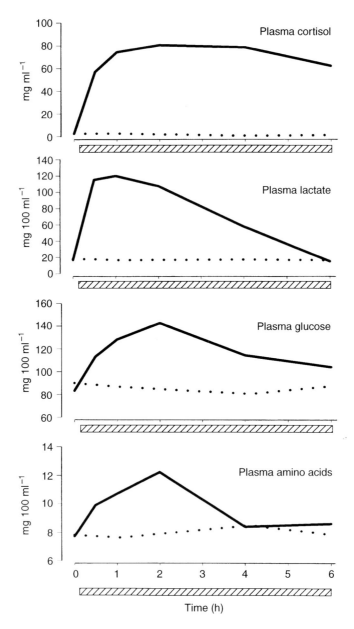

Fig. 15.1   A simplified schematic showing some of the physiological changes which occur in salmonid fish exposed to a short-term. The duration of the stressor is denoted by the horizontal shaded bars. Solid line – stressed fish, dotted line – unstressed fish. Data from T.G. Pottinger and L. Machin, unpublished.

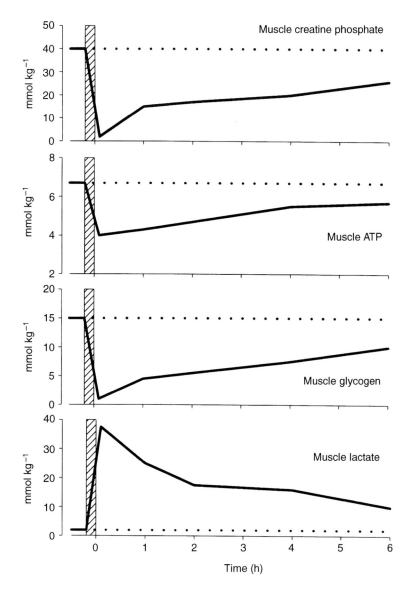

Fig. 15.2 A simplified schematic showing some of the physiological changes which occur in salmonid fish subjected to a period of exhaustive exercise. The duration of the exercise is denoted by the vertical shaded bars. Solid line – exercised fish, dotted line – rested fish. Derived from data in Tufts *et al.* (1991); Kieffer *et al.* (1994); Wang *et al.* (1994).

affects the physicochemical state of the muscle proteins causing changes in appearance and water-holding capacity which are primary markers of DFD meat (Braggins 1996). Spoilage may also be greater in meat with low glycogen/glucose levels because bacterial activity utilises amino acids rather than carbohydrate with the production of ammonia and related odours, and the higher pH favours bacterial growth.

The PSE condition arises as a consequence of an extremely rapid post-mortem glycolytic rate. The ultimate pH (final pH of the muscle post-mortem) may be reached within 15 minutes of slaughter and this rapid change is promoted by high temperature. PSE incidence in pigs is increased when the animals have been stressed prior to slaughter. Published evidence suggests that poultry muscle is affected in a similar way to that of pigs (Solomon *et al.* 1998). Poultry breast meat, like the majority of fish skeletal muscle, is composed primarily of 'white' fibres. In part because of the higher levels of glycogen in white muscle, post-mortem glycolytic rates are greater in white than in red muscle. In poultry, pre-slaughter stress and struggling result in adrenergically accelerated glycogenolysis which causes a reduction in muscle pH and ATP and localised acidosis due to the relatively poor vascularisation of white muscle. There is a consequent increase in denaturation of sarcoplasmic proteins and a reduction in the water-holding capacity of the meat. Accelerated rigor mortis also occurs because of the rapid depletion of muscle ATP (Sosnicki *et al.* 1998).

Clearly, the metabolic status of the animal prior to slaughter is critical in determining the quality of meat. Severe exercise, and stress, both disrupt normal metabolism and therefore potentially impact on meat quality. The rapid mobilisation of metabolic fuel is common to both stress and exercise and it is the depletion of muscle metabolite levels, in particular glycogen levels, and the pH changes in muscle tissue which accompany anaerobic metabolism, that impact on flesh quality.

Effects of stress and severe exercise on flesh quality in fish

Duration, severity, timing
Studies on mammals make it clear that the severity, duration and timing of husbandry-related stressors could be critical in determining how flesh quality in fish will be affected. The likelihood of aquacultured fish being exposed to a severe chronic stressor is limited. Because of the impact of chronic stress on growth, feeding behaviour and the immune system, stress-related problems become apparent to the vigilant grower well in advance of slaughter. Chronic stress can lead to immunosuppression which in turn causes the fish to become more susceptible to parasites and pathogens. Although the proteolytic activity of some parasites, such as the Myxosporean *Kudoa thyrsites*, can cause softened and discoloured flesh (Whitaker & Kent 1991) this is not likely to represent a commonly encountered problem.

Nonetheless, it is worth considering what are the direct effects of long-term, or intermittent acute stress, on flesh quality. Because of the catabolic nature of the stress response, exposure to a prolonged stressor will have effects akin to those of starvation, with the mobilisation of stored energy reserves and possible muscle wasting in extreme cases. However, published data suggest that the effect of prolonged food withdrawal, in the absence of additional stressors, varies in fish according to species.

Long-term stress/starvation

In Atlantic salmon starvation for up to 86 days (at $4°C$) prior to slaughter caused a decrease in the glycogen content of white muscle at slaughter which was proportional to the time of starvation. Fillet fat and protein content were also reduced by starvation. During post-slaughter storage on ice, muscle lactate and glycogen levels were decreased in proportion to the pre-slaughter period of starvation, accompanied by an increase in muscle pH. However, effects of starvation on the texture, freshness and colour of fillets were considered by the authors of this study to be marginal and effects on fat content likely to be masked by large inter- and intra-individual variation in fillet fat content (Einen & Thomassen 1998a, b). In contrast, in rainbow trout deprived of food for up to two months the total lipid content of the fillet was reduced and the eating quality adversely affected, with a putrescent odour and muddy taste being reported (Johansson & Kiessling 1991). Shorter periods of starvation (< 18 days) were not reported to alter the fat content of fillets (Færgemand *et al.* 1995; Ostenfeld *et al.* 1995). It has also been demonstrated that in rainbow trout starvation does not affect the concentration of phosphocreatine and ATP in muscle (Kieffer & Tufts 1998), an observation of some significance when considering effects of pre-slaughter exertion/stress (see below). Ostenfeld *et al.* (1995) examined the effects of a 10.5 hour period of transport in O_2-saturated water following an eight day period of starvation, on flesh quality in rainbow trout. There was no effect of transport or starvation on white muscle lactic acid content glucose content or glycogen content, nor was there any effect of transport on textural parameters (hardness, elasticity, breakpoint).

High stocking densities are considered to be a potent stressor in salmonid fish and thus the effect of stocking density on flesh composition of rainbow trout was examined by Zoccarato *et al.* (1994). These authors found that at densities high enough to cause a reduction in growth rate ($16 \, kg \, m^{-3}$ rising to a maximum of $43 \, kg \, m^{-3}$) relative to fish maintained at lower density ($8 \, kg \, m^{-3}$ rising to a maximum of $25 \, kg \, m^{-3}$) no significant effects were observed on muscle in terms of water content, protein, total lipids and fatty acids, and amino acid composition.

Short-term stress and exercise: effects of method of slaughter on flesh quality indicators

As already discussed, severe exercise or acute stress cause a substantial redistribution and depletion of metabolites in muscle. However, recovery from brief but severe exercise is rapid, requiring a maximum of about eight hours (Milligan 1996). So the exposure of fish to acute stressors is unlikely to impact significantly on flesh quality unless it occurs very close to the time of slaughter. The processes associated with slaughter of fish from capture, through transport, handling and stunning, are undoubtedly stressful. There is some disagreement among published studies as to whether slaughter-related stress affects flesh quality in fish.

Færgamand *et al.* (1995) reported that killing fish by one of two methods, rapid cervical section or asphyxiation, did not affect fat content or texture (determined mechanically and manually) of rainbow trout fillets. Similar conclusions were reached by Azam *et al.* (1989) who examined the effects of three slaughter methods (electroshock, CO_2 narcosis, concussion) on flesh quality in rainbow trout. Overall, these authors found that slaughter method had no significant impact on quality measures (hardness and elasticity, water-holding capacity, bacterial counts) determined during storage. Unsurprisingly, lactate levels were slightly higher and pH slightly lower in the electroshocked and CO_2-killed fish compared to those killed by concussion. However, within five to ten days after slaughter no differences were apparent. A sensory panel, which evaluated toughness, elasticity, firmness and succulence, also failed to discriminate between fillets from fish killed by each of the slaughter methods.

In contrast, the method of killing was found to cause significant differences in flesh quality indicators in horse mackerel (*Trachrus japonicus*; Mochizuki & Sato 1994). Fish allowed to struggle in air until mortality showed more rapid depletion of muscle ATP and more rapid increases in muscle inosine monophosphate (IMP) than fish killed rapidly by pithing or by temperature shock. The onset of rigor was also more rapid in these fish. However, it is arguable that killing the fish by asphyxiation is a prolonged and inhumane slaughter method and the associated struggling will inevitably result in biochemical changes in muscle which resemble those occurring during exhaustive exercise.

Short-term stress and exercise: effects of pre-slaughter disturbance on flesh quality indicators

Jerrett *et al.* (1996) suggested that most studies on the effects of harvesting-related stress on fish muscle quality have failed to control correctly for capture and handling factors and therefore the 'true' condition of 'rested' fish muscle may not have been reported. Using a combination of conditioning, careful handling and anaesthetisation to minimise pre-slaughter stress these authors demonstrated that the tensile strength of chinook salmon (*Oncorhynchus tshawytscha*) white muscle during storage was significantly greater in unstressed than in exhaustively exercised/stressed fish. In this case, loss of tensile strength was equated with 'tenderisation' of the muscle (Ando *et al.* 1993). Accordingly, these authors emphasised the importance of reducing pre-harvest disturbance in producing high quality fish muscle. This assertion is supported by consideration of the results of other similar studies.

The rate of onset of *rigor mortis* is considered to be an important factor in maximising filleting yield which is reduced if fish are processed while in rigor (Azam *et al.* 1990). Because the detachment of actin and myosin filaments is ATP-dependent, rigor is observed when depletion of ATP reaches a critical point. Numerous studies have reported effects of stress on the rate of onset of rigor in fish. In the snapper (*Pagrus auratus*; Lowe *et al.* 1993) the depletion of muscle ATP and thus the onset of rigor was delayed in unstressed fish relative to fish stressed prior to slaughter. Similarly, the onset of rigor was found to be more rapid in severely (although artificially) exercised chinook salmon white muscle (Jerrett & Holland 1998) than in unexercised control tissue. In carp (*Cyprinus carpio*) stressed by severe disturbance prior to slaughter the onset of rigor was significantly more rapid (1.6 h) than in carp killed in the absence of disturbance (21.2 h). The onset of rigor in the stressed fish was reported to occur before muscle ATP depletion and this was ascribed to the occurrence of tetanic contraction in the muscle of stressed fish, related to a high intracellular Ca^{2+} concentration (Nakayama *et al.* 1992).

In Atlantic salmon, the relative concentrations of phosphocreatine, ATP and IMP in white muscle during storage were found to be primarily dependent on the pre-slaughter treatment of the fish (Berg *et al.* 1997). In contrast with other studies, these authors did not find a clear difference in the development of *rigor mortis* in stressed (pumped from the well-boat to slaughter line) and unstressed (killed immediately with a blow to the head) fish but reported that rigor strength was greater in fish stressed prior to slaughter. This result they attributed to the possibility that metabolites were depleted less uniformly within the muscle of unstressed fish than stressed fish, leading to regional differences in the onset of rigor in unstressed fish. Effects of handling on muscle metabolite levels were also reported by Sigholt *et al.* (1997) who examined meat quality in Atlantic salmon killed by CO_2 narcosis and gill bleeding, with or without a pre-slaughter period of confinement stress. Stressed fish displayed lower muscle pH at slaughter, and lower phosphocreatine and ATP levels together with a more rapid onset of rigor than unstressed fish. Fillets from fish stressed prior to slaughter were softer than those from unstressed fish and had a lower breaking strength. These effects were influenced by the temperature at which the fillet was stored.

In a related study carried out under similar conditions no significant differences in muscle metabolite levels were detected in salmon sampled prior to transport in a well-boat, during transport, or immediately after slaughter (Erikson *et al.* 1997). In this case, the authors suggested that the surprising absence of effects of handling stress on muscle biochemistry may have arisen because of differences in the care with which fish were handled under controlled experimental conditions compared to commercial conditions, noting that at two other processing plants where similar studies were carried out (Berg *et al.* 1997 and unpublished) effects of handling stress were more severe. The general trend for exercise and stress prior to slaughter to be reflected in alterations in muscle metabolite levels and rate of onset of rigor was supported by the results of Thomas *et al.* (1999). These authors found that rigor occurred more rapidly in Atlantic salmon which had been subjected to conditions designed to simulate the stress and exercise associated with pre-slaughter disturbances.

Conclusions

Although equivocal and contradictory in some respects, the available data suggest that pre-slaughter husbandry practices can have significant effects on flesh quality in fish. As is the case for mammals and poultry, the adverse changes in the muscle of fish following slaughter appear to be primarily related to the extent to which muscle metabolites are depleted before death and this in turn is directly related to the extent to which the fish are exercised and/or stressed prior to slaughter. The most consistently reported effect of pre-slaughter disturbance is an increase in the rate of onset of *rigor mortis*. However, there appears to be no widespread perception that the quality of farmed fish flesh following slaughter is inadequate. There are, as yet, no reports of the systematic occurrence of flesh quality problems analogous to the DFD and PSE evident in mammalian meat production.

It is unlikely that significant problems exist with current approaches to ongrowing and the general nutrition and husbandry of aquacultured fish which might impact on post-slaughter condition. Any major nutritional imbalances or insufficiencies are likely to become apparent well in advance of slaughter and processing. Instead, a primary goal should be to reduce the extent of pre-slaughter disturbance and thereby reduce the disruption of muscle metabolism which is associated with flesh quality problems. Three main strategies can be identified which may allow the quality of farmed fish flesh to be further improved:

(1)   Handling and disturbance are unavoidable elements of present harvest and slaughter methods and cannot be completely avoided. The development of less stressful and more rapid slaughter methods may lead to improvements in flesh quality by reducing the extent of disruption caused to muscle metabolism.

(2)   Quiescence of the fish prior to slaughter could be achieved using one of a number of anaesthetics, but to be economically viable and practical this approach requires an anaesthetic which does not present residue-related problems and which is affordable. Two possible candidates are sodium bicarbonate/acetic acid (a cheaper alternative to gaseous CO_2 narcosis; Prince *et al.* 1995) and clove oil (Anderson *et al.* 1997).

(3)   Modification of the stress responsiveness or aerobic/anaerobic performance of the fish is possible either genetically, by selectively breeding for a strain with enhanced tolerance of aquacultural procedures (Pottinger & Pickering 1997), or behaviourally, by training (Davison 1997). Both strategies would require a significant investment of time and manpower.

To determine whether the implementation of any of these strategies is justifiable requires further research to establish the degree to which current practices result in unsatisfactory flesh quality, or to assess the extent to which the existing production of flesh of acceptable quality might be enhanced.

References

Anderson, W.G., McKinley, R.S. & Colavecchia, M. (1997) The use of clove oil as an anesthetic for rainbow trout and its effects on swimming performance. *North American Journal of Fisheries Management*, 17, 301–307.

Ando, M., Toyohara, H., Shimizu, Y. & Sakaguchi, M. (1993) Post-mortem tenderization of fish muscle due to weakening of pericellular connective tissue. *Nippon Suisan Gakkaishi*, 59, 1073–1076.

Azam, K., Mackie, I.M. & Smith, J. (1989) The effect of slaughter method on the quality of rainbow trout (*Salmo gairdneri*) during storage on ice. *International Journal of Food Science and Technology*, 24, 69–79.

Azam, K., Strachan, N.J.C., Mackie, I.M., Smith, J. & Nesvadba, P. (1990) Effect of slaughter method on the progress of rigor of rainbow trout (*Salmo gairdneri*) as measured by an image processing system. *International Journal of Food Science and Technology*, 25, 477–82.

Barton, B.A. (1997) Stress in finfish: past, present and future – a historical perspective. In: *Fish Stress and Health in Aquaculture* (eds G.K. Iwama, A.D. Pickering, J.P. Sumpter & C.B. Schreck), pp.1–33. Cambridge University Press, Cambridge.

Becker, B.A., Mayes, H.F., Hahn, G.L., Nienaber, J.A., Jesse, G.W., Anderson, M.E., Heymann, H. & Hedrick, H.B. (1989). Effect of fasting and transportation on various phsyiological parameters and meat quality of slaughter hogs. *Journal of Animal Science*, 67, 334–41.

Berg, T., Erikson, U. & Nordtvedt, T.S. (1997) *Rigor mortis* assessment of Atlantic salmon (*Salmo salar*) and effects of stress. *Journal of Food Science*, 62, 439–46.

Bone, Q. (1978) Locomotor muscle. In: *Fish Physiology. Vol. VII. Locomotion*, (eds W.S. Hoar & D. J. Randall) pp. 361–424. Academic Press, London.

van der Boon, J., Van den Thillart, G.E.E.J.M. & Addink, A.D.F. (1991) The effects of cortisol administration on intermediary metabolism in teleost fish. *Comparative Biochemistry and Physiology*, **100A**, 47–53.

Braggins, T.J. (1996). Effect of stress-related changes in sheepmeat ultimate pH on cooked odour and flavour. *Journal of Agriculture Food Chemistry*, **44**, 2352–60.

Brown, S.N., Warriss, P.D., Nute, G.R., Edwards, J.E. & Knowles, T.G. (1998) Meat quality in pigs subjected to minimal preslaughter stress. *Meat Science*, **49**, 257–65.

Davison, W. (1997) The effects of exercise training on teleost fish, a review of recent literature. *Comparative Biochemistry and Physiology*, **117A**, 67–75.

Dobson, G.P. & Hochachka, P.W. (1987) Role of glycolysis in adenylate depletion and repletion during work and recovery in teleost white muscle. *Journal of Experimental Biology*, **129**, 125–40.

Driedzic, W.R. & Hochachka, P.W. (1978) Metabolism in fish during exercise. In: *Fish Physiology. Vol. VII. Locomotion*, (eds W.S. Hoar & D.J. Randall) pp.503–43. Academic Press, London.

D'Souza, D.N., Dunshea, F.R., Warner, R.D. & Leury, B.J. (1998a) The effect of handling pre-slaughter and carcass processing rate post-slaughter on pork quality. *Meat Science*, **50**, 429–37.

D'Souza, D.N., Warner, R.D., Dunshea, F.R. & Leury, B.J. (1998b) Effect of on-farm and pre-slaughter handling of pigs on meat quality. *Australian Journal of Agricultural Research*, **49**, 1021–5.

Einen, O. & Thomassen, M.S. (1998a) Starvation prior to slaughter in Atlantic salmon (*Salmo salar*). I. Effects on weight loss, body shape, slaughter- and fillet-yield, proximate and fatty acid composition. *Aquaculture*, **166**, 85–104.

Einen, O. & Thomassen, M.S. (1998b) Starvation prior to slaughter in Atlantic salmon (*Salmo salar*). II. White muscle composition and evaluation of freshness, texture and colour characteristics in raw and cooked fillets. *Aquaculture*, **169**, 37–53.

Erikson, U., Sigholt, T. & Seland, A. (1997) Handling stress and water quality during live transportation and slaughter of Atlantic salmon (*Salmo salar*). *Aquaculture*, **149**, 243–52.

Eros, S.K. & Milligan, C.L. (1996) The effect of cortisol on recovery from exhaustive exercise in rainbow trout (*Oncorhynchus mykiss*): potential mechanisms of action. *Physiological Zoology*, **69**, 1196–214.

Færgemand, J., Rønsholdt, B., Alsted, N. & Børresen, T. (1995) Fillet texture of rainbow trout as affected by feeding strategy, slaughtering procedure and storage post mortem. *Water Science and Technology*, **31**, 225–31.

Froning, G.W., Babji, A.S. & Mather, F.B. (1978) The effect of preslaughter temperature, stress, struggle and anesthetization on colour and textural characteristics of turkey muscle. *Poultry Science*, **57**, 630–3.

Gjedrem, T. (1997) Flesh quality improvement in fish through breeding. *Aquaculture International*, **5**, 197–206.

Hochachka, P.W. (1985) Fuels and pathways as designed systems for support of muscle work. *Journal of Experimental Biology*, **115**, 149–64.

Jerrett, A.R. & Holland, A.J. (1998) Rigor tension development in excised 'rested', 'partially exercised' and 'exhausted' chinook salmon white muscle. *Journal of Food Science*, **63**, 48–52.

Jerrett, A.R., Stevens, J. & Holland, A.J. (1996) Tensile properties of white muscle in rested and exhausted chinook salmon (*Oncorhynchus tshawytscha*). *Journal of Food Science*, **61**, 527–32.

Johansson, L. & Kiessling, A. (1991) Effects of starvation on rainbow trout. II. Eating and storage qualities of iced and frozen fish. *Acta Agriculturae Scandinavica*, **41**, 207–16.

Kannan, G., Heath, J.L., Wabeck, C.J. & Mench, J.A. (1997a) Shackling of broilers: effects on stress responses and breast meat quality. *British Poultry Science*, **38**, 323–32.

Kannan, G., Heath, J.L., Wabeck, C.J., Souza, M.C.P., Howe, J.C. & Mench, J.A. (1997b) Effects of crating and transport on stress and meat quality characteristics in broilers. *Poultry Science*, **76**, 523–9.

Kieffer, J.D. & Tufts, B.L. (1998) Effects of food deprivation on white muscle energy reserves in rainbow trout (*Oncorhynchus mykiss*): the relationships with body size and temperature. *Fish Physiology and Biochemistry*, **19**, 239–45.

Kieffer, J.D., Currie, S. & Tufts, B.L. (1994) Effects of environmental temperature on the metabolic and acid-base responses of rainbow trout to exhaustive exercise. *Journal of Experimental Biology*, **194**, 299–317.

Kieffer, J.D., Alsop, D. & Wood, C.M. (1998) A respirometric analysis of fuel use during aerobic swimming at different temperatures in rainbow trout (*Oncorhynchus mykiss*). *Journal of Experimental Biology*, **201**, 3123–33.

Lahucky, R., Palanska, O., Mojto, J., Zaujec, K. & Huba, J. (1998) Effect of preslaughter handling on muscle glycogen level and selected meat quality traits in beef. *Meat Science*, **50**, 389–93.

Lauff, R.F. & Wood, C.M. (1997) Effects of training on respiratory gas exchange, nitrogenous waste excretion, and fuel usage during aerobic swimming in juvenile rainbow trout (*Oncorhynchus mykiss*). *Canadian Journal of Fisheries and Aquatic Sciences*, **54**, 566–71.

Lowe, T.E., Ryder, J.M., Carragher, J.F. & Wells, R.M.G. (1993) Flesh quality in snapper, *Pagrus auratus*, affected by capture stress. *Journal of Food Science*, **58**, 770–73, contd. 796.

Milligan, C.L. (1996) Metabolic recovery from exhaustive exercise in rainbow trout. *Comparative Biochemistry and Physiology*, **113A**, 51–60.

Milligan, C.L. (1997) The role of cortisol in amino acid mobilization and metabolism following exhaustive exercise in rainbow trout (*Oncorhynchus mykiss Walbaum*). *Fish Physiology and Biochemistry*, **16**, 119–28.

Milligan, C.L. & Girard, S.G. (1993) Lactate metabolism in rainbow trout. *Journal of Experimental Biology*, **180**, 175–93.

Mochizuki, S. & Sato, A. (1994) Effects of various killing procedures and storage temperatures on post-mortem changes in the muscle of horse mackerel. *Nippon Suisan Gakkaishi*, **60**, 125–30.

Morales, A.E., GarcPa-Rej, N, L. & de la Higuera, M. (1990) Influence of handling and/or anaesthesia on stress response in rainbow trout. Effects on liver primary metabolism. *Comparative Biochemistry and Physiology*, **95A**, 87–93.

Nakayama, T., Liu, D.-J. & Ooi, A. (1992) Tension change of stressed and unstressed carp muscles in isometric rigor contraction and resolution. *Nippon Suisan Gakkaishi*, **58**, 1517–22.

Ngoka, D.A., Froning, G.W., Lowry, S.R. & Babji, A.S. (1982) Effects of sex, age, pre-slaughter factors, and holding conditions on the quality characteristics and chemical composition of turkey breast muscles. *Poultry Science*, **61**, 1996–2003.

Ostenfeld, T., Thomsen, S., Ingolfdottir, S., Ronsholdt, B. & McLean, E. (1995) Evaluation of the effect of live haulage on metabolites and fillet texture of rainbow trout (*Oncorhynchus mykiss*). *Water Science and Technology*, **31**, 233–7.

Pagnotta, A., Brooks, L. & Milligan, L. (1994) The potential regulatory roles of cortisol in recovery from exhaustive exercise in rainbow trout. *Canadian Journal of Zoology*, **72**, 2136–46.

Pankhurst, N.W. & van der Kraak, G. (1997) Effects of stress on reproduction and growth of fish. In: *Fish Stress and Health in Aquaculture*, (eds G.K. Iwama, A.D. Pickering, J.P. Sumpter & C.B. Schreck), pp.73–93. Cambridge University Press, Cambridge.

Pickering, A.D. & Pottinger, T.G. (1989) Stress responses and disease resistance in salmonid fish: effects of chronic elevation of plasma cortisol. *Fish Physiology and Biochemistry*, **7**, 253–8.

Pickering, A.D. & Pottinger, T.G. (1995) Biochemical effects of stress. In: *Biochemistry and Molecular Biology of Fishes Volume 5. Environmental and Ecological Biochemistry*, (eds P.W. Hochachka and T.P. Mommsen), pp.349–79. Elsevier Science BV, Amsterdam.

Pottinger, T.G. & Pickering, A.D. (1997) Genetic basis to the stress response: selective breeding for stress-tolerant fish. In: *Fish Stress and Health in Aquaculture*, (eds G.K. Iwama, A.D. Pickering, J.P. Sumpter & C.B. Schreck), pp.171–93. Cambridge University Press.

Prince, A.M.J., Low, S.E., Lissimore, T.J., Diewert, R.E. & Hinch, S.G. (1995) Sodium bicarbonate and acetic acid: an effective anaesthetic for field use. *North American Journal of Fisheries Management*, **15**, 170–72.

Randall, D.J. & Perry, S.F. (1992) Catecholamines. In: *Fish Physiology. Vol. XIIB. The Cardiovascular System*, (eds W.S. Hoar, D.J. Randall & A.P. Farrell) pp.255–300. Academic Press, London.

Sigholt, T., Erikson, U., Rustad, T., Johansen, S., Nordtvedt, T.S. & Seland, A. (1997) Handling stress and storage temperature affect meat quality of farm-raised Atlantic salmon (*Salmo salar*). *Journal of Food Science*, **62**, 898–905.

Solomon, M.B., van Laack, R.L.J.M. & Eastridge, J.S. (1998) Biophysical basis of pale, soft, exudative (PSE) pork and poultry muscle: a review. *Journal of Muscle Foods*, **9**, 1–11.

Sosnicki, A.A., Greaser, M.L., Pietrzak, M., Pospiech, E. & Sante, V. (1998) PSE-like syndrome in breast muscle of domestic turkeys: a review. *Journal of Muscle Foods*, **9**, 13–23.

Suarez, R.K. & Mommsen, T.P. (1987) Gluconeogenesis in teleost fishes. *Canadian Journal of Zoology*, **65**, 1869–82.

Sumpter, J.P. (1997) The endocrinology of stress. In: *Fish Stress and Health in Aquaculture*, (eds G.K. Iwama, A.D. Pickering, J.P. Sumpter & C.B. Schreck), pp.95–118. Cambridge University Press, Cambridge.

Thomas, P.M., Pankhurst, N.W. & Bremner, H.A. (1999) The effect of stress and exercise on post-mortem biochemistry of Atlantic salmon and rainbow trout. *Journal of Fish Biology*, **54**, 1177–96.

Tufts, B.L., Tang, Y., Tufts, K. & Boutilier, R.G. (1991) Exhaustive exercise in 'wild' Atlantic salmon (*Salmo salar*): acid-base regulation and blood gas transport. *Canadian Journal of Fisheries and Aquatic Sciences*, **48**, 868–74.

Wang, Y., Heigenhauser, G.J.F. & Wood, C.M. (1994) Integrated responses to exhaustive exercise and recovery in rainbow trout white muscle: acid-base, phosphagen, carbohydrate, lipid, ammonia, fluid volume and electrolyte metabolism. *Journal of Experimental Biology*, **195**, 227–58.

Waring, C.P., Stagg, R.M. & Poxton, M.G. (1992) The effects of handling on flounder (*Platichthys flesus* L.) and Atlantic salmon (*Salmo salar* L.). *Journal of Fish Biology*, **41**, 131–44.

Warriss, P.D. (1990) The handling of cattle pre-slaughter and its effects on carcass and meat quality. *Applied Animal Behaviour Science*, **28**, 171–86.

Weber, J.-M. & Haman, F. (1996) Pathways for metabolic fuels and oxygen in high performance fish. *Comparative Biochemistry and Physiology*, **113A**, 33–8.

Wendelaar Bonga, S.E.W. (1997) The stress response in fish. *Physiological Reviews*, **77**, 591–625.

Whitaker, D.J. & Kent, M.L. (1991) Myxosporean *Kudoa thyrsites*: a cause of soft flesh in farm-reared Atlantic salmon. *Journal of Aquatic Animal Health*, **3**, 291–4.

Zoccarato, I., Benatti, G., Bianchini, M.L., Boccignone, M., Conti, A., Napolitano, R. & Palmegiano, G.B. (1994) Differences in performance, flesh composition and water output quality in relation to density and feeding levels in rainbow trout, *Oncorhynchus mykiss* (Walbaum), farming. *Aquaculture and Fisheries Management*, **25**, 639–47.

Chapter 16

The Effects of Nutrition on the Composition of Farmed Fish

P.C. Morris

BOCH PAULS Fish Feed Group, Renfrew Mill, Wright Street, Renfrew, PA4 8BF, UK

Introduction

While it is not possible to manipulate the composition of wild fish, developments in aquaculture enable those involved to manipulate fish composition to match quality targets. A substantial volume of literature has been produced detailing the effects of nutrition on the composition of farmed fish. This was reviewed by Shearer (1994) who also proposed a more reliable approach to the statistical treatment of proximate composition data. These methods are described in Chapter 4 of this book by Karl Shearer.

Amongst the factors affecting fish composition, nutrition is the most intensively studied and the most easily manipulated under farming conditions. Nutritional effects on fish composition can be classified into two groups: diet formulation and feeding regime. In terms of diet formulation, this chapter will deal with the effects of the major nutrients (protein, lipids, carbohydrate and ash) and their relative proportions/contributions to dietary energy on the proximate composition and slaughter yields of fish. The effects of the micronutrients and pigments on fish quality are discussed elsewhere in this book by Rémi Baker in Chapter 17, and David Nickell and John Springate in Chapter 6.

The advent of high energy diets has focused attention on the effects of feed and feeding on the composition and quality of farmed fish. In this chapter, some of the more recent publications in the field will be reviewed, leading to a number of recommendations to fish farmers who have the unenviable task of optimising the cost of production while matching the ever more rigorous expectations of consumers.

The effects of feed composition

The protein content of fish

The whole body protein content of fish is endogenously regulated such that, for a given developmental stage or size, the protein content is pre-determined by the genetic characteristics of the species or strain (Shearer 1994). Furthermore, when expressed as a percentage of the crude protein, the amino acid profile of fish muscle is very similar both amongst fish from the same genus and between genera (Wilson & Cowey 1985; Cowey 1993). Thus, given the limited number of abundant proteins in muscle and the relatively small size of the free amino acid pool, there is little scope for dietary modification of fish flesh amino acid profiles.

Re-evaluation of historical data using an allometric (size specific) approach clearly showed that whole body protein content was not easily manipulated amongst fed fish (Shearer, 1994). More recent work has served to substantiate this conclusion. Thus, whole body protein level was unaffected by changes in major nutrient composition in diets fed to Mozambique tilapia, *Oreochromis mossambicus* (El-Dahhar & Lovell 1995), hybrid tilapia, *O. niloticus* × *O. aureus* (Chou & Shiau 1996), dentex, *Dentex dentex* (Tibaldi *et al.* 1996), and gilthead seabream, *Sparus aurata* (Company *et al.* 1999).

When fish are further processed, again, there seems to be minimal effect of diet on the protein content of the fish. Thus, for Atlantic halibut, *Hippoglossus hippoglossus*, dietary protein levels ranging from 55 to 72% did not affect the eviscerated carcass protein content (Aksnes *et al.* 1996). Additionally, the fillet and cutlet protein content of Atlantic salmon, *Salmo salar*, was unaffected by dietary protein:energy ratio (Einen & Roem 1997; Einen & Skrede 1998; Hemre & Sandness 1999).

The ash content of fish

As with protein, the whole body ash level in fish is endogenously regulated and therefore can be little modified by the farmer. However, in certain cases of mineral deficiency, the whole body level of the deficient mineral declines and this is often reflected in the bone ash content of the deficient fish (Schwarz 1995; Eya & Lovell 1997). Additionally, imbalance of the mineral content of the diet may result in decreased availability of essential minerals from the feed. This is exemplified by reduced manganese availability to Atlantic salmon parr fed diets containing relatively modest iron supplements (Andersen *et al.* 1996).

It was generally considered that, provided sufficient minerals were available in the correct proportions, the dietary ash content does not directly influence the ash content of farmed fish (Shearer 1994). However, Shearer *et al.* (1992) indicated that excess ash in diets for Atlantic salmon parr in addition to reducing zinc availability also reduced the availability of energy from the diet with adverse affects on parr growth and feed conversion ratio (FCR). This lower energy availability could therefore result in lower levels of whole body fat in fish.

Interestingly, Einen and Roem (1997) recorded significant effects of protein:energy ratio on the ash content of cutlets of large (4.5–5.0 kg) Atlantic salmon. These increases in ash content were not apparent amongst smaller salmon (2.5–3.0 kg) fed the same diets and were not apparent for either size class when the whole body ash content was considered. Amongst the 4.5–5.0 kg salmon examined by Einen and Roem (1997), the condition factor of the fish with the highest level of ash in the cutlet was the lowest and vice versa. Therefore, condition factor may have some relevance to the significant differences in ash content recorded by numerous authors and nearly always ignored.

The carbohydrate content of fish

It is recognised that fish have no qualitative requirement for dietary carbohydrate (Wilson

1994) and that, in its absence, they can match physiological requirements by gluconeo-genesis using amino acid skeletons and selected citric acid cycle intermediates as substrates. However, the carbohydrates still represent the cheapest form of non-protein energy and provide the starch necessary for gelatinisation during the feed manufacturing process. The carbohydrate content of fish is very low, generally accounting for less than 1% whole body weight. Typically, the carbohydrates are present as monosaccharides, glycogen or metabolic intermediates.

In general, changes in the lipid or protein content of the diet do not directly affect the carbohydrate content of farmed fish. The ability of fish to utilise carbohydrate varies on a species by species basis and in accordance with the complexity and quantity of the starch supplied. Cold water and marine fishes are believed to have a poor capacity while warm water species have a high capacity for carbohydrate utilisation (Wilson 1994). The effects of dietary carbohydrate on fish carbohydrate content are discussed below while its effects as a source of non-protein energy in terms of lipid deposition are discussed elsewhere in this chapter.

Typically, fish are able to modulate blood glucose levels and tissues such as the kidney and gills contain low levels of carbohydrate. However, problems with excess carbohydrate or the provision of highly available starch generally manifest themselves in terms of deposition of excess glycogen in the liver associated with increased hepatosomatic index. This has been observed in grey mullet, *Chelon labrosus*, fed diets containing more than 12% corn starch/dextrin (2 : 1) mixture (Ojaveer *et al.* 1996). Similarly, there was a strong correlation between liver glycogen and dietary carbohydrate level in Atlantic salmon fed diets containing 22% oil and between 0 and 31% wheat starch as a substitute for protein (Hemre *et al.* 1995).

The moisture content of fish

Along with fat, the moisture content of fish is easily modified by changes in the diet. Generally, the moisture content declines in response to increased fat in the whole body, fillets and cutlets (Jobling *et al.* 1998; Koskela *et al.* 1998). Fig. 16.1 clearly illustrates this rela-tionship for turbot, *Scophthalmus maximus*, (Morris, unpublished). Having established a reliable calibration curve, the relationship between fat and moisture content can be used to provide an indirect method for the determination of fat levels in farmed fish.

The fat content of fish

The fat content of fish is the topic most likely to cause disagreement between farmers, processors and customers and a point at which views on the acceptability of a product are at their most fractious. The fat content of farmed fish can, to an extent, be manipulated by dietary means in terms not only of its quantity and distribution within the fish but also its perceived quality. This chapter deals mainly with the amount and distribution of fat within the fish. Aspects of perceived lipid quality, e.g. *n*-3 : *n*-6 ratio and HUFA (highly unsaturated fatty acid) content, can be strongly influenced by the type of fat consumed (Morris 1998) and by the feeding regime (Wathne 1995; Einen *et al.* 1998). However, these aspects are outside the remit of this chapter and are dealt with in this book by John Sargent *et al.* in Chapter 2 and Karl Shearer in Chapter 4.

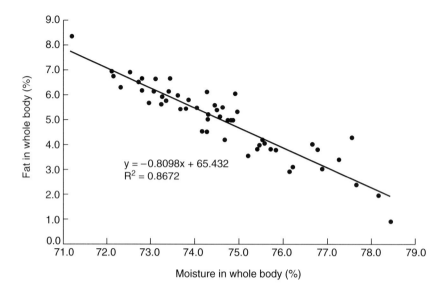

Fig. 16.1   The relationship between whole body moisture and fat content in turbot (unpublished observations).

Fat distribution

The fat content of fish is not uniformly distributed throughout the body but there are regions of the fish which may have very high or low fat content. Using Atlantic salmon as an example, numerous authors have recently focused on the effects of diet on fat localisation (Table 16.1) and on differences in fat content in different regions of the fillet (Fig. 16.2). Generally, the muscle tissue at the posterior end of the fish is lean and the fish becomes fatter when moving in the caudal-cranial direction. Very high levels of fat are found in the tissues surrounding the viscera and in the belly flap region. It is also apparent that there is a cor-

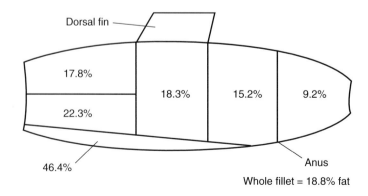

Fig. 16.2   Distribution of fat in a fillet taken from a 6 kg Atlantic salmon (Einen *et al.* 1998).

Table 16.1   Fat content in various portions of Atlantic salmon.

Fish weight (in round, kg)	Portion (as described)	Fat (%)	Reference
Cutlets and steaks			
6.3	NQC	16.7	Einen *et al.* 1998
4.6–5.0	NQC	17.2–18.3	Einen & Roem 1997
4.2	NQC	8.0–13.0	Hillestad *et al.* 1998
2.5–2.9	Cutlet	15.7–18.9	Einen & Roem 1997
2.0–2.2	Cutlet	12.0	Wathne 1995
Fillet			
6.3	Fillet	18.1	Einen *et al.* 1998
4.6–5.0	Fillet	16.1–17.9	Einen & Skrede 1998
3.4–3.9	Muscle	12.3–16.9	Aksnes 1995
2.3–2.5	Muscle	12.0–16.0	Hemre & Sandness 1999
2.0–2.2	White muscle	4.0	Wathne 1995
2.0–2.2	Red muscle	32.8	Wathne 1995
Offal and waste			
6.3	Belly flap	46.4	Einen *et al.* 1998
2.0–2.2	Belly flap adipose	61.1	Wathne 1995
6.3	Viscera	35.1	Einen *et al.* 1998
4.2	Visceral	32.0–46.0	Hillestad *et al.* 1998
2.0–2.2	Visceral adipose	91.9	Wathne 1995
6.3	Liver	4.5	Einen *et al.* 1998
4.6–5.0	Liver	5.0–5.4	Einen & Roem 1997
2.5–2.9	Liver	5.1–5.5	Einen & Roem 1997
2.0–2.2	Liver	4.1	Wathne 1995

relation between fish size and fat content (Table 16.1) and this is discussed in greater detail in this book by Karl Shearer.

Effects of protein, lipid, carbohydrate and energy on fish fat content

Shearer (1994) emphasised the importance of digestible energy in determining the fat content of fish. Fat can accumulate in fish not only as a consequence of the consumption of lipids but, due to lipogenesis, can be derived by the consumption of protein and carbohydrate. Consequently, it is very difficult to isolate the effects of protein, fat or carbohydrate on fish composition individually and therefore it is common to discuss the effects of energy or protein:energy ratio on fish composition. Prompted by a trend towards increasing oil levels and decreasing protein levels in feeds, there have recently been a number of publications investigating the effects of protein:energy ratio on fish composition. Making direct comparisons between different data sets remains difficult as there still is no standardised protocol for application in such studies. Consequently, it is still very common to see energy expressed in terms of gross, digestible and metabolisable values. However, it is possible to draw some conclusions from the available data.

It is now widely accepted that increasing the dietary energy content of fish feeds by increasing the oil content can result in higher levels of fat in the fish. This was clearly demonstrated for rainbow trout, *Oncorhynchus mykiss*, by Jobling *et al.* (1998). Two diets were prepared which were either low fat (59% protein, 13% fat, 21.2 MJ kg^{-1} GE) or high fat (50% protein, 28% fat, 24.4 MJ kg^{-1} GE). The diets were fed in excess to trout of 90 g using automatic feeders for four hours daily. After 11 weeks of feeding, the trout fed the high and low fat diets had grown to 361 and 348 g respectively. Although there were no significant effects of the diets on final weight, the composition of the trout closely reflected the diets fed (Table 16.2). Thus, the trout fed the high fat diet had significantly higher levels of whole body, carcass and visceral fat (and energy) than the trout fed low fat diets. Similarly, the visceral somatic index of the trout fed the high fat diet was higher than that of the fish fed the low fat diet (Table 16.2).

Jobling *et al.* (1998) then continued to feed the remaining trout on a mixture of both low and high fat diets simultaneously. During this second period of feeding, both groups of trout increased in weight to approximately 900 g. The whole body and carcass composition of the fish previously fed the high fat diet was little changed while that of the trout formerly fed the low fat diet changed radically (Table 16.2). Jobling *et al.* (1998) clearly showed that, by comparison to a high fat diet, supplying a low fat feed did not restrict trout growth and

Table 16.2 Effects of feeding low fat (13%) and high fat (28%) diets on the composition of rainbow trout. Fish were fed either a low or high fat diet in phase 1 and were offered both diets simultaneously in phase 2 (Jobling *et al.* 1998).

	Initial	Phase 1 Final High fat		Low fat	Phase 2 Final High fat		Low fat
Whole fish							
Fat (%)	8.9 ± 1.5	15.4 ± 1.6	***	10.5 ± 0.7	15.2 ± 1.2	NS	14.7 ± 1.3
Moisture (%)	75.3 ± 4.4	64.5 ± 2.0	***	69.4 ± 1.2	65.4 ± 1.3	NS	65.9 ± 1.4
Energy (kJ g^{-1})	6.6 ± 1.2	10.1 ± 0.6	***	8.3 ± 0.4	9.6 ± 0.4	NS	9.5 ± 0.5
Viscera							
VSI (%)		11.7 ± 2.0	**	9.1 ± 0.7	13.1 ± 1.6	*	11.6 + 1.3
Fat (%)		38.0 ± 6.7	***	23.2 ± 3.7	28.4 ± 6.2	NS	29.5 + 6.0
Moisture (%)		48.1 ± 5.2	***	59.8 ± 3.8	58.4 + 7.9	NS	55.4 + 6.5
Energy (kJ g^{-1})		17.6 ± 2.3	***	12.6 ± 1.6	13.7 ± 2.7	NS	14.7 + 1.4
Carcass							
Fat (%)		12.4 ± 1.0	***	9.3 ± 0.7	13.2 ± 1.2	NS	12.8 + 1.4
Moisture (%)		67.1 ± 1.8	***	70.5 ± 1.3	66.5 ± 1.5	NS	67.3 + 1.5
Energy (kJ g^{-1})		9.1 ± 0.5	***	7.6 ± 0.4	9.0 ± 0.5	NS	8.8 + 0.5

Differences between treatments are indicated by: * = <0.05; ** = <0.01; *** = <0.001. ns = not significant.
[1] VSI: viscerosomatic index = (viscera weight/whole fish weight) × 100

resulted in a leaner fish. Furthermore, after the fish were offered a choice of feed, it appeared that the already fat fish were capable of maintaining their composition while the lean fish increased in fat content to match that of their counterparts. At both points in this experiment it would have been useful to have used a waste feed collector. Firstly, this would have allowed a determination of FCR in Part 1 and secondly would have established whether the trout actively discriminated between pellet types in order to 'self-regulate' body composition.

Amongst Atlantic salmon there is an additional effect on fish size of the outcome of feeding diets with different protein:energy (P:E) ratios. In the studies of Einen and Roem (1997) and Einen and Skrede (1998) four diets were formulated with the following crude protein:oil ratios: 38/39, 43/35, 48/31 and 52/26. The digestible protein : digestible energy (DP : DE) ratios of these four diets were 14.1, 16.4, 18.8, and 21.9 g MJ^{-1} respectively. The diets were subsequently fed four times daily to satiation for a period of 138 days to Atlantic salmon during which time the salmon grew from approximately 1.1 to 2.7 kg (Group 1) and from approximately 2.5 to 4.8 kg (Group 2). Einen and Roem (1997) observed significant effects of protein : energy ratio on salmon growth and FCR (Table 16.3) such that for fish growing from 1.1 to 3.0 kg and from 2.5 to 5.0 kg DP : DE ratios of 19 and 16–17 g MJ^{-1} respectively were recommended.

Einen and Roem (1997) observed no effect of protein:energy ratio on the fat content of the

Table 16.3 Effects of protein : energy ratio on Atlantic salmon of different sizes (Einen and Roem 1997; Einen & Skrede 1998).

					ANOVA[1]	
Crude protein/oil (%)	38/39	43/35	49/31	52/26		
DP : DE (g MJ^{-1})	14.1	16.4	18.8	21.9	Root MSE	P
Group 1						
Initial weight (kg)	1.08	1.05	1.06	1.06	0.04	0.85
Final weight (kg)	2.50[b]	2.70[ab]	2.92[a]	2.71[ab]	0.11	0.01
Carcass:body weight (%)	90.9[b]	91.1[b]	91.6[ab]	92.1[a]	0.21	0.03
Fat in whole body (%)	17.2[a]	17.3[a]	16.7[a]	15.1[b]	0.55	0.005
Fat in cutlet (%)	18.9	18.3	18.4	15.7	1.53	0.13
Fat in liver (%)	5.5	5.7	5.1	5.1	0.94	0.82
Group 2						
Initial weight (g)	2.51[a]	2.49[a]	2.44[b]	2.49[a]	0.03	0.03
Final weight (g)	4.60[c]	4.95[a]	4.71[b]	4.90[ab]	0.01	0.01
Carcass:whole body (%)	90.0	90.6	91.2	91.0	1.11	0.08
Fat in whole body (%)	16.1	17.2	17.3	16.3	0.71	0.17
Fat in cutlet (%)	17.2	18.3	18.3	17.6	0.87	0.37
Fat in fillet (%)	17.8[a]	17.9[a]	17.0[ab]	16.1[b]	0.02	0.0002
Fat in liver (%)	5.4	5.3	5.4	5.0	0.70	0.88

[1] Results from analysis of variance (ANOVA) where root MSE is the square root of the mean square error and P is the significance level. Mean values marked with different superscripts were found to be significantly different by Duncan's multiple range test at the 5% level.

liver or cutlet amongst the smaller salmon (approximately 2.5 kg at harvest). However, feeding diets with DP : DE ratios of 16.4 g MJ^{-1} and below resulted in a significantly lower slaughter yield than when diets with a higher DP : DE ratio were fed (Table 16.3). Additionally, Einen and Roem (1997) observed a significantly lower whole body fat content amongst salmon fed diets with a DP : DE ratio of 21.9 g MJ^{-1} than when diets with lower DP:DE ratio were consumed. There were no effects of the different DP : DE ratios on slaughter yield, whole body or liver fat content of the larger salmon (5.0 kg at harvest) (Einen & Roem 1997). However, the fillet fat content of the salmon fed the diet with the high DP : DE ratio (21.9 g MJ^{-1}) was significantly lower that that of those fed diets containing 16.4 or 14.1 g MJ^{-1} (Table 16.3) (Einen & Skrede 1998). This data indicated that for larger salmon (> 2–3 kg) it is possible to reduce DP : DE ratios to 16 g MJ^{-1} without adverse effects on growth or fish composition. However, for smaller salmon (1–2 kg) it is necessary to provide diets with DP : DE ratios of 19 g MJ^{-1} or more to maintain optimal growth without adverse effects on yield and composition.

Increased body lipid level in response to feeding diets with decreased protein:energy ratio has been documented for numerous species including Atlantic halibut (Aksens *et al.* 1996), whitefish, *Coregonas lavaretus* (Koskela *et al.* 1998), gilthead seabream, *Sparus aurata* (Company *et al.* 1999) and dentex (Tibaldi *et al.* 1996). Aksnes *et al.* (1996) fed diets with crude protein : gross energy (CP : GE) ratios of 21, 24, 28 or 32 g MJ^{-1} to Atlantic halibut growing from approximately 5 to 550 g. Despite large differences in the protein and gross energy contents of the four feeds, there were no significant effects on the growth and feed efficiency of the halibut (Table 16.4). However, Aksnes *et al.* (1996) documented significant correlations between the decreasing CP:GE ratio and increased gutting loss and lipid and dry matter in the eviscerated fish (Table 16.4).

Although there are numerous examples of increased body fat levels in response to decreasing P : E ratios, Ramseyer and Garling (1998) failed to observe this effect in yellow perch, *Perca flavescens*. Indeed, providing diets with protein:ME ratios of 31 to 14 g MJ^{-1} did not have a significant effect on the whole body lipid content of yellow perch grown from approximately 20 to 50 g. However, when Ramseyer and Garling (1998) fed diets with a fixed protein : ME ratio (22 g MJ^{-1}) but decreasing energy contents, the whole body fat content of yellow perch decreased significantly despite an absence of effect on growth.

Many protein : energy studies in salmonids involve substitution of protein with oil. However, increases in flesh fat content and decreased slaughter yield similar to those recorded for high oil diets have been observed when protein is replaced by carbohydrate in diets for Atlantic salmon. Aksnes (1995) prepared four diets containing a fixed level of lipid (28%) with protein/starch ratios as follows: 60/2, 52/10, 45/17 and 38/23. The diets were fed to satiation to Atlantic salmon which grew from 0.6 to 3.5–3.9 kg during a period of 254 days. Increased starch content in the feeds had significant adverse effects on the apparent digestibility coefficient of starch and lipid from the diets such that the DP : DE ratios of the diets were 24.4, 22.9, 20.7 and 19.6 g MJ^{-1} for the 60/2, 52/10, 45/17 and 38/23 diets respectively. Since the salmon were fed to satiation, the fish were able to increase feed intake to the effect that there were no significant effects of the protein : energy ratios on fish growth though there were adverse effects on FCR. However, Aksnes (1995) recorded low gutting losses amongst salmon fed diets with the two higher DP : DE ratios and lower levels of fillet fat when the feed contained 23.8 g DP MJ^{-1} DE (Table 16.5).

Table 6.4 Influence of protein : energy ratio on the composition of Atlantic halibut (Aksnes *et al.* 1996).

					Regression
Dietary protein (g kg^{-1})	549	611	657	719	
Dietary oil (g kg^{-1})	325	261	198	126	
Dietary gross energy (g kg^{-1})	26.2	25.0	23.7	22.3	
CP : GE ratio (g MJ^{-1})	20.95	24.44	27.72	32.24	
Fish performance					
Initial weight (g)	5.31 ± 0.05	5.73 ± 0.12	5.71 ± 0.70	5.50 ± 0.04	ns
Final weight (g)	539 ± 1	562 ± 100	562 ± 82	558 ± 5	ns
Sgr (% day^{-1})	0.88 ± 0.01	0.87 ± 0.04	0.88 ± 0.05	0.88 ± 0.01	ns
Feed efficiency[1]	1.26 ± 0.03	1.27 ± 0.18	1.31 ± 0.24	1.30 ± 0.01	ns
Fish composition					
Mean weight of fish sampled (g)	496 ± 52	522 ± 56	526 ± 51	564 ± 51	ns
Entrails (% of whole fish weight)	5.31 ± 0.29	4.71 ± 0.33	4.49 ± 0.16	4.11 ± 0.30	$P < 0.01$
Liver (% of whole fish weight)	2.07 ± 0.16	1.98 ± 0.19	1.83 ± 0.15	1.76 ± 0.18	ns
Dry matter in eviscerated fish (g kg^{-1})	314 ± 7	305 ± 4	283 ± 4	272 ± 11	$P < 0.001$
Lipid in eviscerated fish (g kg^{-1})	125 ± 3	112 ± 2	87 ± 3	72.5 ± 10	$P < 0.001$

[1] Feed efficiency = g live weight gain/g feed fed

Table 16.5   Effects of feeding diets with different protein : starch ratios on the growth, feed efficiency and slaughter quality of Atlantic salmon (Aksnes 1995).

Diet	60/2	52/10	45/17	38/23	Linear regression[1]	Quadratic regression
Protein (g kg⁻¹)	602	521	454	384		
Starch (g kg⁻¹)	24	98	165	230		
DP : DE ratio (g MJ⁻¹)	23.4	20.5	19.8	17.8		
Fish performance						
Initial weight (kg)	0.61 ± 0.00	0.61 ± 0.00	0.61 ± 0.00	0.61 ± 0.00	–	–
Final weight (g)	3.40 ± 0.17	3.92 ± 0.30	3.58 ± 0.05	3.61 ± 0.19	ns	ns
Sgr (% day⁻¹)	0.68 ± 0.02	0.74 ± 0.03	0.70 ± 0.01	0.70 ± 0.02	ns	ns
FE[2]	0.91 ± 0.95	0.95 ± 0.06	0.89 ± 0.01	0.80 ± 0.05	$P < 0.05$	$p < 0.01$
Gutting-out loss (%)[4]	11.4 ± 0.9	13.7 ± 2.6	16.0 ± 3.9	15.6 ± 3.7	$P < 0.001$	$P < 0.001$
Fillet dry matter (g kg⁻¹)	341 ± 15	372 ± 16	366 ± 12	369 ± 32	$P < 0.01$	$P < 0.001$
Fillet fat (g kg⁻¹)	123 ± 21	169 ± 22	154 ± 17	160 ± 28	$p < 0.01$	$P < 0.001$

[1] Statistical significance for regression between the levels of carbohydrate in the feed and the various parameters.
ns = not significant ($p > 0.05$)
[3] Feed efficiency (weight gain (kg)/feed fed (kg))
[4] Gutting out loss ((weight viscera/weight fish) × 100%)

Although there is a tendency towards increased body fat levels in response to feeding diets with increasing oil level (decreasing protein : energy ratio), it would appear that there is an upper limit to the accumulation of oil in the flesh of Atlantic salmon. Hemre and Sandness (1999) prepared diets containing a fixed level of protein (39–40%) with three different levels of oil (31, 38 or 47%) substituting for starch. These diets were fed to Atlantic salmon to satiety three times daily for a period of seven months during which time the fish increased in weight from approximately 1.1 to 2.4 kg. FCR improved with each increase in oil level though there were only significant improvements in growth when the oil level in the diet increased from 31 to 38%. When the fillet fat contents of fish of a similar size were compared, Hemre and Sandness (1999) noted that, although there was a significant increase in fillet fat level from 12 to 16% when 31 or 38% fat diets were fed, there was no further increase in fillet fat content when a 47% fat diet was fed (16%).

Feeding regime

Restricted rations

Although there have been a number of studies which examine the effects of feeding regime on the growth and flesh composition of fish, many authors have not considered the effects of the feeding regime on fish weight and thence on composition. Ruohonen *et al.* (1998) documented increased fat content in the gain of rainbow trout in seawater fed either once, twice or four times daily with either dry pelleted feed or herring. However, feed intake increased in proportion to the number of daily feedings which significantly increased growth and the fat content of the trout closely reflected their final weight. Similarly, Johansson *et al.* (1995) fed rainbow trout on rations which were expected to provide 200, 100, 75 or 50% of the feed necessary to support maximum growth according to models for trout produced by Austreng *et al.* (1987). Johansson *et al.* (1995) documented increased fat content in the fillet in response to increased ration level. However, re-evaluation of their data on a size specific basis indicates that the fillet fat content was strongly influenced by fish size.

Over recent years, the feeding of restricted rations has been investigated by salmon farmers as a tool to manipulate flesh fat levels and the efficacy of such a policy was investigated by Einen (1998). Einen fed four groups of 3.9 kg Atlantic salmon with a single diet (45.6% cp, 32.3% oil and 25 MJ kg^{-1} GE) over a period of 110 winter days. Feed was supplied such that, one group of fish was fed to satiation (with waste feed collection) while the remaining four groups were fed a ration equivalent to 75, 50, 25 and 0% of that required to satisfy the appetite of the first group. As shown in Fig. 16.3, the starved fish lost weight while those fed at 25% of satiety were effectively receiving a maintenance ration. Although Einen (1998) recorded a significant increase in slaughter yield in response to decreased ration, this was offset by significant reductions in the fillet yield (as % of gutted weight) from salmon fed at less than 50% satiety. Furthermore, the 2–3% reduction in the fat content of the NQC (Norwegian quality cut) achieved by Einen (1998) (Fig. 16.4) represented only a small decrease given the considerable loss of growth and fillet yield necessary to achieve this effect.

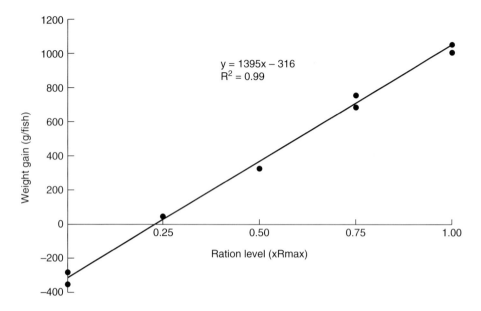

Fig. 16.3 Effects of feeding restricted rations on the pre-harvest weight gain of Atlantic salmon (Einen 1998).

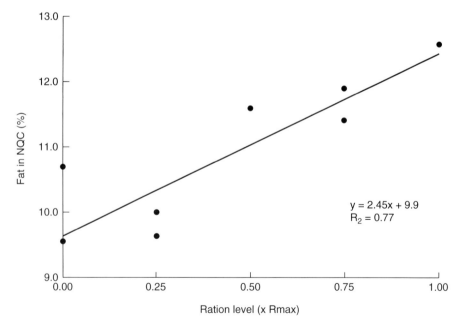

Fig. 16.4 Effects of restricted pre-harvest rations on the fat content of the Norwegian Quality Cut of Atlantic salmon (Einen 1998).

Sveier and Lied (1998) demonstrated that, amongst large Atlantic salmon, the temporal basis of feeding has no effect on slaughter related composition. In their study, Sveier and Lied (1998) fed a single diet (38% cp, 32% oil and 24.6 MJ kg^{-1} GE) in excess for either for one or 22 hours per day with 24 hour illumination. During the three month trial, the salmon grew from approximately 2.9 to 4.9 kg and by using automatic feeders with waste feed collection, accurate estimation of true feed consumption and FCR were made. Sveier and Lied (1998) observed no significant effects of the timing of food availability on the digestibility of the feed, growth, FCR or composition of the salmon (Table 16.6). However, it must be emphasised that this experiment was conducted in tanks where there was a high degree of control over feed availability and distribution between the fish.

Table 16.6   The effects of feeding regime on the growth, FCR and body composition of Atlantic salmon. Values presented as mean ± SEM (Sveier & Lied 1998).

Feeding regime	1 hour	22 hours
Growth and FCR		
Initial weight (g)	2877 ± 30	2854 ± 39
Final weight (g)	4955 ± 54	4922 ± 83
Thermal growth coefficient[1]	3.82 ± 0.07	3.83 ± 0.07
FCR	1.01 ± 0.01	1.03 ± 0.01
Slaughter quality		
Condition factor[2]	1.49 ± 0.01	1.51 ± 0.17
Dressing out percentage (%)[3]	11.5 ± 0.21	11.5 ± 0.73
Protein in whole fish (%)	17.1 ± 0.17	17.2 ± 0.23
Fat in NQC[4] (%)	18.1 ± 0.33	18.1 ± 0.27
Digestibility		
Nitrogen digestibility (%)	87.5 ± 0.67	87.7 ± 0.26
Fat digestibility (%)	94.3 ± 0.26	94.3 ± 0.6

[1] Thermal growth coefficient = $((W_2^{1/3}-W_1^{1/3}/\text{degree days}) *1000)$
[2] Condition factor = (fish weight (g)/fish length^3) × 100
[3] Dressing out percentage = (weight viscera/weight whole fish) × 100%
[4] NQC = Norwegian quality cut

Starvation

A more extreme measure for the reduction of the fat content of Atlantic salmon is starvation (Wathne 1995; Einen *et al.* 1998). Wathne (1995) starved Atlantic salmon of 2.2 kg for five weeks at 8.8°C (308 degree days) with a resultant decrease in ungutted body weight of 4.5%. Dressing-out percentage and visceral fat index were significantly improved by starvation and furthermore, Wathne (1995) demonstrated that this loss of weight was due to the preferential mobilisation of lipids from the adipose tissues though fat levels in the white muscle and NQC were not significantly affected (Table 16.7).

Table 16.7 The effects of starvation on the slaughter traits and the fat content of selected tissues in 2.2 kg Atlantic salmon starved for five weeks (Wathne 1995).

	Initial	Final (after 5 weeks' starvation)	p-value
Slaughter traits			
Ungutted weight (g)	2236 ± 436	2125 ± 422	p<0.01
Dressing out percentage (%)[1]	91.7 ± 1.9	94.1 + 0.8	p<0.01
Visceral fat index (%)	4.77 ± 0.74	3.85 ± 0.46	p<0.01
Fat in selected tissues (g kg^{-1})			
Cutlet	120 ± 8	119 ± 21	p=0.92
Liver	41 ± 1	41 ± 3	p=0.88
White muscle	40 ± 6	40 ± 7	p=0.92
Red muscle	328 ± 24	289 ± 9	p<0.05
Belly flap adipose	611 ± 20	485 ± 19	p<0.01
Visceral adipose	919 ± 55	841 ± 53	p<0.09

[1] Dressing out percentage = ((weight whole fish − weight viscera)/weight whole fish) × 100

Einen *et al.* (1998) conducted an extended starvation trial with salmon of 5.0 kg. Initially, all fish were fed twice daily to satiation on a single diet (48% cp, 32% oil and 24 MJkg^{-1} GE) and feeding was suspended 86, 58, 30, 14, 7, 3 and 0 winter days prior to harvest. During the starvation period the temperature ranged between 3.0° and 6.0°C with an average of 4.1°C. The salmon that were fed continuously until harvest increased in weight by 26% while those that were starved for 86 days lost 11% of their initial weight. As shown in Fig. 16.5, the slaughter yield increased in proportion to the length of starvation for salmon starved for up to 30 days with no further increase in yield for fish starved for between 30 and 86 days. In a similar vein, Einen *et al.* (1998) observed that fillet yield (as % whole body) was not significantly affected for salmon starved up to 30 days though there were significant losses in fillet yield thereafter (Fig. 16.6). Despite the considerable effects on fish weight, Einen *et al.* (1998) recorded only marginal effects of starvation on the fat content of the large salmon (Table 16.8).

Thus, by comparison to fish fed continuously until slaughter, 14 and 58 days of starvation were required to significantly lower the fat content of the visceral tissue and NQC respectively. Furthermore, even after 86 days of starvation at winter temperatures, fat in the whole fillet and belly flaps were not significantly reduced and there appeared to be no preferential loss of fat from any of the fillet regions. Einen *et al.* (1998) concluded that pre-harvest starvation for short periods can be used to increase slaughter yields in large Atlantic salmon though this must be offset against very minor effects on fillet fat levels and decreases in the fillet yield when the starvation period is extended.

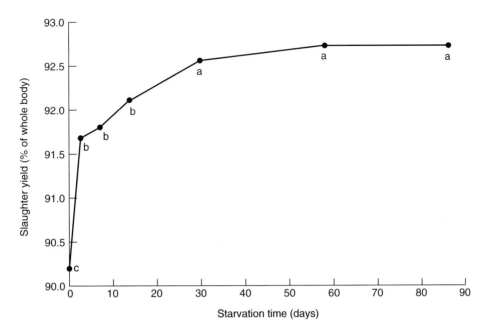

Fig. 16.5   Effect of extended starvation on the slaughter yield of Atlantic salmon (Einen *et al.* 1998).

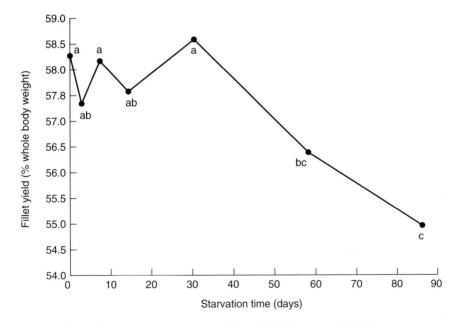

Fig. 16.6   Effect of extended starvation on fillet yield (Einen *et al.* 1998).

Table 16.8   Effects of extended starvation on the fat content of large (> 5.0 kg) Atlantic salmon (Einen *et al*. 1998).

	Starvation group								
								ANOVA[2]	
Starvation time (days)	0	3	7	14	30	58	86		
Starvation period (deg. days)	0	12.3	28.7	57.4	123.0	237.8	352.6		
								SE	P
Fat content (% wet weight)									
Viscera	35.1[bc]	36.6[abc]	39.8[ab]	43.8[a]	42.2[ab]	42.5[ab]	42.3[ab]	2.88	0.028
NQC1	16.7[a]	16.8[a]	16.9[a]	16.2[ab]	16.9[a]	15.3[b]	15.4[b]	0.29	0.0043
Belly flap	46.4	47.5	47.0	45.9	46.9	47.0	47.0	1.27	0.82
Whole fillet	18.8	18.2	18.9	18.2	18.1	18.0	18.4	0.48	0.12

[1] Norwegian quality cut
[2] Results from analysis of variance (ANOVA) where SE is the pooled standard error and P is the significance level. Mean values marked with different superscripts in the same line were found to be significantly different by Duncan's multiple range test at the 5% level.

Summary and recommendations

Much of the material presented in this review has been related to salmonids although many of the principles outlined are applicable to other farmed species. In general, the protein and ash content of fish are very difficult to manipulate and by comparison to fat level, are not necessarily a trait that consumers consider their highest priority.

There is a strong positive and a strong inverse relationship between fish weight and fat content and between moisture and fat content respectively. While there is potential to produce fish with fat levels to meet market demands, these constraints must be achievable at a reasonable cost. For example, high energy diets have served to increase growth rates and reduce FCRs for salmon, although if they are not used judiciously, there is a risk of increased fat content and lower slaughter yields. Certainly, feeding diets with high protein : energy ratios represents a way of producing fish with lower levels of fat. However, the optimum level of protein:energy varies in accordance with fish size and it would appear that in Atlantic salmon, smaller fish (< 2.5 kg) respond more readily to quality-oriented feeding practices than larger fish (> 2.5 kg). However, by comparison with high energy diets, feeds containing high protein:energy ratios raise the cost of production since they can be associated with lower growth rates and poorer FCRs.

The feeding of severely restricted rations and starvation can be used to reduce the fat content of farmed fish. However, depending on the degree of restriction or duration of starvation, such practices can be associated with lower rates of growth and potentially reduced slaughter yields. While the indications are that for large Atlantic salmon ($> 4\,kg$) fed to satiation at each meal, the frequency of feeding has little effect on fish composition, there remains a considerable dearth of information on how feeding practices affect the composition of farmed fish. With the advent of feedback systems to control feed distribution there is now potential to study these effects on a large scale.

This review clearly demonstrates that careful feed selection and optimised feeding practice can be applied to produce fish of a desired composition. Ideally, farmers should select their target market as early in the production cycle as possible. The quality parameters should be agreed on at an early stage allowing the farmer to select a feed type and adopt a feeding regime which will produce fish of the desired quality at harvest. If lower levels of oil in the fish are a prime consideration, the farmer must obtain a premium price for the fish in order to compensate for potentially increased costs of production. It is essential that the composition of the stock is monitored on a regular basis so that the feeding practice can be adjusted to keep the stock within acceptable quality constraints. Growing the fish as fast as possible, only to restrict feeding or starve them prior to harvest, is not best practice.

References

Aksnes, A. (1995) Growth, feed efficiency and slaughter quality of salmon, *Salmo salar* L., given feeds with different ratios of carbohydrate and protein. *Aquaculture Nutrition*, 1, 241–8.

Aksnes, A., Hjertnes, T. & Opstvedt, J. (1996) Effect of dietary protein level and carcass composition in Atlantic halibut (*Hippoglossus hippoglossus* L.). *Aquaculture*, 145, 225–33.

Andersen, F., Maage, A. & Julshamn, K. (1996) An estimation of the dietary iron requirement of Atlantic salmon, *Salmo salar* L., parr. *Aquaculture Nutrition*, 2, 41–7.

Austreng, E., Storebakken, T. & Åsgård, T. (1987) Growth rate estimates for cultured Atlantic salmon and rainbow trout. *Aquaculture*, 60, 157–60.

Chou, B.S. & Shiau, S.Y. (1996) Optimal dietary lipid level for growth of juvenile hybrid tilapia, *Oreochromis niloticus* x *Oreochromis aureus*. *Aquaculture*, 143, 185–95.

Company, R., Calduch-Giner, J.A., Kaushik, S. & Pérez-Sánchez, J. (1999) Growth performance and adiposity in gilthead seabream (*Sparus aurata*): risks and benefits of high energy diets. *Aquaculture*, 171, 279–292.

Cowey, C.B. (1993) Some effects of nutrition on the quality of cultured fish. In: *Fish Nutrition in Practice* (eds S.J. Kaushik & P. Luquet), Proceedings of the IVth International Symposium on Fish Nutrition and Feeding, Biarritz, France, 24–27 June 1991. INRA Editions, Paris. 227–36.

Einen, O. (1998) *Product quality in Atlantic salmon – influence of dietary fat level, feed ration and starvation before slaughter*. Doctor Scientiarum Theses 1998: 27, Agricultural University of Norway, Ås, Norway.

Einen, O. & Roem, A.J. (1997) Dietary protein/energy ratios for Atlantic salmon in relation to fish size: growth, feed utilisation and slaughter quality. *Aquaculture Nutrition*, 3, 115–26.

Einen, O. & Skrede, G. (1998) Quality characteristics in raw and smoked fillets of Atlantic salmon, *Salmo salar*, fed high-energy diets. *Aquaculture Nutrition*, 5, 99–108.

Einen, O., Waagan, B. & Thomassen, M.S. (1998) Starvation prior to slaughter in Atlantic salmon (*Salmo salar*) I. Effects on weight loss, body shape, slaughter- and fillet-yield, proximate and fatty acid composition. *Aquaculture*, 166, 85–104.

El-Dahhar, A.A. & Lovell, R.T. (1995) Effect of protein:energy ratio in purified diets on performance, feed utilisation, and body composition of Mozambique tilapia, *Oreochromis mossambicus* (Peters). *Aquaculture Research*, 26, 451–7.

Eya, J.C. & Lovell, R.T. (1997) Available phosphorous requirements of food sized channel catfish (*Ictalurus punctatus*) fed practical diets in ponds. *Aquaculture*, 154, 283–91.

Hemre, G.I. & Sandness, K. (1999) Effect of dietary lipid level on muscle composition in Atlantic salmon, *Salmo salar*. *Aquaculture Nutrition*, 5, 9–16.

Hemre, G.O., Sandness, K., Lie, Ø. & Waagbø, R. (1995) Blood chemistry and organ nutrient composition in Atlantic salmon, *Salmo salar* L., fed graded amounts of wheat starch. *Aquaculture Nutrition*, 1, 37–42.

Hillestad, M., Johnsen, F., Austreng, E. & Åsgård, T. (1998) Long-term effects of dietary fat level and feeding rate on growth, feed utilisation and carcass quality of Atlantic salmon. *Aquaculture Nutrition*, 4, 89–97.

Jobling, M., Koskela, J. & Savolainen, R. (1998) Influence of dietary fat level and increased adiposity on growth and fat deposition in rainbow trout, *Oncorhynchus mykiss* (Walbaum). *Aquaculture Research*, 29, 601–607.

Johansson, L., Kiesling, A., Åsgård, T. & Berglund, L. (1995) Effects of ration level in rainbow trout, *Oncorhynchus mykiss* (Walbaum), on sensory characteristics, lipid content and fatty acid composition. *Aquaculture Nutrition*, 1, 59–66.

Koskela, J., Jobling, M. & Savolainen, R. (1998) Influence of dietary fat level on feed intake, growth and fat deposition in the whitefish, *Coregonus lavaretus*. *Aquaculture International*, 6, 95–102.

Morris, P.C. (1998) *Fish oil replacement in diets for rainbow trout: partial replacement with vegetable oils*. Paper presented at the British Trout Farming Conference, Winchester, UK, September 1998.

Ojaveer, H., Morris, P.C., Davies, S.J. & Russell, P. (1996) The response of thick-lipped grey mullet, *Chelon labrosus* (Risso), to diets of varied protein-to-energy ratio. *Aquaculture Research*, 27, 603–12.

Ramseyer, L.J. & Garling, D.L. (1998) Effects of dietary protein to metabolisable energy ratios and total protein concentrations on the performance of yellow perch, *Perca flavescens*. *Aquaculture Nutrition*, 4, 217–24.

Ruohonen, K., Vielma, J. & Grove, D.J. (1998) Effects of feeding frequency on growth and food utilisation of rainbow trout (*Oncorhynchus mykiss*) fed low-fat herring or dry pellets. *Aquaculture*, 165, 111–21.

Schwarz, F.J. (1995) Determination of mineral requirements of fish. *Journal of Applied Ichthyology*, 11, 164–74.

Shearer, K.D. (1994) Factors affecting the proximate composition of cultured fishes with emphasis on salmonids. *Aquaculture*, 119, 63–88.

Shearer, K.D., Maage, A., Opstvedt, J. & Mundheim, H. (1992) Effects of high ash diets on growth, feed efficiency and zinc status of Atlantic salmon (*Salmo salar*). *Aquaculture*, 106, 345–55.

Sveier, H. & Lied, E. (1998) The effect of feeding regime on growth, feed utilisation and weight dispersion in large Atlantic salmon (*Salmo salar*) reared in seawater. *Aquaculture*, 165, 333–45.

Tibaldi, E., Beraldo, P., Volpelli, P.A. & Pinosa, M. (1996) Growth response of juvenile dentex (*Dentex dentex* L.) to varying protein level and protein:lipid level in practical diets. *Aquaculture*, 139, 91–9.

Wathne, E. (1995) *Strategies for directing slaughter quality of farmed Atlantic salmon* (Salmo salar) *with emphasis on diet composition and fat deposition.* Doctor Scientiarum theses 1995: 6, Agricultural University of Norway, Ås, Norway.

Wilson, R.P. (1994) Utilisation of carbohydrate by fish. *Aquaculture*, 124, 67–80.

Wilson, R.P. & Cowey, C.B. (1985) Amino acid composition of whole body tissue of rainbow trout and Atlantic salmon. *Aquaculture*, 48, 373–6.

Chapter 17

The Effect of Certain Micronutrients on Fish Flesh Quality

R.T.M. Baker

Marine Harvest McConnell, Craigcrook Castle, Edinburgh, EH4 3TU, UK

Introduction

This review examines the effect of certain dietary micronutrients on the quality of aqua-cultured products. In this chapter, micronutrients are defined as 'dietary components, required in very small quantities, that are either essential, or have nutritionally beneficial roles for fish in terms of growth or health'. Carotenoid pigments may be excluded on this basis, since they are not dietary essential in many life-stages of fish and there is little evidence to support a health role in post-larval fish. Having said this, however, astaxanthin and canthaxanthin are incorporated into aquafeeds for selected species at levels similar to the classical micronutrients. Also, one must not ignore their central role in coloration of white muscle, especially in the salmonids. Given the fact that carotenoid pigments are extensively reviewed elsewhere (Storebakken & No 1992; Torrissen & Christiansen 1995; and see Chapter 6), coverage of these compounds will be brief.

In many instances, fatty acids are omitted from discussions on micronutrients, since these are seen to be packaged together as generic 'lipid', a macronutrient. Not to discuss fatty acids as micronutrients would be an oversight. In fish nutrition there are certain fatty acids that are required at very low levels and these are known as the essential fatty acids (EFAs). EFAs were at one time classified as a vitamin (vitamin F) (Bender 1992). This chapter will discuss the effect of EFAs on product quality and the impact of changing dietary formulation on tissue levels of these important lipids.

The aspects of product quality which are within the scope for control by micronutrients are colour, taste, structure, texture, stability, smell, appearance acceptability, and nutritive value. The focus of this review will be the influence of certain vitamins, fatty acids and carotenoids.

Vitamin E

Vitamin E is important for the tissue's defence against lipid-oxidising free radicals, and is the vitamin with by far the greatest impact on flesh quality. Increased stability of tissues due to vitamin E deposition, as α-tocopherol, leads to a post-mortem protection of tissues against peroxidation, which improves the quality and storage stability of meat and fish flesh. Vitamin E supplementation prolongs the time that beef (Arnold *et al.* 1993), veal (Engeseth *et al.* 1993), pork (Asghar *et al.* 1991), poultry (Sheehy *et al.* 1993) and fish may be stored. Fish are prone to oxidation because their flesh lipids are highly polyunsaturated. Studies on rainbow trout (Frigg *et al.* 1990), channel catfish (Gatlin *et al.* 1992), red seabream (Murata & Yamauchi 1989) and African catfish (Baker & Davies 1996) demonstrate that vitamin E protects tissues in storage, or in *in vitro* stimulated oxidation, against lipid damage. In most species examined, as exemplified by the African catfish (*Clarias gariepinus*) (Baker & Davies 1996), muscle vitamin E level depends on dietary supply, and even at a dietary dose twice as high as currently employed in salmonid diets (200–300 ppm), potential exists for additional α-tocopherol deposition in muscle (Fig. 17.1). Furthermore, increasing muscle vitamin E suppresses forced lipid-oxidation in catfish fillets, as shown in Fig. 17.2. Formation of thiobarbituric acid reactive substances, or TBARS, is lessened as vitamin E deposition increases, since vitamin E is able to prevent the breakdown of lipids to malondialdehyde (MDA), the compound that reacts with thiobarbituric acid under the acidic conditions of the assay. Since laboratory-controlled accelerated peroxidation parallels oxidation of meat products under retail conditions, it is evident that fillets from catfish fed higher doses of vitamin E could have a longer shelf-life.

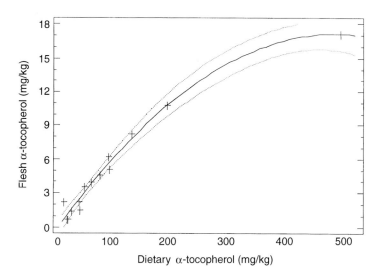

Fig. 17.1 Vitamin E levels in *Clarias* muscle in response to dietary level of α-tocopheryl acetate. Compiled from Baker and Davies (1996, 1997ab). Dashed line shows 95% confidence limit.

Boosting human intake via fish products

Use of modified atmosphere packs and chemical treatment of flesh products decreases oxidative deterioration of meat, but does not actually improve the nutritional quality of muscle-tissues. Although arguably, prevention of oxidation of polyunsaturated fatty acids

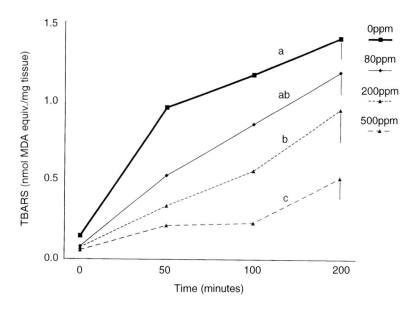

Fig. 17.2   *In vitro* iron-ascorbate induced peroxidation in flesh from *Clarias* fed graded levels of α-tocopheryl acetate. Hanging bars represent −2 sd from the mean. Modified from Baker and Davies (1996).

within foods is achieved, strategies of dietary antioxidant supplementation and fatty acid protection ensure carry-over of these important nutrients to consumers, with potential health benefits. For example, dietary supplementation of vitamin E at 200 ppm into poultry diets increases the contribution of vitamin E from a typical serving of chicken, from 2% to about 15% of the adult recommended daily allowance. Protection of polyunsaturated fats and cholesterol by vitamin E is associated with a reduced risk of coronary heart disease in humans (Hensrud & Heimburger 1994). With fish tissues being a rich source of important unsaturated fats, higher vitamin E levels in cultured fish pass on health benefits to the consumer.

Vitamin E and consumer preference

Appearance as well as taste is an important factor in the perception of food quality. In beef, muscle colour changes to a brown colour from the desirable cherry-red appearance as a result of oxidation of haem-proteins. High levels of vitamin E can drastically retard this pathway, thereby extending the colour display life of beef cuts (Arnold *et al.* 1993). Although these kinds of changes are unlikely to occur in fish fillets, product-colour may be influenced by other mechanisms. Pozo *et al.* (1988) suggest that dietary vitamin E increases the incorporation of canthaxanthin into trout flesh. Since this pigment is one of many responsible for the pink coloration of farmed salmonids, this would improve appearance. Vitamin E is also widely believed to interact with astaxanthin in that both are potent antioxidants *in vitro* and so each may spare the other from free radical action. This will be discussed later.

Although taste and smell measurements are subjective, under controlled conditions, taste-panel evaluations demonstrate an improvement of meat taste by increasing the vitamin E given to poultry and trout (Laksesvela 1960; Frigg *et al.* 1990). Vitamin E exerts its influence by controlling peroxidation processes in lipids. Oxidative rancidity results from the formation of lipid peroxidation products including alcohols, aldehydes and ketones, all of which contribute to off-tastes. As a chain-breaking antioxidant, vitamin E donates a proton to attacking free radicals, thereby terminating the oxidation cascade. Vitamin E is able to exist as a stable free radical due to the nature of the chemical bonds in its chromanol head ring structure.

Maintaining cell structure

Another benefit of vitamin E in flesh is that it lessens moisture loss from meats, especially those that have been frozen. Vitamin E influences cell membrane fluidity since oxidation of membrane-bound fatty acids leads to increased rigidity of the membrane (Dobretsov *et al.* 1977). Evidently, α-tocopherol renders the membrane more flexible, thereby reducing leakage of cell contents. Furthermore, this manifests as protection from the mechanical damage caused when cell-fluids expand during freezing. Upon thawing, increased drip-loss in animal tissues low in vitamin E has been reported in pigs (Asghar *et al.* 1991; Cheah *et al.* 1995), lamb (Monahan *et al.* 1994), and more recently fish (Baker 1997).

Fluidity imparted by vitamin E is unlikely to have direct significant effects on eating quality in terms of texture and softness. Physical softness is more likely due to muscle fibre arrangement, lipid distribution and water content of the flesh. However, during extreme vitamin E deficiency, nutritional muscular dystrophy may occur (Aoe *et al.* 1972) and this affects muscle structure. Also, oxidation of flesh-protein and associated cross-linking may result from acute vitamin E deficiency. How this impacts on texture is unclear.

Current status and concerns

At the time of writing (1999), commercial aquaculture diets for most species are supplemented at about 150–300 ppm α-tocopheryl acetate. It is considered that this figure ensures maximum potential growth and good tissue stability against oxidation by free radicals. This range of values has not been experimentally derived however, and represents an extension from research with feeds containing between 12 and 18 % lipid, as was usual in commercial diets for salmonids in the past. At present, Atlantic salmon diets typically contain in excess of 30% lipid, representing a significantly denser energy source than previously fed. With increased energy metabolism we expect more redox reactions involved in lipid breakdown for energy and this has potential to increase lipid peroxyl formation.

Another concern of today's salmon industry is the fast growth rates that are being genetically selected for. Fast growing strains may be those capable of over-eating and depositing higher levels of tissue lipid and so the vitamin E requirement may be increased. In the UK, this is complicated by the widespread use of early maturing grilse-stocks, which may utilise vitamin E differently. Surprisingly, data on vitamin E levels in developing gonads is limited, but one would imagine that such delicate tissues would be well protected by vitamin E. During this time vitamin E may be diverted or mobilised from muscle tissue to reproductive organs.

Rancid feed compromises vitamin E status

Exogenous free radicals from the oxidation of feeds may affect fish, but this free radical source can be controlled by adding synthetic antioxidants to a formulation (Lauridsen *et al.* 1995). α-tocopheryl acetate is itself not an antioxidant and is not considered to stabilise lipids within feed pellets. Nevertheless, we have demonstrated that dietary vitamin E (supplied as α-tocopheryl acetate) protects tissues of fish fed rancid rations (Baker & Davies 1997a), and that rancid diets cause a drop in the tissue vitamin E level. Could this be due to the use of vitamin E in the tissue as a free radical defence, or perhaps the oxidation of α-tocopherol within the gut, lowering the amount of the vitamin available for uptake? Although esterification of α-tocopherol to the tocopheryl-acetate form stabilises the molecule, within the gut the ester is hydrolysed to a free form. This is then unprotected and open to attack.

Due to its free radical quenching role, feed or culture stresses responsible for increasing oxidation, or perhaps the effects of maturation, could deplete flesh vitamin E stores to the point where health and product quality are affected. Since vitamin E has a very low toxicity, scope exists to greatly increase its supplementation into fish diets. The only constraints may be cost and practical benefit, since it would be unnecessary to extend oxidation resistance beyond the point where spoilage-microbes govern product acceptability.

Vitamin C

Vitamin C (ascorbic acid and derivatives) has been linked to the anti-scorbutic effects (Bender 1992) of increased rate of wound healing, improved stress resistance and immune-potentiation (Roberts *et al.* 1995). Vitamin C is also a water-soluble antioxidant in its own right, as well as functioning to regenerate vitamin E from the tocopheroxyl radical (Machlin 1984). Therefore, this vitamin may have a role to play in the maintenance of flesh quality.

It is clear that in times of stress, vitamin C is used up, and may be mobilised to specific tissues (Henrique *et al.* 1996). High doses of vitamin C may therefore be fed in times of increased stress (such as grading etc.) or during disease challenge, and this may improve the overall health of cultured fish with knock-on benefits to flesh quality. Benefits may be marginal, although research is warranted in this field.

Ascorbic acid is essential for the formation of connective tissue, as a cofactor in the hydroxylation of proline and lysine during collagen formation. Tucker and Halver (1984) stated that ascorbate regulates lysyl oxidase, an enzyme involved in the cross-linking of collagen and elastin, and so deficiency often results in tissue damage. As such, one might expect to see effects of vitamin C status on tissue structure. In fish however, connective tissue is heat unstable (Love 1988) so that any benefits would be negligible in cooked products. Raw products requiring processing, on the other hand, may benefit from improved structural integrity if that can be imparted by dietary vitamin C. With current dietary levels, in practical situations it is likely that factors such as flesh pH, oil and moisture content would predominate as the key factors governing muscle integrity, although one cannot discount vitamin C's impact.

Increased vitamin C status could theoretically also improve the oxidative stability of fish flesh. In broiler chickens, Lauridsen *et al.* (1997) did not see any improvement in chill or freezer storage life in response to increased vitamin C status, but did notice that circulating levels of glutathione peroxidase decreased in high ascorbate fed birds. This points to synergy between these two aqueous phase antioxidants. High levels of vitamin C would protect the tissues to the extent that less synthesis of glutathione peroxidase is necessary to break down free radical oxidation mechanisms. Alternatively, one may expect increases in non-inducible antioxidants since these are being protected by vitamin C through preferential sacrifice.

Although there is a paucity of information in fish, Robb *et al.* (1997) report that the pigment astaxanthin in Atlantic salmon flesh stored on ice for 12 days, was protected by a combined vitamin C/E dietary supplement (42–640 ppm vitamin C + 124–1090 ppm vitamin E). It is quite possible, however, that the vitamin E alone could account for these findings since 120 ppm vitamin E in the low vitamin E diet, relative to the basal diet containing 70 ppm, could have offered considerably improved oxidation resistance in the flesh. The authors' findings should be investigated in rations with combinations of practical and sub-optimal levels of vitamin inclusion. What the authors may have demonstrated however, is that for diets containing that particular oil level and composition, 30 ppm vitamin C and 70 ppm vitamin E is insufficient to guarantee satisfactory protection of flesh pigments.

As an addendum to the antioxidant story, however, recent work has revealed that in some instances, vitamin C and vitamin E may have an antagonistic relationship. Sealey and Gatlin (1999) have presented data suggesting that at high vitamin C levels, plasma tocopherol is slightly depressed. This may be explained by ascorbate driving oxidation via the Fenton reaction, involving the reduction of ferric iron to ferrous iron. Ferrous iron can drive per-oxidation by breaking-down hydrogen peroxide to the highly reactive hydroxyl radical.

Carotenoid pigments

This will be a cursory review of carotenoids in flesh quality since the topic is fully covered in Chapter 6. For information on dietary pigmentation strategies and pigment choice, numerous existing papers may be consulted.

Carotenoid pigments are used extensively in salmonid diets to impart the orange-red colour associated with wild-caught salmon. For these purposes the pigments astaxanthin and canthaxanthin are incorporated into aquafeeds since cultured fish have no access to the natural sources of the pigments. In the wild, fish capable of depositing carotenoids in their white muscle would obtain them from zooplankton and other crustacea that have retained the pigments from ingested algal sources. Carotenoids from feed ingredients, however, may be problematic in situations where a fish is cultured for its white flesh colour. In some instances, efforts have been directed to processing corn gluten meal in such a way that zeaxanthin and lutein are extracted, thus preventing them from imparting a yellow hue to fish receiving rations containing this feedstuff (Park *et al.* 1997).

Commercially synthesised astaxanthin and canthaxanthin are fed at rates from 20 to 80 mg per kg of trout or salmon feed, either singly or in combination, in order to attain the level of flesh pigment that provides the desired visual colour in the final product. Pigment choice really comes down to cost and efficiency, and varies from one producer to the next. Consumer issues may also contribute to the factors to be weighed-up prior to deciding a pigmentation strategy. As an example of this, take the media coverage canthaxanthin received when it was implicated in visual disturbances when taken in very large repeat-doses (Woolf 1998). More recently, assurances have been made that canthaxanthin intake from fish consumption poses no threat to human health.

To obtain a desirable colour in salmon flesh, Torrissen (1995) considers that pigment levels should be above 6 ppm although this also depends on how intensely the colour is perceived. It is noteworthy that salmon destined for smoking should be fairly deeply pigmented, so that for certain markets, a greater concentration of flesh pigment may be aimed for.

Site of pigment binding

Protein structure is of considerable importance because the carotenoid pigments attach to muscle protein, and are not in the lipid associated with the muscle. Henmi *et al.* (1989, 1990) have studied the mechanism of pigment binding to fish muscle and established that astaxanthin and canthaxanthin bind non-specifically to hydrophobic regions of the protein 'actomyosin' within the muscle-fibre cell. They also showed that actomyosin from other fish combined with astaxanthin and canthaxanthin, and that salmonid red-muscle actomyosin also had binding capabilities. So, the fact that it is mainly salmonid white-muscle that retains pigment reflects a selective transport mechanism for these pigments once they have been internalised by the cell. Could the rate of this selective-transport process account for Torrissen and Naevdal's (1988) observation that faster growing salmonids pigment less well? Also, does this explain the observations of Hatlen *et al.* (1995), that older arctic charr are better at depositing astaxanthin in their flesh? Since transport systems may be influenced by water temperature, could a decrease in pigmentation efficiency seen in salmon over winter be due to a slow-down in the pigment transport mechanism during periods of cold sea-temperature, or is it merely due to decreased feed intake? It is clear that a greater research effort is warranted in this area since an ability to optimise the transport mechanisms may lead to improvements in pigment utilisation. Information from such research could also feed into genetic selection programmes in order to allow for the selection of individuals with high pigment binding capabilities.

Carotenoids: antioxidants in vivo?

There has been a great deal of speculation regarding the potential of carotenoids to act as antioxidants in fish tissues. Much of this stems from the fact that *in vitro*, β-carotene, zeaxanthin, canthaxanthin and astaxanthin have all demonstrated potent free radical trapping activity (Palozza & Krinsky 1992). *In vivo*, carotenoids have also been shown to protect rat mitochondria (Kurashige *et al.* 1990), and poultry tissues (Ajuyah *et al.* 1993; Woodall *et al.* 1996) from oxidative damage. Also, in experiments on cell cultures astaxanthin reduced oxidative stress in rat microsomal preparations (Nakagawa *et al.* 1997) and chick cells (Lawlor & O'Brien 1995).

Do we have any evidence to support the antioxidant activity of carotenoids in fish tissues? Nakano *et al.* (1995) demonstrated that serum lipid peroxide levels were decreased in fish fed carotenoids, relative to a control containing none. In blood, carotenoid pigments are transported associated with lipoproteins, where they are free to perform an antioxidant role, along with other antioxidants. Within fish-muscle, the pigment is bound to proteins which may remove the free radical trapping ability. It is perhaps not surprising therefore, that there are few accounts of carotenoid pigments having an impact on flesh stability in fish.

There are two ways in which carotenoids may achieve an antioxidant role: a direct free radical quenching effect, or through synergy with vitamin E. The former mechanism has been reported in frozen rainbow trout fillets where formation of TBARS was suppressed for longer in fillets from very pigmented fish (15 ppm in the flesh) compared to trout pigmented to either 6 or 8 ppm (Bjerkeng & Johnsen 1995). It should be noted that 15 ppm is an unusually high level of flesh pigment and, at present, it would cost a great deal to obtain such a concentration under standard practical culture situations. Additionally, vitamin E levels in those fillets were not reported. Einen and Skrede (1998) also report to have linked fillet astaxanthin concentration to oxidation, since they observed higher smoke-odour and lower off-flavour in smoked salmon with higher astaxanthin contents. An effect of astaxanthin on off-flavour can be explained in terms of retardation of rancidity in line with the antioxidant theory, but effects on smoke-flavour are somewhat less explicable.

In terms of co-operation between nutrients to achieve an improvement in product quality, carotenoids may spare vitamin E by combating free radical mediated tissue oxidation, thereby extending the retail display time of fish products. Despite these hypotheses, studying oxidation in tissues of fish fed varying levels of both of these nutrients has failed to convincingly demonstrate any synergy. The antioxidant role of astaxanthin in fish muscle may only be demonstrable when vitamin E is at critically low levels in the tissue. Even then, other fat-soluble antioxidants may hide any possible benefits of astaxanthin. The best opportunity to see astaxanthin working as an antioxidant may be in fillets from fish fed little or no vitamin E, since post-mortem, antioxidant enzyme activity would decrease, leaving the tissues vulnerable to free radical attack.

The essential fatty acids

Plant oils can successfully replace marine fish-oils in diets for certain fish. Fish growth may not be compromised by feeding diets containing less fish meal and fish oil, but fatty acid composition would undoubtedly be affected. Decreasing the use of feedstuffs containing n-3 fatty acids lessens the n-3 fish flesh content as demonstrated by most workers in the field (reviewed by Steffens 1997). This will lessen the available amount of n-3 fatty acids, such as linolenate and the highly unsaturated fatty acids, available for human consumption.

This would not be beneficial due to the role of n-3s in reducing incidences of human cardiovascular and inflammatory diseases (Hensrud & Heimburger 1994). British Nutrition Foundation (1992) figures state that fish consumption accounts for 13% of human n-3 PUFA intake and that the n-6 : n-3 ratio is approximately 6.7. Even a slight decrease in n-3 contribution from fish products may increase this ratio and this would change the balance existing between these nutrients.

In poultry production, there is much research looking at 'lipid-tailoring' of broilers to yield an ideal profile in terms of n-3 fatty acids and vitamin E (Miller & Huang 1993). Feed formulators replacing fish oils in aquaculture feeds are examining possibilities of employing feedstuffs rich in *n*-3s, so as not to deleteriously affect the fatty acid profiles of cultured fish. In this respect oils from plants such as linseed and rapeseed may be exploited for their fatty acid profiles. Although this would serve to increase *n*-3 content of fish diets over rations based on other vegetable oils, there would be an associated increase in the absolute levels of *n*-6 and a marked lack of HUFA. HUFA are essential for growth and survival in marine fish species. The use of linseed oil in aquafeeds has encountered difficulties, since trials employing this oil have demonstrated poor fish growth (Agradi *et al.* 1993), but recent trials show rapeseed oils to have great potential for future exploitation.

Marine microalgae are naturally abundant in *n*-3 fatty acids, and represent the primary feed source for most cultured marine fish species during larval development and subsequent early growth. It would make sense therefore, that algae could provide fish with the necessary fatty acids for optimum health and tissue quality in aquafeeds. A further benefit of algal use in feeds would be the associated intake of carotenoid pigments from this source and this could decrease the necessity for some of the pigment additives currently used in rations for salmonids. Countering this benefit is the possibility that algal proteins may decrease the bioavailability of vitamin E, as observed in rat studies (Mitchell *et al.* 1989). This would have an impact on fish flesh quality as previously discussed.

So far I have covered the problems of decreasing *n*-3 contents of fish, but there are also distinct advantages. Feedstuffs containing predominantly *n*-6 fatty acids also tend to be more resistant to oxidation due to an overall lower associated HUFA content. Modifying flesh composition by feeding *n*-6 fatty acid rich feedstuffs will decrease the peroxidizability of the flesh, generating a more stable product. Stephan *et al.* (1995) demonstrated that flesh from turbot (*Scophthalmus maximus*) was less oxidised after six months frozen storage when peanut oil was fed in place of cod-liver oil. This is supported by work in poultry, where feeding higher levels of *n*-3 fatty acids from full-fat flax seed (linseed) necessitated the incorporation of antioxidants to maintain the storage stability of meat from broilers (Ajuyah *et al.* 1993).

Role of other micronutrients

This review has covered what may be considered the most important of the micronutrients in flesh quality terms, but evidently the list has not been exhausted. One may add any of the vitamins that are involved in energy yielding pathways, since an imbalance of these would disrupt energy partitioning and thus fat deposition in the flesh. Within this list of micronutrients are pyridoxine, carnitine and thiamine given their roles in the metabolism of proteins, lipid and carbohydrates, respectively. Also, minerals such as selenium and chromium have been identified as having the potential to impact beneficially on flesh quality, whereas pro-oxidant metals may promote lipid deterioration.

What this review has shown however, is that strategies of micronutrient supplementation may offer the culturist the opportunity to make significant improvements in the provision of quality aquacultured products.

References

Agradi, E., Abrami, G., Serrini, G., McKenzie, D., Bolis, C. & Bronzi, P. (1993) The role of dietary *n*-3 fatty acid and vitamin E supplements in growth of sturgeon *(Acipenser naccarii)*. *Comparative Biochemistry and Physiology*, **105A**, 187–95.

Ajuyah, A.O., Ahn, D.U., Hardin, R.T. & Sim, S.J. (1993) Dietary antioxidants and storage affect chemical characteristics of w-3 fatty acid enriched broiler chicken meats. *Journal of Food Science*, **58**, 43–6.

Aoe, H., Abe, I., Saito, T., Fukawa, H. and Koyama, H. (1972). Preventive effects of tocols on muscular dystrophy of young carp. *Bulletin of the Japanese Society of Scientific Fisheries*, **38**, 845–51.

Arnold, R.N., Scheller, K.K., Arp, S.C., Williams, S.N. & Schaeffer, D.M (1993). Dietary alpha-tocopheryl acetate enhances beef quality in holstein and beef breed steers. *Journal of Food Science*, **58**, 28–33.

Asghar, A., Gray, J.I., Booren, A.M., Gomaa, E.A., Abouzied, M.M., Miller, E.R. & Buckley, D.J. (1991) Effects of supranutritional dietary vitamin E on subcellular deposition of α-tocopherol in the muscle and on pork quality. *Journal of the Science of Food and Agriculture*, **57**, 31–41.

Baker, R.T.M. (1997) The effects of dietary α-tocopherol and oxidised lipid on post-thaw drip from catfish muscle. *Animal Feed Science and Technology*, **65**, 35–43.

Baker, R.T.M. & Davies, S.J. (1996). Changes in tissue α-tocopherol status and degree of lipid peroxidation with varying α-tocopheryl acetate inclusion in diets for the African catfish. *Aquaculture Nutrition*, **2**, 71–9.

Baker, R.T.M. & Davies, S.J. (1997a). Modulation of tissue α-tocopherol in African catfish *(Clarias gariepinus)* fed oxidized oils, and the compensatory effect of supplemental dietary vitamin E. *Aquaculture Nutrition*, **3**, 91–7.

Baker, R.T.M. & Davies, S.J. (1997b) The quantitative requirement for α-tocopherol by juvenile African catfish, *Clarias gariepinus* Burchell. *Animal Science*, **65**, 135–42.

Bender, D.A. (1992) *Nutritional Biochemistry of the Vitamins*. 431pp. Cambridge University Press, Cambridge.

Bjerkeng, B. & Johnsen, G. (1995) Frozen storage quality of rainbow trout *(Oncorhynchus mykiss)* as affected by oxygen, illumination, and fillet pigment. *Journal of Food Science*, **60**, 284–8.

British Nutrition Foundation (1992) *Unsaturated Fatty Acids: Nutritional and Physiological Significance*. 211pp. Chapman and Hall, London.

Cheah, K.S., Cheah, A.M. & Krausgrill, D.I. (1995) Effect of dietary supplementation of vitamin E on pig meat quality. *Meat Science*, **39**, 255–64.

Dobretsov, G.E., Borschevskaya, T.A., Petrov, V.A. & Vladimirov, Y.A. (1977) The increase of phospholipid bilayer rigidity after lipid peroxidation. *FEBS letters*, **84**, 125–8.

Einen, O. & Skrede, G. (1998) Quality characteristics in raw and smoked fillets of Atlantic salmon, *Salmo salar*, fed high-energy diets. *Aquaculture Nutrition*, **4**, 99–108.

Engeseth, N.J., Gray, I., Booren, A.M. & Asghar, A. (1993) Improved oxidative stability of veal lipids through dietary vitamin E supplementation. *Meat Science*, **35**, 1–15.

Frigg, M., Prabucki, A.L. & Ruhdel, E.U. (1990) Effect of dietary vitamin E levels on oxidative stability of trout fillets. *Aquaculture*, **84**, 145–58.

Gatlin, D.M. III., Bai, S.C. & Erickson, M.C. (1992) Effects of dietary vitamin E and synthetic antioxidants on composition and storage quality of channel catfish, *Ictalurus punctatus*. *Aquaculture*, 106, 323–32.

Hatlen, G., Aas, G.H., Jorgensen, E.H., Storebakken, T. & Goswami, U.C. (1995) Pigmentation of 1, 2 and 3 year old arctic charr (*Salvelinus alpinus*) fed two different dietary astaxanthin concentrations. *Aquaculture*, 138, 303–12.

Henmi, H., Hata, M. & Hata, M. (1989) Astaxanthin and/or canthaxanthin-actomysin complex in salmon muscle. *Nippon Suisan Gakkaishi*, 55, 1583–9.

Henmi, H., Hata, M. & Hata, M. (1990) Combination of astaxanthin and canthaxanthin with fish muscle actomysins associated with their surface hydrophobicity. *Nippon Suisan Gakkaishi*, 56, 1821–3.

Henrique, M.M.F., Morris, P.C. & Davies, S.J. (1996) Vitamin C status and physiological response of the gilthead seabream, *Sparus aurata* L., to stressors associated with aquaculture. *Aquaculture Research*, 27, 405–12.

Hensrud, D.D. & Heimburger, D.C. (1994) Antioxidant status, fatty acids, and cardiovascular disease. *Nutrition*, 10, 170–75.

Kurashige, M., Okimasu, E., Inoue, M. & Utsumi, K. (1990) Inhibition of oxidative injury of biological membranes by astaxanthin. *Physiological Chemistry and Physics and Medical NMR*, 22, 27–38.

Laksesvela, B. (1960) Supplementation of chick diets with vitamin E to improve meat quality. *Journal of the Science of Food and Agriculture*, 11, 128–33.

Lauridsen, C., Jakobsen, K. & Hansen, T.K. (1995) The influence of dietary ethoxyquin on the vitamin E status in broilers. *Archives of Animal Nutrition*, 47, 245–54.

Lauridsen, C., Jensen, C., Jakobsen, K., Engberg, R.M., Andersen, J.O., Jensen, S.K., Sorensen, P., Henckel, P., Skibsted, L.H. & Bertelsen, G. (1997) The influence of vitamin C on the antioxidative status of chickens *in vitro*, at slaughter and on the oxidative stability of broiler meat products. *Acta Agriculturae Scandinavica Section A, Animal Science*, 47, 187–96.

Lawlor, S.M. & O'Brien, N.M. (1995) Astaxanthin: antioxidant effects in chicken embryo fibroblasts. *Nutrition Research*, 15, 1695–704.

Love, R.M. (1988) *The Food Fishes: Their Intrinsic Variation and Practical Implications.* 276pp. Farrand Press, London.

Machlin, L.J. (1984) *Handbook of Vitamins: Nutritional, Biochemical and Clinical Aspects.* Marcel Dekker Inc., New York.

Miller, E.L. & Huang, Y.X. (1993) Improving the nutritional value of broiler meat through increased n-3 fatty acid and vitamin E content. In: *Proceedings of the 11th European Symposium on the Quality of Poultry Meat.* (eds Colin, Culioli & Ricard), pp. 404–11.

Mitchell, G.V., Jenkins, M.Y. & Grundel, E. (1989) Tissue α-tocopherol, thiobarbituric acid-reactive substances (TBA-RS), and glutathione levels in rats fed algal proteins. In: *Vitamin E: Biochemistry and Health Implications* (eds A.T. Diplock, L.J. Machlin, L. Packer & W.A. Pryor). *Annals of the New York Academy of Sciences*, 570, 478–9.

Monahan, F.J., Gray, J.I., Asghar, A., Haug, A., Strasburg, G.M., Buckley, D.J. & Morrissey, P.A. (1994) Influence of diet on lipid oxidation and membrane structure in porcine muscle microsomes. *Journal of Agricultural and Food Chemistry*, 42, 59–63.

Murata, H. & Yamauchi, K. (1989) Relationship between the 2-thiobarbituric acid values of some tissues from cultured red sea bream and its dietary α-tocopherol levels. *Nippon Suisan Gakkaishi*, 55, 1435–9.

Nakagawa, K., Kang, S-D., Park, D-K., Handelman, G.J. & Miyazawa, T. (1997) Inhibition by α-carotene and astaxanthin of NADPH-dependent microsomal phospholipid peroxidation. *Journal of Nutritional Science and Vitaminology*, **43**, 345–55.

Nakano, T., Tosa, M. & Takeuchi, M. (1995) Improvement of biochemical features in fish health by red yeast and synthetic astaxanthin. *Journal of Agriculture and Food Chemistry*, **43**, 1570–73.

Palozza, P. & Krinsky, N.I. (1992) Antioxidant effects of carotenoids *in vitro* and *in vivo*: an overview. *Methods in Enzymology*, **213**, 403–20.

Park, H., Flores, R.A. & Johnson, L.A. (1997) Preparation of fish feed ingredients: reduction of carotenoids in corn gluten meal. *Journal of Agriculture and Food Chemistry*, **45**, 2088–92.

Pozo, R., Lavety, J. & Love, R.M. (1988) The role of dietary α-tocopherol (vitamin E) in stabilising the canthaxanthin and lipids of rainbow trout muscle. *Aquaculture*, **73**, 165–75.

Robb, D., Volker, L., Warriss, P. & Kestin, S. (1997) Effect of vitamin C and vitamin E on astaxanthin loss during storage of fillets of Atlantic salmon (*Salmo salar*). *Abstract presented at World Aquaculture '97*, 19–23 February, 1997, Seattle, Washington, USA, pp. 392–3.

Roberts, M.L., Davies, S.J. & Pulsford, A.L. (1995) The influence of ascorbic acid (vitamin C) on non-specific immunity in the turbot (*Scophthalmus maximus* L.). *Fish and Shellfish Immunology*, **5**, 27–38.

Sealey, W.M. & Gatlin, D.M. (1999) Interactions between vitamin C and vitamin E in hybrid striped bass (*Morone chrysops* × *M. saxatilis*). *Presentation at the Annual International Conference and Exposition of the World Aquaculture Society*, 26 April–2 May 1999, Sydney, Australia, abstract p.683.

Sheehy, P.J.A., Morrissey, P.A. & Flynn, A. (1993) Increased storage stability of chicken muscle by dietary alpha-tocopherol supplementation. *Irish Journal of Agricultural and Food Research*, **32**, 67–73.

Steffens, W. (1997) Effects of variation in essential fatty acids in fish feeds on the nutritive value of freshwater fish for humans. *Aquaculture*, **151**, 97–119.

Stephan, G., Guillaume, J. & Lamour, F. (1995) Lipid peroxidation in turbot (*Scophthalmus maximus*) tissue:effect of dietary vitamin E and dietary n-6 or n-3 polyunsaturated fatty acids. *Aquaculture*, **130**, 251–68.

Storebakken, T. & No, H.K. (1992) Pigmentation of rainbow trout. *Aquaculture*, **100**, 209–29.

Torrissen, O.J. (1995) Strategies for salmonid pigmentation. *Journal of Applied Ichthyology*, **11**, 276–81.

Torrissen, O.J. & Christiansen, R. (1995) Requirements for carotenoids in fish diets. *Journal of Applied Ichthyology*, **11**, 225–30.

Torrissen, O.J. & Naevdal, G. (1988) Pigmentation of salmonids–variation in flesh carotenoids of Atlantic salmon. *Aquaculture*, **68**, 305–10.

Tucker, B.W. & Halver, J.E. (1984) Ascorbate-2-sulfate metabolism in fish. *Nutrition Reviews*, **5**, 173–9.

Woodall, A.A., Britton, G. & Jackson, M.J. (1996) Dietary supplementation with carotenoids: effects on α-tocopherol levels and susceptibility of tissues to oxidative stress. *British Journal of Nutrition*, **76**, 307–17.

Woolf, M. (1998) Egg yolk dye could damage children's eyes. *Observer* newspaper (UK), Sunday 8 February 1998.

Chapter 18

Tainting of Aquaculture Products by Natural and Anthropogenic Contaminants

P. Howgate

26 Lavender Row, Stedham, Midhurst, West Sussex, GU20 0NS, UK

Introduction

The International Standards Organisation (ISO) in its glossary of terms used in the sensory evaluation of foods defines a 'taint' as 'an odour or flavour foreign to the product' (ISO 1992). The glossary distinguishes between taints and off-flavours or off-odours, which are defined as atypical flavours or odours associated with deterioration or transformation of the product. The distinction is that taints are derived from substances present in the surroundings of the product; off-odours and off-flavours are produced by processes within the product. It is common in North America to apply the word 'off-flavour' both to taints and to off-flavours as defined above, and to assume any taint adversely affects the flavour of the product. However, it is useful when investigating atypical flavours to distinguish between taints and off-flavours. While it is true that taints induced in fish by chemicals, natural or anthropogenic, in ambient water are often unpleasant, the ISO definition does not require that a taint be necessarily unpleasant, or that tainted fish is necessarily unfit for consumption.

Fish, including crustacean and bivalve shellfish, readily take up chemicals from the ambient water and if a chemical has a detectable odour at its concentration in the water, then any aquatic product exposed to the water is very likely to take up the chemical and become tainted. The general principle is that if a body of water has an odour, fish harvested from that water will have a similar odour at a similar intensity. The causative chemicals can be in the water as a result of natural processes or as a result of pollution.

Tainting arising from natural processes

Muddy/earthy flavours

Muddy flavours have long been reported in freshwater and brackish water fish from both natural and farmed environments (Persson 1995). These muddy odours also affect drinking water supplies. The causative chemicals in both cases have often been shown to be geosmin or 2-methylisoborneol released by micro-organisms in the water, and there is now quite a large literature on the subject, including reviews (Arganosa & Flick 1992; Maga 1987;

Tucker & Martin 1991). These two chemicals have extremely low detection thresholds in water; reported values differ somewhat among publications, but are in the order of 1 ng/l (Maga 1987). Most published reports of muddy flavours in aquatic food products refer to the taint in vertebrate fish, but it has been reported in shrimp (Lovell & Broce 1985), and in brackish water clams (Hsieh *et al.* 1988). Though muddy flavours in fish and shellfish have a world-wide occurrence, the literature on the taint in farmed fish is dominated by investigations into its presence in channel catfish farmed in the USA.

The presence of muddy, and other, taints which render the product unacceptable to consumers is a serious problem for catfish farmers in the USA, costing them millions of dollars annually (Tucker & Martin 1991). The presence and intensity of the taint varies seasonally and is associated with blooms of cyanobacteria in the phytoplankton in the warmer months (Lovell *et al.* 1986; Lorio *et al.* 1992; Tucker & van der Ploeg 1993; van der Ploeg & Tucker 1993). Pond conditions affect both the intensity and the composition of the phytoplankton flora (Paerl & Tucker 1995; Tucker 1996). Nutrient levels, enhanced by high feed rates (Brown & Boyd 1982), influence the growth of the phytoplankton, and factors like hardness and pH affect the composition. In Alabama, where the ponds are constructed in acidic soils, the bloom is dominated by *Anabaena* species which produce geosmin; and in Mississippi, where ponds are constructed on alkaline soils, by *Oscillatoria* species, which produce 2-methylisoborneol (Hariyadi *et al.* 1994; van der Ploeg *et al.* 1992).

Though the presence of muddy taints in the product is troublesome and costly for catfish farmers, there seems to be little the farmer can do by cultural practices to mitigate the problem (Johnsen & Diongi 1993). The quality assurance practice is to monitor the product for the presence of taint, and to transfer tainted fish to clean water for a few days to allow the taint to depurate. The organisms responsible for producing the tainting chemicals can grow in brackish water, but not in pure seawater (Paerl & Tucker 1995), and aquaculture production systems in brackish waters should keep the salt concentration above 1.0% to prevent blooms of the causative species.

As was pointed out above, the literature on muddy taints in aquacultured products comes almost exclusively from experience in catfish production in the USA. According to FAO statistics (FAO 1998), 57% of world aquaculture production in 1996 came from freshwater, and judging from the species grown, and the countries involved, most of this will have come from earth ponds systems in warm climates fertilised to encourage phytoplankton production. It is to be expected that muddy taints must be present, if not prevalent, in products from these systems, and it is unfortunate that information on muddy taints elsewhere than from the USA is lacking.

Other natural taints

Though the incidence of musty taints in catfish culture is most frequent in the summer months, taints of various sorts can be detected in fish throughout the year (Lovell 1983; van der Ploeg & Tucker 1993). Martin *et al.* (1988a,b) reported the presence of 2-methylene-bornane and 2-methyl-2-bornene, dehydration products of 2-methylisoborneol, in chronically tainted fish and considered that these could be the cause of muddy taints in winter months. Mills *et al.* (1993) confirmed the presence of these chemicals in both tainted and untainted catfish, but did not consider they induced the taint. Experienced assessors can discriminate different characters of taints and even in summer months flavours other than muddy can be detected along with the muddy flavours. The chemicals responsible for these other flavours, and their sources, have not been identified in the context of catfish ponds, but they are almost certain to originate from micro-organisms. Cotsaris *et al.* (1995) have reported on odiferous chemicals produced by monocultures of four species of algae found in freshwater and the sensory characteristics of the chemicals. Only one of the species investigated, *Scenedesmus subspicatus*, has been reported in catfish ponds, but this species alone produces a wide range of odiferous chemicals with characters the same as some of those described in tainted catfish.

An interesting example of a taint that can both enhance and detract from the acceptability of fishery products is provided by the flavour of bromophenols. Those who have had the opportunity of tasting fish of the one species, usually a salmonid, harvested from both freshwater and marine environments will remark on the greater intensity of, and more pleasant character of, the flavour in the product from the marine environment. Whitfield *et al.* (1997) assessed the flavours of wild-caught and of farmed shrimp and also measured their bromophenol contents. They found that the wild-caught samples had briny, prawn-like flavours whereas the cultivated prawns were bland. Bromophenol content in the wild specimens was higher than in the farmed and the authors considered the bromophenols could have contributed to the characteristic marine flavours of the wild specimens. Boyle *et al.* (1992a,b) measured the bromophenol contents of salmon and other species of fish harvested from marine and from freshwater environments and found that those from the latter environment had only trace amounts of bromophenols whereas salmon from the marine environment contained between 5.1 and 33.2 µg/kg of 2,4,6 tribromophenol. The authors studied the sensory properties of bromophenols and concluded that these substances at the concentrations found in fish could impart the seaweedy, briny flavours characteristic of seafoods compared with freshwater fish. Fish in the wild can pick up bromophenols through their diet (Boyle *et al.* 1993; Whitfield *et al.* 1997), but, as with most chemicals, bromophenols can be taken up from solution through the gills. Boyle *et al.* (1992a) found that unrefined sea salt contained 18.2 µg/kg of 2,4,6 tribromophenol so this compound is presumably a normal constituent of seawater.

If the bromophenol content in product is high, its flavour is unpleasant and the product is unacceptable to the consumer (Bemelmans & den Braber 1983; Anthoni *et al.* 1990; Münker (1995), but the work just described shows that at low concentrations, bromophenols contribute to the pleasant flavours of marine fish, and this raises the possibility of improving the normally bland flavours of freshwater-raised fish by adding small amounts of tribromophenol to the ambient water. The concept of influencing the flavour of farmed fish by adding chemicals to the water has been explored by Maligalig *et al.* (1975).

Tainting from anthropogenic sources

The aquatic environment is a sink for a wide variety of chemical pollutants of various origins. Some of these have strong flavours and if they enter the environment of an aquaculture installation, can taint the fish. Most reports of tainting of aquatic food products by pollutants concern petroleum as the chemical substance concerned, and there are few reports of tainting by other chemicals.

Tainting by petroleum

Petroleum includes crude oils and substance like fuel oils derived from them. They are notorious for having strong odours and clearly will taint fish if present in the ambient water. Accounts of tainting of aquatic foods by petroleum have been included in reviews by Connell and Miller (1981), Howgate (1999), and GESAMP (1977, 1993), though the examples therein are predominately of tainting of wild stocks. Moller *et al.* (1989) have reported on impacts of oil spills on mariculture and refer to tainting.

Tainting of fish following oil spills is rarely systematically investigated – though it is often perceived as a consequence – and reports of tainting are hardly more than anecdotal. The best recorded case of tainting of aquaculture products following an oil spill is that of the *Braer* incident (Whittle *et al.* 1997). Farmed salmon from 21 sites within the impact zone of the spread of the oil were sampled and tested for oil taint by an expert panel of assessors, and fish from nine of these were found to be tainted. The stocks were destroyed in some farms, but in others loss of taint was monitored and taint was still detectable up to 30 weeks after the spill. It is to be hoped that the *Braer* incident with its impact on mariculture will be a rare occurrence – there are few places in the world with such a high intensity of mariculture sites in proximity to tanker traffic – but mariculture sites, including sites for bivalve culture, are at risk of tainting by spills of oils from ships.

I am aware from my own experience of advising on quality of aquatic products of several incidents in Britain of tainting of trout by spills of vehicle fuel oil and of heating fuel oil into ponds. In these cases the farmer has decided he could not market the fish, or it has been condemned as unfit for consumption by local public health officials, and the stock has been destroyed. I am also aware of a case where trout in cages in an estuary were tainted by a spill of oil which drifted over the site. These small cases do not get reported in the technical literature and it is not possible to estimate how often they occur.

Tainting by industrial and agricultural chemicals

There are reports of stocks of wild fish being tainted by industrial chemicals discharged into waterways (Howgate 1999), but none concern farmed fish. Rimkus and Wolf (1993) have reported the presence of musk xylols in aquaculture products – trout and mussels. Musk xylols have low detection thresholds in water and are used as fragrances in detergents, but the authors did not assess the flavour of the products. Though there are no specific reports on tainting of farmed fish by chemical contaminants in feed water, it must be an element of the quality assurance programme for fish farms to check that any water used in aquaculture installations is free from odours.

Aquaculture in pond systems is at risk from contamination by agricultural chemicals in run-off draining into ponds, or by drifting of sprays. In general, but not always, insecticides are odourless, but sometimes petroleum fractions are used as solvents and there is a potential for tainting of fish in impacted ponds. Faust and Aly (1964) have reported the odour detection thresholds of some pesticide formulations, but the authors acknowledge it is likely the odour impact came from the solvents. Folmar (1980) found that two herbicides, acrolein and 2,4-D (DMA), the dimethylamine salt of 2,4-dichlorophenoxyacetic acid, tainted trout under experimental conditions. It is possible that the tainting properties of 2,4-dichlorophenoxyacetic acid preparations is due to residual amounts of 2,4-dichlorophenol, which taints fish at 0.024 mg/l in the ambient water (Howgate 1987). Catfish have been tainted by molinate used as a herbicide for weed control on levees (Martin *et al.* 1992).

Quality assurance

Discernible taints are unacceptable to consumers. Apart from any dislike of the flavours, taints suggest to a consumer that the fish has been contaminated and therefore might be harmful. In fact, chemicals like geosmin and 2-methylisoborneol and other chemicals produced by phytoplankton are not known to be toxic to humans, or are not likely to be at the concentrations that can impart taint. It is also unlikely that industrial and agricultural chemicals that could impart taint would be harmful at the effective concentrations. However, the customer would not be expected to know that and is justified in being suspicious of any product with an unusual flavour, particularly if that flavour is 'chemical'. It must be part of any quality assurance plan for aquaculture products that procedures are in place to ensure that the products are free of taints. If taints are found in the product, the quality controllers will need to trace the source of the taint, take necessary steps to correct the problem, and perhaps monitor the affected batch for loss of taint during any remedial depuration stage.

Assessment of taint

The author has recently discussed in some detail the assessment of taints in foods (Howgate 1999), and there are other reviews for food in general (Kilcast 1992) and for fish in particular (Botta 1994).

Taint is a sensory experience and the primary means of evaluating taints must be by sensory methods. The sensory procedures need not be elaborate for quality control purposes; it will usually be sufficient to record if the product is tainted or not, that is, it has or has not an atypical flavour. It is not realistic to expect an assessor to be able to recognise all possible flavours that might occur from chemical contamination, and control has to be based on absence of atypical flavours rather than presence of specific taint flavours. It follows that assessors must be familiar by appropriate training and experience with the typical flavours of the product. A slight degree of a natural taint like muddiness might be acceptable as a typical flavour in freshwater fish, but the extent must be clearly defined in the quality specification. Any suggestion of 'chemical' taint would not be acceptable.

An extension of the simple pass/fail criterion is to register the strength of the taint on an intensity scale. A typical scale used in sensory evaluation is a numeric intensity scale with $0 =$ absence of taint, $1 =$ slight, $2 =$ moderate, $3 =$ strong, $4 =$ very strong, and $5 =$ extremely strong. Such scales are used in the American catfish industry to record the intensity of muddy taints. A scale which is the reverse of this, $10 =$ absence of taint, $2 =$ extreme, has also been used in the catfish industry (Lovell *et al.* 1986). It might seem counter-intuitive at first to use a scale like this because it feels more natural to assign 0 to absence and to use increasing numbers to denote increasing strength, but the scale makes sense in the quality control context. Where taint is common, as it is in the American catfish industry, and presence of taint reduces quality, then a high score, meaning absence of taint, denotes high quality, and a low score given with high intensities of taint denotes low quality.

More elaborate sensory procedures can be used in investigation of taints and tainting episodes. An appropriate one is quantitative descriptive analysis (QDA). Descriptive terms which characterise the flavour or odour of a product are derived by a panel of assessors evaluating samples of the product with a wide range of qualities. Those terms which best differentiate among qualities and which panel members can consistently use are selected to give a lexicon of terms. Samples are then assessed and the perceived flavour is described by terms from the lexicon, and the intensities of the various flavour notes are rated on an intensity scale (Johnsen & Kelly 1990). The procedure generates large amounts of data that can become unwieldy to examine and evaluate, and multivariate statistical methods are typically used to reduce the data. QDA is a powerful method for investigation of flavours, but is not suitable for use in quality control in a typical processing plant, and is probably best left for research institutes.

If the causative chemical is known, then chemical analysis can be used to determine the presence of the chemical and its amount. These data perhaps can be used to predict the presence and intensity of taint, but chemical tests are not good substitutes for sensory evaluation.

Uptake and depuration of taints

There is no doubt from experience with actual tainting incidents, and from laboratory studies, that taint is taken up very rapidly, within minutes, from the water into the fish, though equilibrium might take a day or two (Johnsen & Lloyd 1992; Johnsen *et al.* 1996). Uptake is primarily through the gills, though there can be some absorption through the skin and from the gut (Lovell & Sackey 1973; Maligalig *et al.* 1975; From & Hørlyck 1984). In contrast, loss of taint when the fish is transferred to clean water is slow. Clearance of geosmin and 2-methylisoborneol taints can take several days, and in the case of petroleum taints can take weeks.

Uptake and distribution can be considered as passive processes and governed by the thermodynamic properties of the chemical. When the chemical potential (fugacity) of the chemical outside the fish is greater than it is within the fish, the chemical will pass through the gills into the fish – that is uptake occurs. When it is higher in the fish than in the ambient water the chemical will flow out of the fish – that is depuration occurs. The chemical will circulate though the fish and will partition between the water and lipid phases in the tissues until the chemical potentials in the phases are equal. At equilibrium, the ratio of the concentration of the chemical in the lipid and water phases is the partition coefficient of the chemical in those two phases and, which is predicted by the octanol/water partition coefficient, a parameter of considerable importance in predicting the fate of pollutants in the environment (Mackay 1991). When the partition coefficient is high the concentration of chemical in the muscle tissue of the fish is greater than that in the ambient water – bioconcentration – and depends on the lipid content of the muscle. This accumulation of chemical in the lipid of the muscle complicates the relationship between concentration of the tainting chemical in the muscle tissue and the intensity of odour and flavour. The detection threshold of a chemical like geosmin or 2-methylisoborneol is much higher in lipid than in water, perhaps by 2–4 orders of magnitude. Thus, for a given concentration of a chemical in muscle tissue, the perceived odour will be of lower intensity in a fish with a high lipid content that in one with a low lipid content. Another factor that bears on the relationship of concentration as determined by chemical analysis and perception is that the perceived intensity of odour of a chemical is proportional to the logarithm of its concentration (Rashash *et al.* 1997). This means that for very odiferous chemicals like geosmin and 2-methylisoborneol, their odours can still be perceived at concentrations close to the sensitivity of chemical analysis.

References

Anthoni, U., Larsen, C., Nielsen, P.H. & Christophersen, C. (1990) Off-flavour from commercial crustaceans from the North Atlantic zone. *Biochemical Systematics and Ecology*, 18, 377–9.

Arganosa, G.C. & Flick, G.J. (1992) Off-flavours in fish and shellfish. In: *Off-Flavours in Foods and Beverages*. (ed G. Charalambous), pp.103–26. Elsevier Science Publishers, Amsterdam.

Bemelmans, J.M.H. & den Braber, H.J.A. (1983) Investigation of an iodine-like taste in herring from the Baltic Sea. *Water Science and Technology*, 15, 105–13.

Botta, J.R. (1994) Sensory evaluation of tainted aquatic resources. In: *Analysis of Contaminants in Edible Aquatic Resources: General Considerations, Metals, Organometallics, Tainting and Organics* (eds J.W. Kiceniuk & S. Ray), pp.257–73. VCH Publishers, New York.

Boyle, J.L., Lindsay, R.C. & Stuiber, D.A. (1992a) Bromophenol distribution in salmon and selected seafoods of fresh- and saltwater origin. *Journal of Food Science*, 57, 918–22.

Boyle, J.L., Lindsay, R.L. & Stuiber, D.A. (1992b) Contributions of bromophenols to marine-associated flavors of fish and seafoods. *Journal of Aquatic Food Product Technology*, 1, 43–63.

Boyle, J.L., Lindsay, R.L. & Stuiber, D.A. (1993) Occurrence and properties of flavor-related bromophenols found in the marine environment: a review. *Journal of Aquatic Food Product Technology*, 2, 75–81.

Brown, S.W. & Boyd, C.E. (1982) Off-flavour in channel catfish from commercial ponds. *Transactions of the American Fisheries Society*, 111, 379–83.

Connell, D.W. & Miller, G.J. (1981) Petroleum hydrocarbons in aquatic ecosystems – behaviour and effects of sublethal concentrations. Part 2. *CRC Critical Reviews in Environmental Control*, 11, 105–62.

Cotsaris, E., Bruchet, A., Mallevialle, J. & Bursill, D.B. (1995) The identification of odorous metabolites produced from algal monocultures. *Water Science and Technology*, 31, 251–8.

Faust, S.D. & Aly, O.M. (1964) Water pollution by organic pesticides. *Journal of the American Water Works Association*, 56, 267–79.

Folmar, L.C. (1980) Effects of short-term field applications of acrolein and 2,4-(DMA) on flavor of the flesh of rainbow trout. *Bulletin of Environmental Contamination and Toxicology*, 24, 217–24.

FAO (1998) *Fishstat plus*. An electronic data base of fisheries statistics. Food and Agriculture Organization, Rome, Italy.

From, J. & Hørlyck, V. (1984) Sites of uptake of geosmin, a cause of earthy-flavor in rainbow trout (*Salmo gairdneri*). *Canadian Journal of Fisheries and Aquatic Sciences*, **41**, 1224–6.

GESAMP (1977) IMO/FAO/UNESCO/WHO/IAEA/UN/UNEP Joint Group of Experts on the Scientific Aspects of Marine Pollution. *Impact of oil on the marine environment.* GESAMP Reports and Studies No. 6, International Maritime Organization, London.

GESAMP (1993) IMO/FAO/UNESCO/WMO/IAEA/UN/UNEP Joint Group of Experts on the Scientific Aspects of Marine Pollution. *Impact of oil and related chemicals and wastes on the marine environment.* Reports and Studies GESAMP 50, International Maritime Organization, London.

Hariyadi, S., Tucker, C.S., Steeby, J.A., van der Ploeg, M. & Boyd, C.E. (1994) Environmental conditions and channel catfish, *Ictalurus punctatus*, production under similar pond management regimes in Alabama and Mississippi. *Journal of the World Aquaculture Society*, **25**, 236–49.

Howgate, P. (1987) Measurement of tainting in seafoods. In: *Seafood Quality Determination* (eds D.E. Kramer & J. Liston), pp.63–72. Elsevier Science Publishers, Amsterdam.

Howgate, P. (1999) Tainting of food by chemical contaminants. In: *Environmental Contaminants in Food* (eds C.F. Moffat & K.J. Whittle), pp.430–70. Sheffield Press, Sheffield, England.

Hsieh, T.C.-Y., Tanchotikul, U. & Matiella, J.E. (1988) Identification of geosmin as the major muddy off-flavour of Louisiana brackish water clam (*Rangia cuneata*). *Journal of Food Science*, **53**, 1228–9.

ISO (1992) *Sensory analysis – Methodology – Vocabulary*, International Standards Organization, ISO 5492–1992. International Standards Organization, Geneva, Switzerland.

Johnsen, P.B. & Diongi, C.P. (1993) Physiological approaches to the management of off-flavours in farmed-raised Channel catfish, *Ictalurus punctatus*. *Journal of Applied Aquaculture*, **3**, 141–61.

Johnsen, P.B. & Kelly, C.A. (1990) A technique for the quantitative sensory evaluation of farm-raised catfish. *Journal of Sensory Studies*, **4**, 189–99.

Johnsen, P.B. & Lloyd, S.W. (1992) Influence of fat content on uptake and depuration of the off-flavor 2-methylisoborneol by channel catfish (*Ictalurus punctatus*). *Canadian Journal of Fisheries and Aquatic Science*, **49**, 2406–11.

Johnsen, P.B., Lloyd, S.W., Vinyard, B.T. & Diongi, C.P. (1996) Effects of temperature on the uptake and depuration of 2-methylisoborneol (MIB) in channel catfish *Ictalurus punctatus Journal of the World Aquaculture Society*, **27**, 15–20.

Kilcast, D. (1992) Sensory evaluation of taints and off-flavours. In: *Food Taints and Off-Flavours* (ed M.J. Saxby) pp.1–34. Blackie Academic and Professional, London.

Lorio, W.J., Perschbacher, P.W. & Johnsen, P.B. (1992) Relationship between water quality, phytoplankton community and off-flavors in channel catfish (*Ictalurus punctatus*) production ponds. *Aquaculture*, **106**, 285–92.

Lovell, R.T. (1983) Off-flavors in pond-cultured channel catfish. *Water Science and Technology*, **15**, 67–73.

Lovell, R.T. & Broce, D. (1985) Cause of musty flavor in pond-cultured penaeid shrimp. *Aquaculture*, **50**, 169–74.

Lovell, R.T. & Sackey, L.A. (1973) Absorption by channel catfish of earthy-musty flavor compounds synthesized by cultures of blue-green algae. *Transactions of the American Fisheries Society*, **102**, 774–7.

Lovell, R.T., Lelana, I.Y., Boyd, C.E. & Armstrong, M.S. (1986) Geosmin and musty-muddy flavors in pond-raised channel catfish. *Transactions of the American Fisheries Society*, **115**, 485–9.

Mackay, D. (1991) *Multimedia Environmental Models: the Fugacity Approach*. Lewis Publishers, Michigan, USA.

Maga, J.A. (1987) Musty/earthy aromas. *Food Reviews International*, **3**, 269–84.

Maligalig, L.L., Caul, J.F., Bassette, R. & Tiermeier, O.W. (1975) Flavoring live channel catfish (*Ictalurus punctatus*) experimentally. Effects of concentration and exposure time. *Journal of Food Science*, **40**, 1242–5.

Martin, J.F., Bennett, L.W. & Graham, W.H. (1988a) Off-flavor in the channel catfish (*Ictalurus punctatus*) due to 2-methylisoborneol and its dehydration products. *Water Science and Technology*, **20**, 99–105.

Martin, J.F., Fisher, T.H. & Bennett, L.W. (1988b) Musty odour in chronically off-flavoured channel catfish: isolation of 2-methylenebornane and 2-methyl-2-bornene. *Journal of Agriculture and Food Chemistry*, **36**, 1257–60.

Martin, J.F., Bennett, L.W. & Anderson, W. (1992) Off-flavor in commercial fish ponds resulting from molinate contamination. *The Science of the Total Environment*, **119**, 281–7.

Mills, O.E., Chung, S-Y. and Johnsen, P.B. (1993) Dehydration products of 2-methylisoborneol are not responsible for off-flavour in the channel catfish. *Journal of Agricultural and Food Chemistry*, **41**, 1690–92.

Moller, T.H., Dicks, B. & Goodman, C.N. (1989) Fisheries and mariculture affected by oil spills. *Proceedings of the 1989 Oil Spill Conference, San Antonio, Texas*. American Petroleum Institute, Washington, DC. Report no. 4479, pp. 389–94.

Münker, W. (1995) Occurrence of off-flavour (tainting) in spring herring (*Clupea harengus*) from the Baltic Sea. *Informationen Fischwirtschaft*, **42**, 202–209.

Paerl, H.W. & Tucker, C.S. (1995) Ecology of blue-green algae in aquaculture ponds. *Journal of the World Aquaculture Society*, **26**, 109–131.

Persson, P-E. (1995) 19th century and early 20th century studies on aquatic off-flavours – a historical review. *Water Science and Technology*, **31**, 9–13.

van der Ploeg, M. & Tucker, C.S. (1993) Seasonal trends in flavour quality of channel catfish, *Ictalurus punctatus*, from commercial ponds in Mississippi. *Journal of Applied Aquaculture*, **3**, 121–40.

van der Ploeg, M., Tucker, C.S. & Boyd, C.E. (1992) Geosmin and 2-methylisoborneol production by the cyanobacteria in fish ponds in the southeastern United States. *Water Science and Technology*, **25**, 283–90.

Rashash, D.M.C., Dietrich, A.M. & Hoehn, R.C. (1997) FPA of selected odorous compounds. *Journal of the American Water Works Association*, **89**, 131–41.

Rimkus, G. & Wolf, M. (1993) Rückstände und Verunreinigungen in Fischen aus Aquakultur. 2 Mitteilung: Nachweis von Moschus Xylol und Moschus Keton in Fischchen. *Deutsche Lebensmittel-Rundschau*, **89**, 171–4.

Tucker, C.S. (1996) The ecology of channel catfish culture ponds in northwest Mississippi. *Reviews in Fisheries Science*, **5**, 1–55.

Tucker, C.S. & Martin, J.F. (1991) Environment-related off-flavours in fish. In: *Water*

Quality in Aquaculture (eds J.R. Tomasa & D. Brune) pp.133–79. World Aquaculture Books, Baton Rouge, USA.

Tucker, C.S. & van der Ploeg, M. (1993) Seasonal changes in water quality in commercial channel catfish in Mississippi. *Journal of the World Aquaculture Society*, 24, 473–81.

Whitfield, F.B., Helidoniotis, F., Shaw, K.J. & Svornos, D. (1997) Distribution of bromophenols in Australian wild-harvested and cultivated prawns (shrimp). *Journal of Agricultural and Food Chemistry*, 45, 4398–4405.

Whittle, K.J., Anderson, D.A., Mackie, P.R., Moffat, C.F., Shepherd, N.J. & McVicar, A.H. (1997) The impact of the *Braer* oil on caged salmon. In: *The Impact of an Oil Spill in Turbulent Waters: The Braer*. Proceedings of a Symposium held at the Royal Society of Edinburgh, 7–8 September 1995 (eds J.M. Davies & G. Topping), pp.144–160. The Stationery Office, Edinburgh.

Chapter 19

Potential Effects of Preslaughter Fasting, Handling and Transport

U. Erikson

SINTEF Fisheries and Aquaculture, N-7465 Trondheim, Norway

Introduction

Under natural circumstances several fish species are able to survive long periods without food. In fish farming the fish should be fasted for a certain period before harvesting to clean the digestive tracts, so minimizing the risk of flesh contamination during gutting and processing. Also, long-term fasting (months) has been applied to reduce or stabilise salmon production. The common fasting period used in the production of salmonids in Norway is about two weeks in the winter at seawater temperatures about 3–8°C and one week in the summer at 12–18°C.

After fasting in the sea cages, fish are less active and can more readily be transported. The fish are slaughtered either at the cages or they may be transported live to a plant for processing. This is done by towing the entire cage or by using specially designed well-boats where the fish are netted or pumped into the hold. Depending on biomass and temperature, oxygenation of the transport water may be necessary. In well-boats, the fish are normally transported in an open system with good water exchange.

Another aspect of live fish transport is the idea of bringing the fish directly to the market for consumption. In such cases transport usually takes place using other means of transport, often in closed systems demanding closer attention to water quality to avoid mortality or, potentially, to avoid a reduced quality of the product. Since the edible part of the fish mainly consists of white muscle, the current review focuses on the potential effects of fasting and transport on the flesh quality.

Fasting

During fasting of poikilothermic fish, the energy turnover and energy loss rise with an increase in water temperature. This loss can be related to body mass and nutritional status (Beck & Gropp 1995). Fish species able to store fat in the muscle depend more readily on fat as an energy reserve during long periods of low food intake compared with lean fish which mobilise tissue protein (Love 1980). In the muscle, the quantitatively most important sources of energy during fasting are protein or lipid (species dependent), whereas glycogen is of less importance.

Appearance, weight, length and yield

Prolonged starvation is characterised by loss of weight and change in shape where the fish become thinner and the head becomes relatively larger in proportion causing a decreasing condition factor (Akiyama & Nose 1980; Black & Skinner 1986; Lie & Huse 1992). As the general appearance in changed, this effect may be taken into consideration regarding the commercial value of the fish in different markets. Also, the skin colour has been reported to become darker as a result of starvation (Roberts & Bullock 1989). For example, during winter-time fasting of Atlantic salmon (*Salmo salar*) (2.1–2.6 kg), the weight loss of the whole fish was most pronounced in the first 35 days (7–8%) and thereafter became less (loss 3–4% from 35 to 78 days). In the first period, the reduction in weight of the viscera accounted for more than 50% of the weight loss. The fish length, however, continued to increase throughout the fasting period, reducing the condition factor from initial mean values (two groups) of 1.23–1.35 to 1.10–1.19 and 1.02–1.14 after 35 and 78 days, respectively (Lie & Huse 1992).

The percentage yield at slaughter usually increases during fasting due to the fact that fasting results in a higher weight loss in the viscera than in the gutted body. However, Akse and Midling (1997) found no effect on filleting yield when Atlantic cod (*Gadus morhua*) caught by seine net and transported to storage netpens were fasted for 73 days. Einen *et al.* (1998) studied Atlantic salmon with initial weight about 5 kg fasted for 0, 3, 7, 14, 30, 58 and 86 days prior to slaughter. No residual feed was found in the gastrointestinal tract of fish fasted for 3 days or more. At 86 days the weight loss was 11.3% and the rate of weight loss decreased as starvation time increased. The body shape systematically got leaner thus reducing the condition factor. Slaughter yield increased up to 30 days of starvation whereas fillet yield decreased after 58 and 86 days of starvation. The most important sources of fuel were both protein and fat. Material in the fillet was used most, followed by viscera and liver. It was concluded that long-term starvation at an average temperature of 4.1°C caused only marginal changes in body composition. The loss of body mass, lower fillet-yield and a leaner body shape were the most important effects of starvation.

Lipids and fatty acids

In fish, lipid is regarded as the most important source of energy during fasting (Love 1980). A general trend in several species seems to be that initially the lipids in the liver are mobilised before the lipids in the muscle (Cowey & Walton 1989). The triacylglycerols, as the predominant form of lipid reserves, are always mobilised before phospholipids. As the lipids from the muscle (fatty fish) are catabolised, the relative amount of water in tissues increases equally (Sargent *et al.* 1989).

During a fast for 48 days, the visceral fat depots in rainbow trout (*Oncorhynchus mykiss*) were the first to be mobilised, before the fat in liver and muscle. The lipid content of the muscle remained nearly constant during the first 27 days of fasting before dropping significantly during the next 21 days. The saturated fatty acids were preferentially mobilised from the viscera resulting in a rise in the percentage of monoenes and polyunsaturates. In the muscle, a substantial increase in the percentage of saturates was caused by a decline in monoenes whereas the polyunsaturates remained constant (Jezierska *et al.* 1982).

In the same species, no depletion of the total lipid content was found after a fast for 38 days (Choubert 1985) in contrast to Johansson and Kiessling (1991) who found after a fast for one and two months, a decrease in the total lipid content of the abdominal wall only. A change in the relative distribution of fatty acids where the monounsaturates decreased and polyunsaturates increased was also found. Lie and Huse (1992) reported that the fillet fat from two groups of Atlantic salmon decreased during 78 days of fasting from 11.3 and 14.0% initially, to 9.3 and 11.9%, respectively, whereas the fat content of 4–5 % in rainbow trout fillet was unaffected after a fast for 18 days (Færgemand *et al.* 1995). Einen *et al.* (1998) found a high variation in fat content among different fillet segments independent of starvation times up to 86 days. Generally, only marginal differences in fatty acid composition were observed in muscle, belly flap and liver.

During a fast for one month, the lipids in *Tilapia mossambica* muscle lost a higher proportion of C20 and C22 polyunsaturates than the visceral lipids (Nair & Gopakumar 1981) while Satoh *et al.* (1984) found that only 18:2 *n*-6 decreased to any extent in *T. nilotica* fasted for 82 days. In carp (*Cyprinus carpio*) the percentage of 16 : 0 decreased during a fast for 93 days whereas 20 : 4 *n*-6 and 22 : 6 *n*-3 increased (Takeuchi & Watanabe 1982) although it has also been suggested that 18 : 1 *n*-9 and 18 : 2 *n*-6 were primarily mobilised in this species (Murata & Higashi 1980). In sea bass (*Dicentrarchus labrax* L.) fasted for eight weeks, the general pattern was that the saturates in the liver and muscle decreased while the polyunsaturates were selectively retained, especially in muscle phospholipids (Delgado *et al.* 1994).

Proteins and amino acids

During prolonged starvation, the proteolytic activity in the muscle increases leading to mobilisation of amino acids as a major source of energy. The white muscle is mobilised before red muscle and at later stages the connective tissues (see Cowey & Walton 1989). Créach and Serfaty (1974) reported that during fasting up to 15 days, there was little change in the amino acid concentrations in carp muscle and liver. Magnitude of nitrogen loss was in the order: intestine > kidney > liver > spleen > muscle. But, after eight months of starvation, the nitrogen loss from the muscle was greatest with the amino acid concentrations decreasing, 10-fold compared with initial concentrations. A similar pattern has been reported for other species (see Cowey & Walton 1989).

In cod, after the liver lipids were exhausted, the proteins in the red and white muscles were mobilised. This occurred after about 20 weeks of starvation and the proteins were then the only source of energy (Black & Love 1986) whereas in Pyrenean brown trout juveniles (*Salmo trutta fario*) muscle proteins were significantly mobilised after 50 days of fasting (Navarro *et al.* 1992). However, fasting Atlantic salmon for 78 days (Lie & Huse 1992) or 86 days (Einen *et al.* 1998) did not affect the fillet protein content. The protein content in white muscle of rainbow trout increased slightly up to two months of fasting, but after three months a significantly lower content was observed (Kiessling *et al.* 1990).

Glycogen

In a review of endogenous fuels, van den Thillart and van Raaij (1995) concluded that during starvation fish species may be divided in two categories: species which do (common carp; roach, *Rutilus rutilus*; killifish, *Fundulus heteroclitus*; rainbow trout; brown trout) or do not (migrating sockeye salmon, *Oncorhynchus nerka*; European eel, *Anguilla anguilla*; American eel, *A. rostrata*) mobilise glycogen during the initial phase of starvation. In the first group, the glycogen utilisation may be of transient importance before lipids and proteins are mobilised while, in the latter group, glycogen is only used as energy source during prolonged periods of starvation after the availability of other fuels is reduced. For instance, European eel could be starved for at least 96 and 164 days before initial glycogen levels were reduced in liver and muscle, respectively (Dave *et al.* 1975). In Pyrenean brown trout juveniles a significant decrease in muscle glycogen was observed after 30 days of fasting (Navarro *et al.* 1992) whereas in rainbow trout the glycogen content in red and white muscles during the first 1–2 months of fasting increased, followed by a rapid decrease between two and three months (Kiessling *et al.* 1990). Einen and Thomassen (1998) showed that the glycogen content in Atlantic salmon white muscle decreased progressively with the length of starvation. Consequently, after 4 and 12 days of ice-storage the lactate levels also decreased, resulting in a higher ultimate pH. The effect was observed after 58 days of starvation at an average seawater temperature of 4.1°C.

The glycogen content in the muscle at the point of death may affect product quality. In Atlantic cod, it has been reported that heavy feeding was related to soft textured muscle. This was associated with a high initial glycogen content with a corresponding low ultimate pH affecting denaturation of myofibrillar protein (Ang & Haard 1985). When this species was fasted for 73 days, Akse and Midling (1997) found no significant differences in the ultimate pH.

Water

Since there is an inverse relationship between the relative proportions of lipid and water, in fatty fish catabolised lipid is replaced during fasting by equal amounts of water whereas in lean fish the protein will partially be replaced by water (Love 1980).

Carotenoids

Fasting for eight weeks had no effect on the astaxanthin or canthaxanthin levels in rainbow trout muscle (Foss *et al.* 1984) while Choubert (1985) reported a significant decrease in canthaxanthin content after 24 days. When starving Atlantic salmon for up to 86 days the muscle astaxanthin content and Roche colour card readings were not significantly affected. However, instrumental colour analyses (CIE 1976) of raw fillets showed changes in lightness (L*) and yellowness (b*) (Einen & Thomassen 1998). A reduction in the fat content during fasting, may lead to a more reddish visual colour impression (Torrissen *et al.* 1989).

Enzymes

Overall enzyme levels, muscle metabolism and muscle protein synthesis decrease during fasting (see Cowey & Walton 1989), but during long-term starvation increased proteolytic activity has been reported (Mommsen *et al.* 1980; Ando *et al.* 1986). For example, starvation of rainbow trout caused a decrease in the activity of certain enzymes involved in energy metabolism and the glycolytic activity was reduced by 70% after three months of starvation. In red muscle the oxidative capacity and fatty acid oxidation decreased during short-term starvation but were partly restored during prolonged starvation (Kiessling *et al.* 1990).

Structure and texture

As a result of muscle proteolytic activity during prolonged fasting, it has been shown in carp and cod that the muscle extracellular space increased in addition to a thickening of the myosomes (see Cowey & Walton 1989). A shrinking of white muscle fibres has been reported to occur in cod (Greer-Walker 1971), carp (Love 1980) and in plaice (*Pleuronectes platessa*) (Johnston 1981). Kiessling *et al.* (1990) showed that rainbow trout muscle fibres did not respond to starvation for three months in a consistent way. In the red muscle, a narrowing of the range of fibre sizes occurred whereas in the white muscle starvation had no effect on fibre sizes. The pink muscle showed an intermediate response. A narrowing of the range of the red muscle fibre area has also been observed in starved plaice (Johnston 1981).

Færgemand *et al.* (1995) reported that a fasting period of 18 days had no effect on rainbow trout fillet texture as assessed by sensory evaluations and Instron measurements. After longer periods without food, Einen and Thomassen (1998) demonstrated using instrumental texture analysis, measured after four days of ice-storage, that the Atlantic salmon raw fillet texture was harder in groups starved for 58 or 86 days than in groups starved for a shorter time.

Storage and eating quality

Johansson and Kiessling (1991) studied the effect of fasting rainbow trout for one and two months on sensory characteristics. The results indicated that eating and storage quality would be changed as total odour, total taste and juiciness were reduced. However, the changes were considered moderate and no drastic action concerning altering the fasting period was considered necessary.

Trimethylamine was not detected in Atlantic salmon muscle of fish starved up to 86 days. Sensory analyses of cooked fillets performed after ice-storage of gutted fish for 13–16 days revealed that fish starved for 86 days had less fresh flavour than fish starved for 30 days or less. Fillets from the latter group had higher acidulous flavour compared with groups starved for 0–14 days, whereas groups starved for 58 or 86 days had less acidulous flavour than all other groups. Sensory assessed hardness was lowest in the fish starved for 86 days compared with those starved for 0–30 days. It was concluded that although starvation did induce some changes in the composition of the white muscle, the differences were considered rather marginal. Thus, starvation seems to be a rather weak tool for changing fillet quality in this species (Einen & Thomassen, 1998).

Akse and Midling (1997) found almost no effect on sensory properties (colour, odour, taste and texture) as well as in processing quality when wild-caught Atlantic cod were fasted in storage sea cages for 73 days. It was concluded that live storage and fasting of capelin fed cod neither improved nor reduced eating quality.

Metabolic rates

Since fasting reduces the physical activity of the fish (Love 1980), the fish should preferably be fasted prior to transport. During fasting, the rates of oxygen consumption, the excretion of carbon dioxide and ammonia, as well as the amount of faeces produced, are decreased (Westers 1984). Falconer (1964) suggested that salmonids should be fasted for at least three days to reduce metabolic rates before transport. Moreover, Dickson and Kramer (1971) reported that the standard metabolic rate of rainbow trout decreased by about 38% from one to two days of fasting. Thereafter, a slight increase was observed until the experiment was terminated after nine days of fasting. Beamish (1964) found that 72 h without food was required to decrease standard oxygen consumption of brook trout (*Salvelinus fontinalis*) and that continued fasting had no further practical effect. In rainbow trout, at least 63 h of fasting is necessary to reduce ammonia excretion by 50% (Phillips & Brockway 1954). Altogether, the recommended practice is to fast salmonids for 48–72 h before transport to have any material effect on the metabolic rate (Wedemeyer 1996).

Conclusions

During fasting the energy metabolism is primarily fuelled by fatty acids and amino acids whereas, depending on species, glycogen may be an additional substrate either during the first stage of fasting or only during prolonged periods of starvation. However, only long-term fasting may affect flesh quality. Therefore, the common fasting period of 1–2 weeks for salmonids can be regarded as adequate. However, to clean digestive tracts and to reduce metabolic rate before transport and slaughter, the fasting period should not be less than 2–3 days.

Transport

A good review of basic information regarding water quality and physiologial effects, fish handling techniques and transport recommendations, is that of Wedemeyer (1996) and some transport designs are reviewed by Berka (1986).

Different fish species may have different tolerance of any pressure variations during transport as some species, such as salmonids, have a duct from the swim-bladder to the oesophagus permitting gas escape (physostome fish). This means if salmonids are exposed to pressure changes, the swim-bladder may be emptied and the fish lose their buoyancy. Such species may be observed filling their swim-bladders by gulping air at the water surface. In fish with no channel connecting the swim-bladder and the digestive tract (physoclist fish) such as cod, the swim-bladder will burst when the surrounding pressure is reduced by more than 60% (Tytler & Blaxter 1973). Indeed, this has been reported to occur after cod were caught by seine net and transported to storing sea cages. Some fish were unable to dive due to expanded or ruptured swim-bladders. Consequently, they were observed floating belly up on the surface (Akse & Midling 1997).

A considerable number of studies have been conducted concerning transport and stress as defined by changes in hormone levels or changes in blood chemistry of fish. On the other hand, available information is scarce with respect to potential changes in the muscle as a result of transport, and how this might affect flesh quality.

In closed systems, accumulation of un-ionised ammonia (NH_3) and carbon dioxide can be potential stressors during transport. However, when fish are to be slaughtered directly after transport, many of the long-term effects of handling or possible adverse water quality may be disregarded. Consequently, water quality requirements may be less important than what is commonly regarded necessary for live fish transport. The water quality recommendations for intensive cultures are intended to provide optimal conditions for growth, good health and low mortality. When the fish are not slaughtered immediately after transport, the water quality must obviously be such that the fish are able to perform essential life functions also after transport. Tomasso and Carmichael (1988) showed that when red drum fingerlings (*Sciaenops ocellatus*) were transported by truck up to about 9 h, the mortality at the end of transport was typically less than 1%. However, the cumulative mortality for the next 10 days ranged from 12% to 51%.

During limitations in water exchange when the water is not aerated during transport, it is important that the carbon dioxide levels do not reach toxic levels. In the case of European sea bass (*Dicentrarus labrax*), the LC_{50} levels of carbon dioxide were about 100 mg/L (Grøttum & Sigholt 1996).

Stress as defined by changes in hormone levels and blood chemistry

A truck transport for 30 min of yearling coho salmon (*Oncorhynchus kitutch*) caused a marked physiological stress response as judged by elevated plasma cortisol levels. Depending on season, recovery to near resting levels took from 3 h to more than 24 h (Schreck *et al.* 1989).

Similarly, transport of red drum (*Sciaenops ocellatus*) induced rapid but transient plasma cortisol, glucose and osmolality responses but caused no mortality (Robertson *et al.* 1987). McDonald *et al.* (1993) reported that loading densities of 69–170 kg/m^3 and trip durations of 3.5–11 h had only minor effects on physiological responses as judged by plasma constituents in hatchery-reared salmonids. Barton and Peter (1982) showed that plasma cortisol levels in rainbow trout fingerlings increased markedly after 0.5 h of transport. The increase was maintained over the next 4 h before a significant decrease occurred after 8 h of transport. The dissolved oxygen (DO) saturation levels did not significantly affect plasma cortisol values. During and following transport by road for 8 h, juvenile white suckers (*Catostomus commersoni*) had elevated plasma cortisol concentrations and required several days to recover before they reached baseline values (Bandeen & Leatherland 1997). Compared with well-fed groups, a fast for four weeks did not affect the plasma cortisol respose pattern when the fish were subjected to a 5 min chase challenge. However, the recovery period was prolonged in fasted fish indicating that prolonged fasting seems to reduce the ability to recover from handling stress.

Provided adequate water quality can be maintained, the processes of loading and unloading seem to be the most stressful events rather than the actual transport per se (Specker & Schreck 1980; Robertson *et al.* 1988).

Stress as defined by biochemistry of the muscle

As outlined above, transport may induce changes in blood constituents. Consequently, we may ask if there are any links between (adverse) water quality, change in blood chemistry (stress) and flesh quality. An obvious answer to this question is that any stressor (e.g. low levels of DO) that makes the fish struggle, will also adversely affect flesh quality (Erikson *et al.* 1997). In other cases, the fish may exhibit a sluggish behaviour with no vigorous muscle activity when subjected to adverse water quality. Using blood-related parameters as stress indicators, the fish may be defined as stressed, but it is not clear to what extent this might affect the muscle.

When rainbow trout were exposed to an elevated ammonia concentration of 0.72 mg/L for 24 h, the body content of ammonia also increased (Paley *et al.* 1993). However, Grøttum *et al.* (1998) showed using non-invasive ^{31}P-NMR, that the whole fish (predominantly white muscle) high-energy phosphate balance, as well as the intracellular pH, were not significantly affected when goby (*Pomatoschistus* sp.) were exposed to sublethal levels of ammonia.

Butler and Day (1993) found no significant effects on plasma, red or white muscle pH when brown trout were exposed to sublethal pH 4.5. However, a significantly higher total water content of the white muscle was found. Although acute exposure of adult rainbow trout to pH 4.0 caused significant ionoregulatory disturbances, no significant change in water content was observed in epaxial white muscle (Audet & Wood 1988). Furthermore, it was concluded that the fish did not acclimate to acid stress. A non-invasive ^{31}P-NMR study of tilapia exposed to pH 4.0 induced a transient pH drop in gills and plasma which recovered during the next 10 h exposure to acid water. Notably in the muscle, the intracellular pH was only slightly affected and the high-energy phosphate stores were not affected (van Waarde *et al.* 1990). However, acid-stressed juvenile rainbow trout seemed to be more sensitive to additional stressors than unstressed fish (Barton *et al.* 1985). By contrast, Haya *et al.* (1985) suggested that the control fish were more responsive to handling stress than acid-exposed juvenile Atlantic salmon. Exposure at pH 4.7 for up to 76 days seemed to increase gluconeogenesis as well as stressing the energy metabolism.

A simulated open system transport for 2 days at 8°C of Atlantic cod weighing 0.8–2.5 kg at a fish density of 540 kg/m^3 caused a general stress response, as indicated by an increase in plasma cortisol and glucose levels. The degree of iono-osmoregulatory disturbance and mortality were negligible. Notably, transport did not affect muscle contents of water, glycogen and lactate as well as liver glycogen (Staurnes *et al.* 1994). Ostenfeld *et al.* (1995) followed a commercial truck transport of fasted (eight days) rainbow trout (0.3–0.5 kg) at a fish density of 167 kg/m^3 at 6.1–8.2°C for 10.5 h. Water pH decreased from 6.9 to 6.5 whereas NH$_4^+$ rose from 0.431 to 3.625 mg/L after 6.5 h transport. Upon arrival the fish were in excellent condition. Analysis of white muscle revealed no differences in glycogen, glucose or lactate when the fish were compared before and after transport. Nor did the dry matter or the textural parameters hardness, elasticity and breakpoint of the raw fillet (Instron assessments) differ. Post rigor muscle pH was slightly higher after transport compared with control fish. In dark muscle glycogen and lactate decreased after transport. Altogether, the results suggested that transport had limited effects on the flesh quality. Erikson *et al.* (1997) studied a well-boat transport (90 min) of Atlantic salmon (5.1 $\pm$ 1.1 kg fasted for 12 days) at a fish density of 125 kg/m^3 at about 7°C. During transport (open system) and at the quay, the transport water quality largely resembled typical seawater values. Indicators of muscle handling stress were high-energy phosphates, inosine mono-phosphate, adenylate energy charge, pH and redox potential. It was concluded that salmon can be loaded using a water-filled net, transported and delivered to the quay ready for slaughter without any dramatic effects of handling stress which in turn would otherwise have resulted in early rigor and reduced freshness. In fact, the fish were less stressed after transport, indicating that the fish actually recovered after moderate loading stress.

Effect of water quality on post rigor fillet quality with special attention to oxygen supersaturation

To minimise transport costs, there is currently a tendency to increase the transported bio-masses either by increasing the loading capacity of the vessel or the fish density. In the latter case, water quality in open systems may also have to be improved by aeration or oxyge-nation. During transport of high biomasses there may be considerable gradients in DO from the inlet to the outlet side of the hold. When using effective equipment for oxygenation, oxygen supersaturation may occur during transport. A study of the effect of water quality, and in particular the effect of oxygen supersaturation, on the flesh quality of Atlantic salmon transported by well-boats was conducted in three parts (U. Erikson, T. Rosten, A. Sverdrup, J.E. Steen, T. Ordemann & G. Aanderud unpublished results), as outlined here.

Survey of commercial well-boat transports
Results from a survey of 150 commercial transports with nine different well-boats showed that Atlantic salmon can be transported in open systems under various conditions without negative feedback from the processing plants with regard to product quality. The survey was based on data from questionnaires given to well-boat and processing plant personnel. Although such data are subjective and may be considerably biased, they do indicate that no dramatic effects of transports were observed. In 39% of the transports, additional oxygen was supplied to the holds. The transport conditions were as follows: water flow-through (based on maximum pump capacity used at quay) 1500–2500 m^3/h; fish density 41–255 kg/m^3; seawater temperature 3–17°C; DO 70–120% saturation; transport time 0.3–8.5 h; time at quay before unloading (slaughter) 0–13 h.

Laboratory studies of simulated transport in open systems, with reduced water-flow, and in closed systems
In this part, we studied simulated transports with fish densities 227–329 kg/m^3 for 5 h under laboratory conditions. The fish (3.9–5.5 kg) were commercially farmed before being accli-mated in the laboratory to the experimental temperature of 15°C. Total fasting time was 15 days. The effects of water quality on behaviour, stress and post rigor muscle quality in open and closed systems as well as with reduced seawater-flow were studied. Different equipment (prototypes) for oxygenation and aeration were also tested. After simulated transport all

groups were anaesthetized with benzocaine (68 mg/L) and subsequently killed with a blow to the head except for two groups of fish exposed to oxygen supersaturated water prior to either chasing to exhaustion or carbon dioxide anaesthesia in separate tanks to simulate commercial slaughter procedures. These fish were ultimately killed by a blow to the head. Plasma chloride and muscle-pH were used as indicators of stress. The fish were immediately gutted, but not bled, before transfer to boxes with ice. Muscle-pH and rigor developments were then followed. After four days, the fish were filleted and the water content and colour (Roche colour card) were determined. Gaping and fillet softness were evaluated by trained personnel from the industry.

Both in the open system (good water quality) and closed system (adverse water quality), high DO concentrations (200–250% saturation) within a few minutes affected the fish behaviour. The fish became sluggish, keeping their mouth above the water surface exhibiting a gulping or coughing behaviour. After about 3 h, the fish exhibited short (5 s) sporadic bursts of activity with 'the tail on the water surface' and at about 4.5 h most fish tended to lie quietly at the bottom of the tank until the experiment was terminated. Notably, almost *no* vigorous swimming activity took place during the experiments. The opercular amplitude gradually diminished and at the end of the experiment there were scarcely any gill movements.

It is well-known that at high DO levels, the gill ventilation rate is reduced. As a result, carbon dioxide accumulates in the blood (hypercapnia) and eventually the blood pH is reduced (acidosis) (Hobe *et al.* 1984). In turn, this may affect the brain activity and thus behaviour. At high fish densities in closed systems with normal oxygen levels (about 80% saturation) the fish also exhibited abnormal behaviour but this was less extreme. In the closed system (with the most adverse water quality) carbon dioxide and total ammonia (TA = $NH_4^+ + NH_3$) gradually accumulated during 'transport' to 10 mg/L and 687 µg/L after 5 h, respectively. However, the toxic NH_3 never exceeded 2.4 µg/L which is below the recommended value for fish farming. The total gas saturation level never exceeded 105% and the alkalinity essentially stayed constant at 2.3 mmol/L. Particularly in the closed system, heavy mucus formation and foaming was observed. However, when the fish were subjected to handling stress after simulated transport, the initial muscle pH-values of pH 7.3–7.4, showed that the muscle was in a rested state and the mean plasma chloride values (control: 140–145 mmol/L; treatments: 155–170 mmol/L) indicated a mild stress response as assessed in the blood of fish subjected to adverse water quality.

Compared with the control groups, no effects on the product quality were observed except in the fish subjected to oxygen supersaturated water. During ice storage of the fish from this group, the pH-drop was slower and rigor lasted somewhat longer than in all other groups. The fillets did not differ from those in the other groups with respect to water content (68%) or colour (Roche colour card: 14–15) and the gaping score was also low (almost no gaping). However, the fillets from the oxygen supersaturated fish were soft and had more mucus on the skin side. Notably, this effect was reduced when the fish were subjected to handling stress after simulated transport. The results strongly suggest that high levels of oxygen in early post mortem muscle tissues delay post-mortem degradations. However, during transport, the Root and Bohr effects would normally have reduced the oxygen delivery capacity to the tissues, but it is unclear to what extent the simultaneous high DO levels may have affected this process. It is suggested that, due to a slow pH-reduction, certain autolytic enzymes are active causing a softening of the muscle tissues. It might be mentioned that long-term rearing of rainbow trout in oxygen-supersaturated waters up to 183% saturation had no effect on growth or mortality (Edsall & Smith 1990).

Evaluation of three commercial transports

Subsequently, we studied three commercial well-boat transports all using open systems. The experimental conditions are shown in Table 19.1. As pointed out previously (Erikson *et al.* 1997), the fish in this study also seemed somewhat quieter during unloading than during loading at the sea cage. In all cases evaluation of handling stress and fillets was done before and after transport as well as after slaughter. In all cases, we found that transport did not affect the flesh quality. During transport with the highest fish density (177 kg/m^3) some fish showed signs of adverse behaviour (swimming with mouth open at the water surface). At the quay, these fish were selectively netted individually. It turned out that these fish had elevated plasma chloride values (164 $\pm$ 11 mmol/L) compared with pre-transport values (147 $\pm$ 4 mmol/L) whereas muscle pH values, corresponding to rested fish, were not different (pH 7.2 $\pm$ 0.2).

Table 19.1   Three commercial well-boat transports (open systems) of live Atlantic salmon (4–6 kg) with good water quality. Stress and fillet quality were evaluated.

	Transport 1	Transport 2	Transport 3
Fasting period (days)	14	6	5
Biomass (tons)	31	56[1]	31
Fish density (kg/m^3)	119	162	177
Transport time (h)	5	2	2
Time at quay (h)[2]	5	13.5	7.5
SW-temperature (°C)	15	6	13
DO saturation (%)	80–100	90–120[3,4]; 60–90[4]	60–100[3]
Muscle quality[5]	A	B	C

[1] Divided equally in two similar holds
[2] Until slaughter commenced
[3] oxygenation of hold
[4] Simultaneous transport of fish with and without oxygenation.
[5] Muscle quality: A = pH and rigor; B = pH, rigor, gaping, colour; water content, water holding capacity and texture (Instron); C = pH, rigor, colour and texture (Instron) (n = 6 12).

Conclusions

Overall, we concluded that commercial well-boat transport using modern technology is not detrimental in terms of affecting flesh quality. However, once the fish were transferred to the slaughter line, they were subjected to considerable stress during carbon dioxide anaesthesia. Thus, the current 'bottleneck' regarding the detrimental effects of handling stress is the slaughter procedure. Instantaneous stunning of the fish is therefore highly recommended (rested harvesting).

Even if fish showed abnormal behaviour, this did not necessarily seem to affect flesh quality as long as no vigorous muscle activity took place. In terms of water quality, only the levels of DO seemed to affect flesh quality. Thus, recommended DO levels should be in the range of 70–100% saturation.

Transport in closed systems with chilling (hypothermia, hibernation)

As an alternative transport concept, the fish can be held in a state of hibernation by reducing the body temperature to a point where fish movement is practically eliminated (hypothermia). Due to reduced metabolism and water fouling, the loading densities can be increased. For example, in case of salmonids the Q_{10} values for oxygen consumption and excretion of metabolic waste products are in the range of 1.4–1.9 (Smith 1977). Hypothermia may also be considered as a method for partially anaesthetizing the fish to reduce the potential detrimental effects of handling stress during transport. By lowering the body temperature, the reflexes are blocked as the nerve conduction depends on acclimation temperature (Roots & Prosser 1962). Mittal and Whitear (1978) studied cold anaesthesia of several poikilotherms. For instance, when *Monopterus cuchia* acclimated to 23°C were chilled to 10°C, the fish became stiff and immobile and could be killed without a struggle. It was concluded that cold-adapted fish can be made relatively torpid by reducing temperature to 0°C, whereas warm-adapted fish appeared to be completely anaesthetized at temperatures around 0°C. Considerable individual and species variations were observed though.

Considering transport economy, the use of chilling is mostly confined to closed systems. Strict control of temperature and water quality are therefore essential requirements. Also, light sedation (rather than anaesthesia) should be used as the fish should not lose equilibrium and sink to the bottom of the transport tank where the risks of injuries are higher. Temperature changes of less than 10°C can be tolerated physiologically by fish in good health. Otherwise, temperature shock may occur which can be stressful to the fish. For example, lowering the water temperature from 10° to 5°C reduced hyperglycemia by about 30% in smolts transported by truck at $120 \, kg/m^3$. However, chilled water offers little protection against blood electrolyte depletion which is a major factor in the delayed mortality which can occur following transport (Wedemeyer 1996).

Fingerling rainbow trout adapted to 10–11°C showed a large increase in plasma cortisol levels when they were exposed to chilled water (1°C). When the fish were acclimated to chilled water for 24 h and then transported in the chilled water, the plasma cortisol values nevertheless increased during transport (Barton & Peter 1982).

Dissolved carbon dioxide concentrations greater than 15–20 mg/L are normally considered undesirable because oxygen transport to the tissues begins to be impaired by the Bohr and Root effects. In addition, the DO required to provide sufficient oxygen to the tissues increases. To prevent hypercapnia, the dissolved carbon dioxide should therefore be removed from the water. However, another strategy may be used when transporting fish to be slaughtered. Unlike many chemical anaesthetics, carbon dioxide leaves no toxic residues in the fish. It may be used in combination with low temperature to improve overall effectiveness of the anaesthetic treatment. In fact, Yokoyama *et al.* (1989) showed that initial carbon dioxide anaesthesia ($pCO_2 = 80 \, mmHg$) combined with low temperature (4°C) ('cold-CO_2') could be used successfully prior to adult carp (acclimated at 23°C) transport for

10 h at 14°C in a closed system. Somewhat surprisingly, the combination appeared to be stressful as judged by ATP-related compounds, glycogen and pH in muscle. If the carp were allowed to recover after 'cold-CO_2' anaesthesia, the meat quality was however unaffected (Yokoyama *et al.* 1993).

Commercial live transport of Atlantic salmon using hypothermia

Recently we studied the commercial well-boat transport of Atlantic salmon (3–6 kg, fasted for 12 days) in a closed system with refrigerated sea water (RSW) chilling (U. Erikson, T. Rosten and J.E. Steen unpublished results). The objectives were to evaluate water quality, behaviour and stress during transport at low temperature using a new well-boat with two separate holds (250 m^2 each). The potential benefits from transport in closed systems would be reducing stress during transport, sedating the fish prior to subsequent slaughter, more rapid chilling post mortem to preserve freshness, and lowering core temperature at packing.

In one of the holds, the fish were transported traditionally, i.e. in an open system with no oxygenation and at seawater temperature (8°C). In the other hold, the fish were transported in pre-chilled seawater at 1°C in a closed system with oxygenation but without aeration. Fish densities in both holds were 90 kg/m^3 and transport time was 2 h. The fish were kept for not more than 4 h in the closed system due to loading and unloading processes. The oxygen levels in both holds varied within 70–110% saturation. In the open system, typical seawater conditions prevailed. In the closed system, the carbon dioxide levels increased steadily up to 45 mg/L during transport causing a pH drop of 1.3 units from pH 8.0. The alkalinity increased from 2.25 to 2.45 μmol/L and the salinity was constant at 33.3 parts per thousand. The TA increased to 2520 μg/L, but the toxic NH_3 showed only a moderate increase up to 2.0 μg/L which is below the proposed safety level for fish farming (20 μg/L). The water gradually became less transparent with some foaming at the surface as reflected by increases in colour from 2–3 to 8 mg Pt/L (distilled water = 0 mg Pt/L) and in total organic carbon from 1.4 to 4.8 mg/L. Notably, an increase in Fe^{3+} content from about 10 to 108 μg/L was observed. The source of the iron is unknown, but it may originate from haemoglobin. Thus, it may be speculated that transfer of fish to chilled water induced bleeding from the gills.

Upon arrival at the quay, only the gills of the chilled fish had a bright red colour, which may be regarded as a support for this idea. During transport, the chilled fish were more torpid than in the open system and they were close to the bottom of the hold. When netted individually after transport, these fish hardly struggled at all. By this time, the body temperature was equilibriated with that of the transport water (1°C). Plasma chloride values in the chilled fish were 155 ± 5 mmol/L. Compared with the values at the cage, 149 ± 8 mmol/L, and after transport in open system, 141 ± 8 mmol/L, a mild stress effect was indicated in the chilled fish. White muscle pH values at the cage were pH 7.3 ± 0.1 and after transport in the open system pH 7.5 ± 0.1, corresponding to a rested fish. The flesh quality was not assessed in this study, but since the fish were very torpid, they could potentially be slaughtered with minimal handling stress. If the low body temperature could be maintained through the processing line, this may be an efficient method for rapid chilling. However, from an ethical point of view, the possibility of bleeding from the gills as a result of rapid transfer of fish to low temperatures should be investigated further.

Conclusions

Using modern technology, the transport of live fish does not seem to have adverse effects on flesh quality. Vigorous muscle activity is the major potential contributor to reduced flesh quality. Also, mechanical damage such as bruising and scale loss due to crowding may occur during transport. Tentatively, the levels of dissolved oxygen seem to be the only water quality parameter linked to flesh quality. For assessment of potential effects on flesh quality, the choice of suitable stress indicators (blood versus muscle) is important.

Acknowledgements

The author wishes to thank the Norwegian Research Council, Hydro Seafood, Brønnbåteiernes Forening, Flatsetsund Slipp AS, Lakstrans as well as the crew on M/S Caroline and M/S Frøytrans (Water quality project). Also, thanks to Frøyfisk AS and the crew on M/S Roy Kristian (chilled transport project).

References

Akiyama, T. & Nose, T. (1980) Changes in body weight, condition factor, and body composition of fingerling chum salmon with various sizes during starvation. *Bulletin of the National Research Institute of Aquaculture*, **1**, 71–8.

Akse, L. & Midling, K. (1997) Live capture and starvation of capelin cod (*Gadus morhua* L.) in order to improve the quality. In: *Seafood from Producer to Consumer, Integrated Approach to Quality* (eds J.B. Luten, T. Børresen & J. Oehlenschläger), pp.47–58. Elsevier Science BV, Amsterdam.

Ando, S., Hatano, M. & Zama, K. (1986) Protein degradation and protease activity of chum salmon (*Oncorhynchus keta*) muscle during spawning migration. *Fish Physiology and Biochemistry*, **1**, 17–26.

Ang, J.F. & Haard, N.F. (1985) Chemical composition and postmortem changes in soft textured muscle from intensely feeding Atlantic cod (*Gadus morhua*, L.). *Journal of Food Biochemistry*, **9**, 49–64.

Audet, C. & Wood, C.M. (1988) Do rainbow trout (*Salmo gairdneri*) acclimate to low pH? *Canadian Journal of Fisheries and Aquatic Sciences*, **45**, 1399–1405.

Bandeen, J. & Leatherland, J.F. (1997) Transportation and handling stress of white suckers raised in cages. *Aquaculture International*, **5**, 385–96.

Barton, B.A. & Peter, R.E. (1982) Plasma cortisol stress response in fingerling rainbow trout, *Salmo gairdneri* Richardson, to various transport conditions, anaesthesia, and cold shock. *Journal of Fish Biology*, **20**, 39–51.

Barton, B.A., Weiner, G.S. & Schreck, C.B. (1985) Effect of prior acid exposure on physiological responses of juvenile rainbow trout (*Salmo gairdneri*) to acute handling stress. *Canadian Journal of Fisheries and Aquatic Sciences*, **42**, 710–17.

Beamish, F.H.W. (1964) Influence of starvation on standard and routine oxygen consumption. *Transactions of the American Fisheries Society*, **93**, 103–107.

Beck, F. & Gropp, J. (1995) Estimation of starvation losses of nitrogen and energy in rainbow trout (*Oncorhynchus mykiss*) with special regard to protein and energy maintenance requirements. *Journal of Applied Ichthyology*, 11, 263–75.

Berka, R. (1986) *The transport of live fish. A review*. EIFAC Technical Paper (48), p.52.

Black, D. & Love, R.M. (1986) The sequential mobilization and restoration of energy reserves in tissues of Atlantic cod during starvation and refeeding. *Journal of Comparative Physiology B*, 156, 469–79.

Black, D. & Skinner, E.R. (1986) Features of the lipid transport system of fish as demonstrated by studies on starvation in the rainbow trout. *Journal of Comparative Physiology B*, 156, 497–502.

Butler, P.J. & Day, N. (1993) The relationship between intracellular pH and swimming performance of brown trout exposed to neutral and sublethal pH. *Journal of Experimental Biology*, 176, 271–84.

Choubert, G. (1985) Effects of starvation and feeding on canthaxanthin depletion in the muscle of rainbow trout (*Salmo gairdneri* Rich.). *Aquaculture*, 46, 293–8.

CIE (1976) 18th session, London, UK, September 1975. CIE Publication 36. Commission International d'Eclairage, Paris, France.

Cowey, C.B. & Walton, M.J. (1989) Intermediary metabolism. In: *Fish Nutrition* (ed. J.E. Halver), 2nd edn, pp.259–329. Academic Press, San Diego.

Créach, Y. & Serfaty, A. (1974) Le jeûne et la réalimentation chez la carpe (*Cyprinus carpio* L.). *Journal of Physiology (Paris)*, 68, 245–60.

Dave, G., Johansson-Sjöbeck, M.L., Larsson, A., Lewander, K. & Lidman, U. (1975) Metabolic and hematological effects of starvation in the European eel, *Anguilla anguilla* L. I. Carbohydrate, lipid, protein and inorganic ion metabolism. *Comparative Biochemistry and Physiology*, 52, 423–30.

Delgado, A., Estevez, P., Hortelano, P. & Alejandre, M.J. (1994) Analyses of fatty acids from different lipids in liver and muscle of sea bass (*Dicentrarchus labrax* L.). Influence of temperature and fasting. *Comparative Biochemistry and Physiology*, 108A, 673–80.

Dickson, I.W. and Kramer, R.H. (1971) Factors influencing scope for activity and active and standard metabolism of rainbow trout (*Salmo gairdneri*). *Journal of the Fisheries Research Board of Canada*, 28, 587–96.

Edsall, D.E. & Smith, C.E. (1990) Performance of rainbow trout and Snake River cutthroat trout reared in oxygen-supersaturated water. *Aquaculture*, 90, 251–9.

Einen, O. & Thomassen, M.S. (1998) Starvation prior to slaughter in Atlantic salmon (*Salmo salar*). II. White muscle composition and evaluation of freshness, texture and colour characteristics in raw and cooked fillets. *Aquaculture*, 169, 37–53.

Einen, O., Waagan, B. & Thomassen, M.S. (1998) Starvation prior to slaughter in Atlantic salmon (*Salmo salar*). I. Effects on weight loss, body shape, slaughter- and fillet-yield, proximate and fatty acid composition. *Aquaculture*, 166, 85–104.

Erikson, U., Sigholt, T. & Seland, A. (1997) Handling stress and water quality during live transportation and slaughter of Atlantic salmon (*Salmo salar*). *Aquaculture*, 149, 243–52.

Falconer, D.D. (1964) Practical trout transport techniques. *The Progressive Fish-Culturist*, 26, 51–8.

Foss, P., Storebakken, T., Schiedt, K., Liaaen-Jensen, S., Austreng, E. & Streiff, K. (1984) Carotenoids in diets for salmonids. I. Pigmentation of rainbow trout with the individual optical isomers of astaxanthin in comparison with canthaxanthin. *Aquaculture*, **41**, 213–36.

Færgemand, J., Rønsholdt, B., Alsted, N. & Børresen, T. (1995) Fillet texture of rainbow trout as affected by feeding strategy, slaughtering procedure and storage post mortem. *Water Science and Technology*, **31**, 225–31.

Greer-Walker, M.G. (1971) Effect of starvation and exercise on the skeletal muscle fibres on the cod (*Gadus morhua* L.) and the coalfish (*Gadus virens* L.) respectively. *Journal du Conseil, Conseil International pour l'Exploration de la Mer*, **33**, 421–6.

Grøttum, J.A. & Sigholt, T. (1996) Acute toxicity of carbon dioxide on European seabass (*Dicentrarchus labrax*): Mortality and effects on plasma ions. *Comparative Biochemistry and Physiology*, **115A**, 323–7.

Grøttum, J.A., Erikson, U. Grasdalen, H. & Staurnes, M. (1998) In vivo ^{31}P-NMR spectroscopy and respiration measurements of anaesthetized goby (*Pomatoschistus* sp.) pre-exposed to ammonia. *Comparative Biochemistry and Physiology*, **120A**, 469–75.

Haya, K., Waiwood, B.A. & van Eeckhaute (1985) Disruption of energy metabolism and smoltification during exposure of juvenile Atlantic salmon (*Salmo salar*) to low pH. *Comparative Biochemistry and Physiology*, **82C**, 323–9.

Hobe, H., Wood, C.M. & Wheatly, M.G. (1984) The mechanisms of acid-base and ionoregulation in the freshwater rainbow trout during environmental hyperoxia and subsequent normoxia. I. Extra- and intracellular acid-base status. *Respiration Physiology*, **55**, 139–55.

Jezierska, B., Hazel, J.R. & Gerking, S.D. (1982) Lipid mobilization during starvation in the rainbow trout, *Salmo gairdneri* Richardson, with attention to fatty acids. *Journal of Fish Biology*, **21**, 681–92.

Johansson, L. & Kiessling, A. (1991) Effects of starvation on rainbow trout. II. Eating and storage qualities of iced and frozen fish. *Acta Agriculturae Scandinavica*, **41**, 207–16.

Johnston, I.A. (1981) Quantitative analysis of muscle breakdown during starvation in the marine flatfish (*Pleuronectes platessa*). *Cell and Tissue Research*, **214**, 369–86.

Kiessling, A., Johansson, L. & Kiessling, K.-H (1990) Effects of starvation on rainbow trout muscle. I. Histochemistry, metabolism and composition of white and red muscle in mature and immature fish. *Acta Agriculturae Scandinavica*, **40**, 309–24.

Lie, Ø. & Huse, I. (1992) The effect of starvation on the composition of Atlantic salmon (*Salmo salar*). *Fiskeridirektoratets Skrifter, Serie Ernæring*, **5**, 11–16.

Love, R.M. (1980) *The Chemical Biology of Fishes*, Vol. 2. Academic Press, New York.

McDonald, D.G., Goldstein, M.D. & Mitton, C. (1993) Responses of hatchery-reared brook trout, lake trout, and splake to transport stress. *Transactions of the American Fisheries Society*, **122**, 1127–38.

Mittal, A.K. & Whitear, M. (1978) A note on cold anaesthesia of poikilotherms. *Journal of Fish Biology*, **13**, 519–20.

Mommsen, T.P., French, C.J. & Hochachka, P.W. (1980) Sites and patterns of protein and amino acid utilization during the spawning migration of salmon. *Canadian Journal of Zoology*, **58**, 1785–99.

Murata, H. & Higashi, T. (1980) Selective utilization of fatty acid as energy source in carp. *Bulletin of the Japanese Society of Scientific Fisheries*, **46**, 1333–8.

Nair, K.G.R. and Gopakumar, K. (1981) *Fishery Technology*, **18**, 123–7.

Navarro, I., Gutiérrez, J. & Planas, J. (1992) Changes in plasma glucagon, insulin and tissue metabolites associated with prolonged fasting in brown trout (*Salmo trutta fario*) during two different seasons of the year. *Comparative Biochemistry and Physiology*, 102A, 401–7.

Ostenfeld, T., Thomsen, S., Ingólfdóttir, S., Rønsholdt, B. & McLean, E. (1995) Evaluation on the effect of live haulage on metabolites and fillet texture of rainbow trout (*Oncorhynchus mykiss*). *Water Science and Technology*, 31, 233–7.

Paley, R.K., Twitchen, I.D. & Eddy, F.B. (1993) Ammonia, Na$^+$, K$^+$ and Cl$^-$ levels in rainbow trout yolk-sac fry in response to external ammonia. *Journal of Experimental Biology*, 180, 273–84.

Phillips, A.M. & Brockway, D.R. (1954) Effect of starvation, water temperature and sodium amytal on the metabolic rate of brook trout. *The Progressive Fish-Culturist*, 16, 65–8.

Roberts, R.J. & Bullock, A.M. (1989) Nutritional pathology. In: *Fish Nutrition* (ed. J.E. Halver), 2nd edn, pp.423–73. Academic Press, San Diego.

Robertson, L., Thomas, P., Arnold, C.R. & Trant, J.M. (1987) Plasma cortisol and secondary stress responses of red drum to handling, transport, rearing density, and a disease outbreak. *The Progressive Fish-Culturist*, 49, 1–12.

Robertson, L., Thomas, P. and Arnold, C.R. (1988) Plasma cortisol and secondary stress responses of cultured red drum (*Sciaenops ocellatus*) to several transportation procedures. *Aquaculture*, 68, 115–30.

Roots, B.I. & Prosser, C.L. (1962) Temperature acclimation and the nervous system in fish. *Journal of Experimental Biology*, 39, 617–29.

Sargent, J., Henderson, R.J. & Tocher, D.R. (1989) The lipids. In: *Fish Nutrition* (ed. J.E. Halver), 2nd edn, pp.153–218. Academic Press, San Diego.

Satoh, S., Takeuchi, T. & Watanabe, T. (1984) *Bulletin of the Japanese Society of Scientific Fisheries*, 50, 79–84.

Schreck, C.B., Solazzi, M.F., Johnson, S.L. & Nickelson, T.E. (1989) Transportation stress affects performance of coho salmon, *Oncorhynchus kisutch*. *Aquaculture*, 82, 15–20.

Smith, R.R. (1977) *Studies on the energy metabolism of cultured fishes*. 59p. PhD dissertation, Cornell University, Cortland, NY.

Specker, J.L. & Schreck, C.B. (1980) Stress responses to transportation and fitness for marine survival in coho salmon (*Oncorhynchus kisutch*) smolts. *Canadian Journal of Fisheries and Aquatic Sciences*, 37, 765–9.

Staurnes, M., Sigholt, T., Pedersen, H.P. & Rustad, T. (1994) Physiological effects of simulated high-density transport of Atlantic cod (*Gadus morhua*). *Aquaculture*, 119, 381–91.

Takeuchi, T. & Watanabe, T. (1982) The effects of starvation and environmental temperature on proximate and fatty acid compositions on carp and rainbow trout. *Bulletin of the Japanese Society of Scientific Fisheries*, 48, 1307–16.

van den Thillart, G. & Van Raaij, M. (1995) Endogenous fuels; non-invasive versus invasive approaches. In: *Biochemistry and Molecular Biology of Fishes*, Vol. 4. (eds T.P. Mommsen & P.W. Hochachka), pp.33–63. Elsevier Science BV, Amsterdam.

Tomasso, J.R. & Carmichael, G.J. (1988) Handling and transport-induced stress in red drum fingerlings (*Sciaenops ocellatus*). *Contributions in Marine Science*, 30, 133–8.

Torrissen, O.J., Hardy, R.W. & Shearer, K.D. (1989) Pigmentation in salmonids – carotenoid deposition and metabolism. *CRC Critical Reviews in Aquatic Sciences*, 1, 209–25.

Tytler, P. & Blaxter, J.H.S. (1973) Adaption by cod and saithe to pressure changes. *Netherlands Journal of Sea Research*, 7, 31–45.

van Waarde, A., van Dijk, P., van den Thillart, G., Verhagen, M., Erkelens, C., Bonga, S.W, Addink, A., & Lugtenburg J. (1990) ^{31}P-NMR studies on acid-base balance and energy metabolism of acid-exposed fish. *Journal of Experimental Biology*, 154, 223–36.

Wedemeyer, G.A. (1996) Transportation and handling. In: *Principles of Salmonid Culture* (eds W. Pennell & B.A. Barton), pp.727–58. Elsevier, Amsterdam.

Westers, H. (1984) *Principles of Intensive Fish Culture* (a manual for Michigan's state fish hatcheries). 109 pp. Michigan Department of Natural Resources, Lansing, MI.

Yokoyama, Y., Yoshikawa, H. Ueno, S. & Mitsuda, H. (1989). Application of CO_2-anesthesia combined with low temperature for long-term anesthesia in carp. *Nippon Suisan Gakkaishi*, 55, 1203–1209.

Yokoyama, Y., Kawai, F. & Kanamori, M. (1993) Effect of cold-CO_2 anesthesia on postmortem levels of ATP-related compounds, pH, and glycogen in carp muscle. *Nippon Suisan Gakkaishi*, 59, 2047–52.

Chapter 20

The Relationship Between Killing Methods and Quality

D.H.F. Robb

Department of Clinical Veterinary Science, University of Bristol, Bristol, BS40 5DU, UK

Introduction

Slaughter is the end point of the life of a food animal. It is the point where the living animal is turned into meat. Many quality characteristics are fixed at this point. For instance in salmonids the colour of the flesh, a very important flesh quality attribute, is derived from carotenoid pigments fed to the fish in their diets. After slaughter the quality of the flesh may slowly show deleterious changes. This is well accepted and there are a range of procedures carried out to minimise this, such as chilling or freezing which slow the rate of degradation. However, little is known about the effects that slaughter itself can have on the quality of the flesh. This chapter will summarise the work carried out in this area. Initially quality will be defined and then current commercial slaughter methods will be briefly described and the potential effects discussed. The chapter concludes with some general recommendations.

Definition of quality

Quality can mean many things to many people. However, for the purposes of this chapter it will relate to two things: ethical and flesh quality. Both are subjective terms, and their meaning depends on the experiences of the individual. Ethical quality refers to the perception of how humanely the fish has been treated during the process. In general, consumers are poorly informed about this, especially where the slaughter of fish is concerned. However, attitudes are changing and in Germany this has led to legislation to phase out a current slaughter method used for eels (*Anguilla anguilla*) which causes a great deal of suffering to the fish.

Flesh quality, for the purpose of this paper, is defined as the appearance, texture and flavour of the meat. The demands of the consumer vary from country to country, depending on custom and previous experience. Colour and the visual structure of the flesh are important for appearance. Texture is important for flesh processing and also during eating.

Current slaughter methods

There are a variety of killing methods used for fish, depending on the needs of the market. Some species, such as Atlantic salmon (*Salmo salar*) need to be bled out and so are stunned to stop them moving and facilitate this. Other large fish, such as tuna, are also killed outright to stop them from moving on the boat and potentially damaging themselves. Smaller fish tend not to be individually processed as the numbers are too great for the procedure to be commercially viable. The range of killing methods used is discussed below.

Death in air

The oldest method of killing fish, death in air relies on the fact that when fish are lifted out of water their gills collapse and are less efficient at gaseous exchange. The fish eventually die of anoxia. Death is speeded up by the fish struggling as it tries to escape. This uses up the oxygen faster, leading to faster brain death. The body temperature of the fish also affects the rate of death. At higher temperatures the rate of respiration is higher, leading to a faster death once the fish is removed from water (Kestin *et al.* 1991). Thus at 20°C rainbow trout took 2.6 min for brain death to occur and 11.5 min for the fish to stop all movement. At 14°C these processes took 3.0 min and 27.7 min respectively and at 2°C the times were increased to 9.6 min for brain death to occur and 197.6 min for the cessation of all movement.

Death in ice slurry

Death in ice slurry is a variation on death in air. Fish are added to an ice/water slurry. The water is then drained off leaving the fish mixed in with the ice. This cools the fish, resulting in much slower movements. However, they still struggle and eventually die of anoxia. The time to death is much longer than that in air as the metabolic rate of the fish is lower at lower temperature, so less oxygen is used in a given time (Kestin *et al.* 1991). However, the total activity of the fish from the start of slaughter to the point of death may be the same if the fish starts with the same level of oxygen.

Exsanguination

Large fish need to have the blood removed from their flesh. To carry out this procedure with the minimum work, the gills are either cut with a sharp knife or broken by a pull from the operators' fingers. At its simplest, the fish are not stunned and die from anoxia caused by blood loss. The bleed-out following a cut with a sharp knife may be faster than with a blunter knife, or by breaking, as the blood vessels will tend to occlude following a poor cut or pull, slowing the loss of blood. This will slow the rate of death. During death, the fish can move vigorously which uses up their oxygen reserves more quickly and may help to force the blood out of their body. This can promote exsanguination but it still takes about five minutes for the fish to die (Robb *et al.* 2000).

Evisceration as a slaughter method

The viscera of the fish contain proteases and other enzymes required for digestion and metabolism. After death these enzymes start autolysis and can cause severe damage to the flesh. It is therefore desirable to get rid of the viscera as soon after death as possible. In some fish, evisceration is used as part of the slaughter method. The viscera are removed from the live fish, causing loss of blood. This contributes to the onset of anoxia which also results from the fish being out of water. This procedure is used for a lot of wild-caught fish, including flat fish which can take over 6 hours to die after evisceration. It also forms part of the current procedure for the slaughter of farmed eels in Holland and Germany and is the only part of the procedure which actually causes eventual death. The eels are placed fully conscious in a salt or ammonia bath to deslime them. After approximately ten minutes of intense struggling the fish are removed and rinsed in freshwater. They are then exhausted and can be handled easily for evisceration.

Anaesthesia

As will be seen later in this chapter, stress and muscle activity at slaughter are detrimental to quality – both ethical and flesh. By adding an anaesthetic to the water it is possible to minimise handling and the subsequent stress and muscle activity. In New Zealand and Australia it is possible to add the anaesthetic AQUI-S (AQUI-S Ltd, New Zealand) to the water around loosely crowded fish. After ten to thirty minutes the fish become anaesthetised, having gone gently through the stages of induction of anaesthesia. The fish are then netted from the pen or tank and killed using a percussive blow or a spike (see below).

This method is not more widely used as there are concerns about the addition of chemicals to the fish immediately prior to slaughter. Such a procedure does not give the fish a chance to eliminate the substances from their flesh. However, the concentrations of the chemicals are very low and they are food grade ingredients. Ethically the procedure seems very humane and the flesh quality of the product is high (A.R. Jerrett, pers. comm.; Robb 1998; Robb & Frost 1999).

Carbon dioxide narcosis

Carbon dioxide dissolves in water to form an acid. The blood pH of any fish in the water is consequently lowered and the drop causes disruption of processes in the brain and the eventual death. This process is used commercially to stun or kill fish. Water in a tank is saturated with carbon dioxide gas, resulting in a pH of approximately 5.0 (Anon 1995). This compares with the normal rested pH of the fish of 7.8. When salmon are added to this water their blood takes up the carbon dioxide. Initially this causes the fish to show a large number of aversive reactions. The fish swim repeatedly round the tank, trying to escape. After about two minutes salmon become quiet, but are still conscious (Kestin *et al.* 1995). Marx *et al.* (1997) found that trout took 3.2 min to be anaesthetised and carp took 9.2 min, while eels took 109.7 min. All three species showed strong aversive reactions during the induction of anaesthesia, as described in salmon.

According to guidelines issued for salmon, the fish should be removed from the tank and bled after four minutes (Anon. 1995). Gill cutting will result in the death of the fish. However, the whole procedure, from immersion in the carbon dioxide bath to the point of brain death, takes over 5 min in salmon (Robb, unpublished results), and 4.5 min in rainbow trout (Kestin *et al.* 1995).

Electrical stunning and electrocution

Electricity is widely used for the stunning of red meat animals and poultry. Potentially, it is a very useful tool for the stunning and slaughter of fish as water is a good conductor of electricity. Therefore it should be possible to stun fish in water and without imposing high levels of stress on them. Initial investigations into electrical stunning imply that it has high potential as a slaughter method. Kestin *et al.* (1995) found that electrical stunning rendered fish instantly insensible when applied across the head for 1 s. The duration of insensibility is increased with increasing current and duration of current application. Current applications longer than 30 s can kill the fish (Robb, unpublished observations).

However, in some fish, such as trout, electrical stunning at 50 Hz has resulted in a high proportion of carcass haemorrhages and broken vertebrae (Kestin *et al.* 1995). This has led to lower currents being investigated. It has been found that lower currents can cause less carcass damage, but that up to 15 min is required to 'render the fish insensible'. However, it is unlikely that this method does render the fish insensible. It is more likely that the fish are immobilised by the current stimulating the muscles to tetany. After 15 min the muscles are probably exhausted of energy and the fish are unable to move, thus appearing insensible. This method is probably inhumane, as anyone who has touched a live electric wire will confirm.

More research is required into electrical stunning of fish. It has potential to be a very humane method as it can be instantaneous and the fish can be killed in water with little pre-slaughter handling. However, problems of carcass damage need to be addressed.

Percussive stunning

If an animal is struck hard on the head the brain is accelerated relative to the skull. This causes shearing within the brain and disruption of the neural processes. As the force imparted by the blow increases, the degree of damage to the brain increases. At lower energies the blow renders the animal insensible and at higher energies the animal is killed. This method is commonly used to kill fish. The fish are lifted out of water, restrained and hit on the head with a lightweight and fast moving club (often termed a priest). The process in salmon is completed by cutting the gills of the fish to bleed them out (Anon. 1995). This ensures the death of the fish if the blow was not sufficient to do so. Insensibility is instantaneous (Kestin *et al.* 1995; Robb *et al.* 2000), with the only potential stress to the fish occurring when they are lifted from the water on to the killing table.

Spiking

Physical destruction of the brain causes death. This killing method is used in tuna and salmon (Taniguchi 1977; A. Jerrett pers. comm.). A sharp spike is inserted through the skull of salmon into the brain. The size of the spike is determined by the size of the brain. By moving the spike once it is inserted into the skull the brain is destroyed. This is likely to be rapid provided the application is accurate. However, it is possible to miss the brain as the target is small and the fish struggle during the process (Robb *et al.* 2000).

With tuna, a similar procedure is carried out, but to ensure destruction of the central nervous system a further step is taken. First, a large diameter (approximately 3 cm) hollow pipe is pushed into the skull. Known as coring, this aims to completely destroy the brain. Then a flexible rod approximately 5 mm in diameter is pushed down the backbone. This prevents any further movement by the fish.

Summary of slaughter methods

The slaughter methods currently used fall into two categories, fast and slow. The slow methods result in a lot of movement by the fish and are generally thought aversive. The fast methods, including percussive stunning and spiking, kill the fish immediately and the only movement occurs when the fish are removed from water and handled prior to the blow or spike. The effects of the amount of movement at slaughter are discussed in the following sections.

Muscle activity

Two types of striated muscle are recognised, red and white. Red muscle is used for relatively slow and repetitive movements and white muscle for faster and more sporadic activity. The structure and anatomy of muscle has been described by Ian Johnston in Chapter 3. The major contractile proteins are myosin and actin, together with the associated tropomyosin and troponin. When the muscle is at rest the tropomyosin blocks link between the actin and the myosin myofilaments. However, when the muscle cells are stimulated by motor neurones the tropomyosin-troponin complexes are pulled away from active sites on the actin. The heads of the myosin molecules then bind to these sites. As the myosin heads rotate during binding they expose sites which have ATPase activity, breaking down adenosine triphosphate (ATP) to adenosine diphosphate (ADP) and releasing energy. This energy causes the myosin heads to release from the actin, straighten and reattach to the next actin binding site. Thus the actin myofilament is pulled over the myosin filament using a series of ratchet-like movements. When the muscle stimulation by the motor neurone ceases, further ATP is required to release the myosin heads and allow the tropomyosin-troponin complexes to cover the receptor sites on the actin myofilaments.

The ATP molecules required for the processes are generated by respiration. This uses energy derived from the oxidation of glycogen to synthesise ATP from ADP (Fig. 20.1). Respiration can be aerobic or anaerobic. In white muscle the respiration is predominantly anaerobic, but in red muscle it is predominantly aerobic. The products of aerobic respiration are carbon dioxide and water, in contrast to the lactic acid produced from anaerobic respiration. Under normal conditions the lactic acid is later oxidised when aerobic conditions return. For instance, when an animal has to move quickly white muscle is used, producing

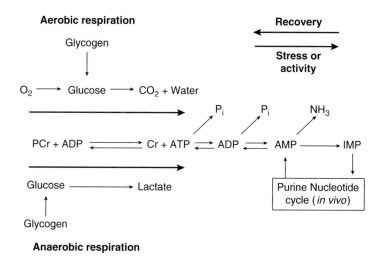

Fig. 20.1   Energy metabolism in fish white muscle during rest (aerobic), stress or activity (anaerobic) and recovery (after Erikson 1997). PCr = phosphocreatine, Cr = creatine, ATP = adenosine triphosphate, ADP = adenosine diphosphate, AMP = adenosine mono-phosphate, IMP = inosine monophosphate, P_i = inorganic phosphate.

lactic acid. When the muscle stops moving oxygen can be transported to that muscle and the lactic acid removed in the blood stream for oxidation in the liver.

During normal swimming at slow speeds, fish predominantly use red muscle for move-ment. However, as the level of activity increases, white muscle is used more, so that at very high levels of activity white muscle will be predominantly used. This results in a large amount of anaerobic respiration and the production of large amounts of lactic acid. If the animal is allowed to recover from the activity the lactic acid will be cleared from the blood and muscle (Milligan 1996). However, if the animal is killed immediately after the activity this will not happen and the pH of the flesh will remain low. This could have effects on the flesh quality of the fish.

Effect of muscle activity on flesh quality

The muscle activity prior to slaughter can affect the quality of the fish in two ways: phy-sically and biochemically. When the fish are swimming vigorously during crowding, they are likely to hit other fish or solid objects. This can cause them to lose scales, become cut or bruised. Some slaughter methods, such as electrical stunning with 50 Hz ac, result in such a degree of muscle activity that vertebrae may be crushed and blood vessels ruptured (Kestin *et al.* 1995), causing significantly more damage than is observed with fish slaughtered by other methods. Obviously such damage degrades the value of the fish. Rough handling also causes physical damage. Excellent pictorial examples of such physical damage are given by Doyle (1995).

For a time after blood circulation stops, the muscle cells continue to respire, first using up oxygen remaining and then respiring anaerobically, producing lactic acid which lowers the pH (this process is described in detail by Jeacocke 1996). This gives a characteristic drop in the muscle pH post-mortem. When the pH reaches a certain level it will interfere with the synthesis of ATP, the energy supply for the muscle cells, causing the cessation of this chemical pathway. When the muscle cells use up their remaining ATP they go into rigor, where the actin and myosin myofilaments are unable to slide over each other. As more individual cells enter rigor the body of the fish becomes stiffer.

If a fish is killed after muscle activity the cells will contain a certain level of lactic acid from their previous anaerobic respiration. This brings them closer to the pH which prevents the synthesis of ATP and so they enter rigor earlier (Fig. 20.2). The more strenuous the muscular activity prior to slaughter the higher the muscle lactic acid concentration (Azam *et al.* 1989; Lowe *et al.* 1993). This lowers muscle pH and so increases the rate of the onset of rigor (Korhonen *et al.* 1990; Robb 1998). The use of spiking to immediately destroy the brain, and hence prevent muscular activity, has been shown to increase the time to the onset of rigor compared to immersion in ice slurry (Boyd *et al.* 1984). However, the ultimate muscle pH post-slaughter for stressed and unstressed fish of the same species is not different, despite the significant differences immediately post-mortem (Korhonen *et al.* 1990; Robb 1998).

Watabe *et al.* (1989) found the rate of the onset of rigor in plaice (*Paralichthys olicaceus*) to also be temperature dependent. The storage temperature of the fish immediately after

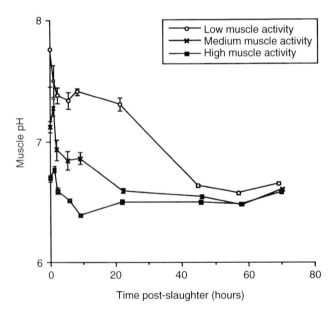

Fig. 20.2   The post-mortem drop in muscle pH following three different levels of muscle activity at slaughter. The low activity group were anaesthetised, the medium group were chased with a net and the high activity group were electrostimulated for two minutes. (After Robb 1998)

slaughter will affect the rate of the various reactions occurring post-mortem. This is caused by the reduction in ATPase activity with the reduction in temperature and by the reduction in the uptake of Ca^{2+} ions from the myofibrils. The increased Ca^{2+} ions trigger the activation of the ATPase in the actin-myosin complex, increasing the rate of ATP breakdown and reducing the time to rigor. At $0°C$ the rate of uptake of Ca^{2+} ions from the myofibrils is reduced compared to that at $10°C$. Therefore, ATPase activity is triggered more readily at the lower temperature, leading to a faster onset of rigor. However, this is not consistent with the results of Mochizuki and Sato (1994) who found almost no difference in the rate of progress of rigor at $0°C$ and $8°C$ in horse mackerel.

The early cessation of the synthesis of ATP and its metabolism (shown in Fig. 20.1) in fish which showed a high level of muscle activity prior to slaughter results in a more rapid build up of the metabolites of this molecule (Korhonen *et al.* 1990; Mochizuki & Sato 1994; Berg *et al.* 1997). This is manifested as a faster increase in the K values of the fish post-mortem (Lowe *et al.* 1993; Erikson *et al.* 1995; Erikson *et al.* 1997), although Sigholt *et al.* (1997) found no effect on K values. The K value is an index of freshness determined by the levels of the high energy phosphates. In contrast, fish which were killed instantly by spiking the brain showed the presence of ATP in the white muscle for a longer time post-mortem, than fish which were killed by immersion in ice slurry or by dying in air (Boyd *et al.* 1984 and Mochizuki & Sato 1994 respectively). This shows that the cessation of the activity of the central nervous system and its control of the muscles results in a longer time to the loss of ATP and the corresponding onset of rigor. Boyd *et al.* (1984) also suggested that this may promote high quality flesh.

The degree of muscle activity prior to slaughter also affects how firm the fish becomes during rigor (Berg *et al.* 1997; Robb 1998). As discussed above, the fish uses increasing amounts of muscle as the level of activity increases. At low levels of activity, very few muscles are used. When the animal is killed, the muscle cells which had been active will go into rigor first. The result is that the fish remains more flexible even when in 'full' rigor. In contrast the fish which showed most muscle activity have a very different rigor profile (Fig. 20.3). As most of the muscle cells were active they went into rigor first and all entered rigor at approximately the same time. This leads to a very stiff fish during rigor.

When the fish is in rigor it should not be handled or this will cause damage to the flesh. However, post-slaughter it is often necessary for fish to be eviscerated as early as possible. With a rapid onset of rigor, this often leads to fish being eviscerated and packed on ice while entering or already in rigor. By reducing pre-slaughter stress and activity the fish will take longer to go into rigor. This will allow more time to process and pack the fish before they enter rigor, reducing the possibility of damage to the flesh.

The 'strength' of the rigor may also have a direct impact on flesh quality. Fish which show a high degree of activity pre-slaughter show a greater degree of 'gaping'. 'Gaping' occurs when the myotomes separate at the myosepta. This causes a large amount of damage to the fillet, looking very unattractive to the consumer. It also makes processing the flesh more difficult, especially for smoked salmon where thin slices along the length of the fish are required. Gaping is a major problem in the salmon industry, especially for smoked salmon, as it spoils the appearance of the fish and makes it harder to slice (A. Dingwall, pers. comm.). Gaping may be a result of fish having a high proportion of their muscle fibres in rigor at the same time, placing a large force on the connective tissue in the myosepta and pulling it apart

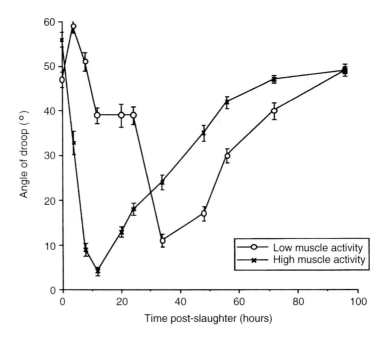

Fig. 20.3   The development of rigor after slaughter as measured by the angle of droop from the horizontal of the posterior half of a fish clamped horizontally. The levels of muscle activity are as for Fig. 20.2. (After Robb 1998)

(Robb 1998). Fletcher *et al.* (1997) found no effect of exercise on gaping at the myo-commata-muscle fibre junction, but from their previous results (Jerrett *et al.* 1996) they doubted the usefulness of their gaping scores as an indicator of propensity of the flesh to gape.

Muscle fibre strength is also directly affected by the degree of muscle activity pre-slaughter. Increasing activity results in greatly decreased strength of the individual fibres (Mochizuki & Sato 1994; Jerrett *et al.* 1996). The whole fillets of stressed fish were also weaker than rested fish (Sigholt *et al.* 1997). These results of increased muscle activity are probably the cause of an increased degree of gaping (Fig. 20.4) appearing after handling post-rigor which was observed by Robb (1998).

Flesh colour is also affected by the rapid drop in muscle pH post-slaughter. With a more rapid decrease in pH being observed in fish which showed more muscle activity prior to slaughter, it has been shown that there are significant changes in the colour of the fillet, in both salmonids (Robb & Frost 1999) and tuna (A. Smart, pers. comm.). Increasing the rate of pH drop resulted in a significantly higher L*, angle of hue and chroma for rainbow trout flesh (Robb 1998) and in higher L* for hybrid striped bass (Eifert *et al.* 1992). This was also reflected in a lower Roche colour card score for the trout (Fig. 20.5).

The cause of the loss of colour is that fish showing a high level of muscle activity at slaughter have a significantly lower soluble muscle protein content (Rustad & Løken 1995; Robb unpubl. obs.). This is very similar to an effect which occurs in pigs that are acutely

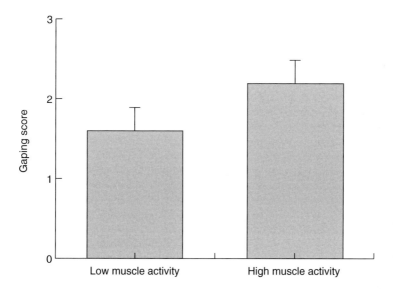

Fig. 20.4 Fillet gaping scores following the muscle activity levels generated as for Fig. 20.2. A score of 0 indicates no gaping and a score of 3 indicates gaping at most myosetpa and within the myomeres. (After Robb 1998)

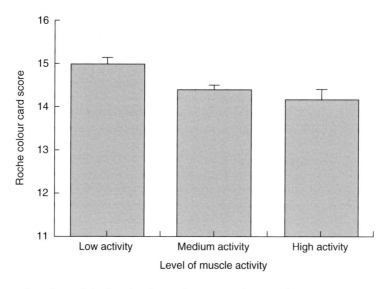

Fig. 20.5 The effect of the levels of muscle activity discussed in Fig. 20.2 on the Roche colour card scores. (After Robb 1998)

stressed at slaughter, which results in paler flesh (Warriss 1996). Known as pale, soft and exudative (PSE) flesh, the muscle of such animals has lower levels of soluble muscle proteins. The rapid change in the muscle pH post-slaughter causes the muscle proteins to denature, becoming insoluble and releasing water (Penny 1977). The change in conformation of the muscle results in a change in the reflection of incident light. The light becomes scattered, causing interference and thus making the flesh appear whiter to an observer. The pigments within the flesh are masked by this interference and the flesh appears more opaque. This finding is very similar to the results of Robb (1998, and unpubl. obs.) and therefore it appears probable that PSE can occur in fish.

Sensory effects

There have been few investigations reporting on the effects of slaughter methods on sensory properties of fish. Sigholt *et al.* (1997) found that in a triangle test, taste panelists differentiated stressed (killed after carbon dioxide stunning) from unstressed control salmon stored at 0.4°C. The stressed fish were also found to have lower texture and lower odour at 3.3°C storage. The reduced texture is detrimental to the properties of salmon and most other fish species. It may also make it harder to process, as processing often requires that the fish be firm to allow slicing. However, Eifert *et al.* (1992) and N.K. Sørensen (pers. comm.) found no effects on sensory parameters following different slaughter treatments.

   A good case could be made for the collection of more sensory information from samples following different slaughter methods. It is obvious that slaughter methods have an effect on the physical properties of the flesh. It is likely that texture will be affected as found by Sigholt *et al.* (1997). Flavour may also be affected in fish during storage. During the metabolism of ATP post-mortem, inosine monophosphate (IMP) is formed, which is a flavour enhancer, and then hypoxanthine (Hx) which is bitter and is regarded as a contributor to off-flavours (Erikson 1997). It is therefore possible that, as stressed fish produce Hx earlier than unstressed fish, they will be regarded as spoiled earlier.

Conclusions

Slaughter methods can have an important effect on the flesh quality of fish. The more a fish struggles or is stressed during the slaughter process, the faster its rate of pH falls post-slaughter. This changes the state of soluble muscle proteins, which become insoluble. The change in state affects the appearance of the fish, making the flesh lighter and more opaque.

   Increased muscle activity or stress also results in weaker muscle fibres and reduced overall fillet strength. This has an effect on the sensory qualities of the flesh, making it softer, which is regarded as detrimental in fish. It may also have an effect on the technological properties of the fish, making it more difficult to process. Increased muscle activity and stress at slaughter also have a direct effect on the appearance of the fish. Physical damage is easily caused, especially in some species, resulting in the downgrading of the whole fish.

   Some slaughter methods have been shown to have a direct effect on the biochemical properties of the fish. Rapid killing, for instance by spiking the brain, has been shown to lead to a slower drop in pH and the longer presence of ATP post-mortem. This may promote flesh quality. In contrast, slow acting slaughter methods, such as death in air or in ice, or processes causing greater muscle activity, such as carbon dioxide stunning, have large effects on the rate of pH fall post-mortem and have detrimental effects on flesh quality. It is therefore preferable to use methods which kill the animals rapidly or in a stress free manner with the least amount of muscle activity. This will help achieve good flesh quality. It will also improve the ethical quality of the fish.

Slaughter methods and pre-slaughter handling (discussed by Ulf Erikson in Chapter 19) are important topics for research. The effects of the two may have masked many real, but more subtle effects, in previous work. For the purposes of research, care should be taken to minimise and control pre-slaughter and slaughter stresses. For instance, by slowly anaesthetising the fish and then killing them by spiking it is possible to minimise variation in muscle pH post-slaughter.

Pre-slaughter handling and slaughter methods are the start of an irreversible process of degradation of the fish flesh. By adopting improved methods, the aquaculture industries can provide high quality flesh to processors and consumers.

References

Anon (1995) *Operating manual for the product certification schemes for Scottish quality farmed salmon and smoked Scottish quality salmon.* Scottish Quality Salmon Ltd. Inverness, Scotland.

Azam, K., Mackie, I.M. & Smith, J. (1989) The effect of slaughter method on the quality of rainbow trout (*Salmo gairdneri*) during storage on ice. *International Journal of Food Science and Technology*, **24**, 69–79.

Berg, T., Erikson, U. & Nordtvedt, T.S. (1997) Rigor mortis assessment of Atlantic salmon (*Salmo salar*) and effects of stress. *Journal of Food Science*, 439–46.

Boyd, N.S., Wilson, N.D., Jerrett, A.R. & Hall, B.I. (1984) Effects of brain destruction on post harvest muscle metabolism in the fish kahawai (*Arripis trutta*). *Journal of Food Science*, **49**, 177–9.

Doyle, J.P. (1995) *Care and Handling of Salmon: The Key to Quality*. Marine Advisory Bulletin No. 45, 66pp. University of Alaska Fairbanks, USA.

Eifert, J.D., Hackney, C.R., Libey, G.S. & Flick, G.J. Jr (1992) Aquacultured hybrid striped bass fillet quality resulting from post-harvest cooling or CO_2 treatments. *Journal of Food Science*, **57**, 1099–1102.

Erikson, U. (1997) *Muscle quality of Atlantic salmon (Salmo salar) as affected by handling stress.* Doctoral thesis, Norwegian University of Science and Technology, Trondheim, Norway.

Erikson, U., Beyer, A.R., Sigholt, T. & Jørgensen, L. (1995) Effect of handling stress on the phosphate metabolites and k-value of Atlantic salmon. In: *Quality in Aquaculture*, Special Publication 23, European Aquaculture Society, Ghent, Belgium.

Erikson, U., Beyer, A.R. & Sigholt, T. (1997) Muscle high-energy phosphates and stress affect *K*-values during ice storage of Atlantic salmon (*Salmo salar*). *Journal of Food Science*, **62**, 43–7.

Fletcher, G.C., Hallett, I.C., Jerrett, A.R. & Holland, A.J. (1997) Changes in the fine structure of the myocommata-muscle fibre junction related to gaping in rested and exercised muscle from king salmon (*Oncorhynchus tshawytscha*). *Lebensmittel Wissenschaft und Technologie*, **30**, 246–52.

Jeacocke, R.E. (1996) The control of post-mortem metabolism and the onset of *rigor mortis*. In: *Recent Advances in the Chemistry of Meat* (ed. A.J. Bailey) Special Publication 47, pp.41–57. The Royal Society of Chemistry, London, England.

Jerrett, A.J., Stevens, J. & Holland, A.J. (1996) Tensile properties of white muscle in rested and exhausted king salmon (*Oncorhynchus tshawytscha*). *Journal of Food Science*, **61**, 527–32.

Kestin, S.C., Wotton, S.B. & Gregory, N.G. (1991) Effect of slaughter by removal from water on visual evoked activity in the brain and reflex movement of rainbow trout (*Oncorhynchus mykiss*). *Veterinary Record*, **128**, 443–6.

Kestin, S., Wotton, S. & Adams, S. (1995) The effect of CO_2, concussion or electrical stunning of rainbow trout (*Oncorhynchus mykiss*) on fish welfare. In: *Quality in Aquaculture*, Special Publication 23, pp.380–81. European Aquaculture Society, Ghent, Belgium.

Korhonen, R.W., Lanier, T.C. & Giesbrecht, F. (1990) An evaluation for simple methods for following rigor development in fish. *Journal of Food Science*, **55**, 346–8, 368.

Lowe, T.E., Ryder, J.M., Carragher, J.F. & Wells, R.M.G. (1993) Flesh quality in snapper, *Pagrus auratus*, affected by capture stress. *Journal of Food Science*, **58**, 770–73, 796.

Marx, H., Brunner, B., Weinzierl, W., Hoffmann, R. & Stolle, A. (1997) Methods of stunning freshwater fish: impact on meat quality and aspects of animal welfare. *Zeitschrift für Lebensmittel und Untersuschring Forsch A*, **204**, 282–6.

Milligan, C.L. (1996) Metabolic recovery from exhaustive exercise in rainbow trout. *Comparative Biochemistry and Physiology*, **113A**, 51–60.

Mochizuki, S. & Sato, A. (1994) Effects of various killing procedures and storage temperatures on post-mortem changes in the muscle of horse mackerel. *Nippon Suisan Gakkaishi*, **60**, 125–30.

Penny, I.F. (1977) The effect of temperature on the drip, denaturation and extracellular space of pork *Longissimus dorsi* muscle. *Journal of the Science of Food and Agriculture*, **28**, 329–38.

Robb, D.H.F. (1998) *Some factors affecting the flesh quality of salmonids: pigmentation, composition and eating quality.* Doctoral thesis, University of Bristol, UK.

Robb, D.H.F. & Frost, S. (1999) Welfare and quality. What is the relationship? *Presentation at Innovations for Seafood '99*, Surfer's Paradise, Queensland, Australia, 21–23 April 1999.

Robb, D.H.F., Wotton, S.B., McKinstry, J., Sørensen, N-K. and Kestin, S.C. (2000) Commercial slaughter methods used on Atlantic salmon: determination of the onset of brain failure by electroencephalography. *Veterinary Record* (in press).

Rustad, T. & Løken, J.-T. (1995) Effect of handling on the muscle chemistry and quality of salmon. In: *Quality in Aquaculture*, Special Publication 23, pp.385–6. European Aquaculture Society, Ghent, Belgium.

Sigholt, T., Erikson, U., Rustad, T., Johansen, S., Nordvedt, T. & Seland, A. (1997) Handling stress and storage temperature affect meat quality of farm-raised Atlantic salmon (*Salmo salar*). *Journal of Food Science*, **62**, 898–905.

Taniguchi, H. (1977) How to effectively kill tunas in order to maintain quality and higher prices. *Suisan Sekai*, **9**, 52. Quoted by Boyd, N.S., Wilson, N.D., Jerrett, A.R., & Hall, B.I. (1984). Effects of brain destruction on post harvest muscle metabolism in the fish kahawai (*Arripis trutta*). *Journal of Food Science*, **49**, 177–9.

Warriss, P.D. (1996) Instrumental Measurement of Colour. In: *Meat Quality and Meat Packaging.* (eds S.A. Taylor, A. Raimundo, M. Severini & F.J.M. Smulders), pp.221–31. ECCEAMST, Utrecht, The Netherlands.

Watabe, S., Ushio, H., Iwamoto, M., Yamanaka, H. & Hashimoto, K. (1989) Temperature-dependency of rigor-mortis of fish muscle: myofibrillar Mg^{2+}-ATPase activity and Ca^{2+} uptake by sarcoplasmic reticulum. *Journal of Food Science*, **54**, 1107–10, 1115.

Chapter 21

Effect of the Commercial and Experimental Slaughter of Eels (*Anguilla anguilla* L.) on Quality and Welfare

*J.W. van de Vis[1], J. Oehlenschläger[2],
H. Kuhlmann[3], W. Münkner[2], D.H.F. Robb[4] and
A.A.M. Schelvis-Smit[1]*

[1] *Netherlands Institute for Fisheries Research (RIVO), P.O. Box 68, 1970 AB IJmuiden,
The Netherlands*
[2] *Bundesforschungsanstalt für Fischerei, Institut für Biochemie und Technologie,
Palmaille 9, D-22767 Hamburg, Germany*
[3] *Bundesforschungsanstalt für Fischerei, Institut für Fischereiökologie, Aussenstelle
Ahrensburg, Wulffsdorfer Weg 204, D-22926, Hamburg, Germany*
[4] *Department of Clinical Veterinary Science, University of Bristol, Langford, Bristol,
BS40 5DU, UK.*

Introduction

Harvest procedures, quality and welfare

In fish the welfare ante-mortem and the quality of the flesh post-mortem can be adversely affected by farming and harvest conditions. Farming conditions comprise, amongst others, water quality and stocking density. Harvest consists of crowding, catching, transport, lairage and slaughter, and is one of the most intense stressors in fish farming (Thomas *et al.* 1999). There are similarities in anatomy, (neuro) physiology and behaviour between fish, mammals and birds (Kestin 1994; FAWC 1996; Wendelaar Bonga 1997; Wiepkema 1997). Therefore, it is likely that the welfare of fish can potentially be similarly adversely affected by husbandry and harvesting conditions.

In red and white meat animals handling and slaughter procedures may have profound effects on the course of chemical changes post-mortem and consequently on the quality of the fresh and processed meat. There is some evidence that in fish a similar relationship exists between harvest procedures and aspects of quality such as water holding capacity, texture and keeping quality (Azam *et al.* 1989; Iwamoto *et al.* 1990; Proctor *et al.* 1992a,b; Lowe *et al.* 1993; Kals *et al.* 1995; Templeton 1996; Marx *et al.* 1997; Sigholt *et al.* 1997; Thomas *et al.* 1999). However, not all researchers have observed effects on sensory parameters attributable to harvest methods used (see Chapter 20 by Dave Robb). Nevertheless, it is known that both handling and slaughter methods used may affect post-mortem biochemical changes (discussed by Ulf Erikson in Chapter 19). It is therefore possible that any potential improvement of product quality from applying an optimised slaughter method may be masked by effects of the handling method used.

Humane slaughter

Humane slaughter is required by the EU Council Directive 93/119EC (1993) for mammals and poultry. According to this Directive, horses, ruminants, pigs, rabbits and poultry brought into abattoirs for slaughter shall be (a) moved and if necessary lairaged, (b) restrained and (c) stunned before slaughter. Animals must be restrained in an appropriate manner, so as to spare them any avoidable pain, suffering, agitation, injury or contusions. Slaughter is defined as 'Causing the death of an animal by bleeding' (EU Council Directive 1993). The term slaughter is also used in practice to designate the process, which is applied to turn a living animal into food. Within the scope of this study the term slaughter is used as a description of this process.

In the case of fish there is no EU legislation, but for red and white meat animals it is generally stated that unconsciousness and insensitivity should be induced as soon as possible without a detrimental effect on the welfare of the animal and meat quality (Blackmore & Delany 1988). This approach can be used as a basis to propose a general term of reference and criteria for the evaluation and development of slaughter methods for fish. The criteria are met when stunning induces instantaneous unconsciousness and insensitivity which lasts until death or, when the achievement of instantaneous induction is not possible, the animal is rendered unconscious and insensitive without avoidable stress, pain or suffering.

Evaluation of welfare during slaughter

Slaughter methods can be evaluated by assessment of stress, the state of unconsciousness and insensitivity and trauma. Trauma comprises damage (e.g. haemorrhages) of the skin, muscle tissue, joints and bones. Handling stress may occur prior to the induction of unconsciousness and insensitivity. The term stress is used when control systems in an animal are overtaxed and there is likely to be a reduction of biological fitness (Broom 1988; Broom & Johnson 1993). Handling stress can be assessed by analysis of levels of cortisol in the blood and analysis of the white muscle tissue with respect to nucleotides, creatine phosphate and glycogen (Erikson 1997).

Stress may cause changes in behaviour. In many cases the behaviour can be adaptive (e.g. avoidance) and thereby increases the probability of survival. Adaptive behaviour mitigates the exposure to a stressor. However, if behavioural mitigation is not possible during pre-slaughter handling, changes in behaviour may reflect deleterious changes in how the fish senses and responds to its environment (Schreck *et al.* 1997). From the observation of the behaviour of fish at slaughter, indications may be obtained about whether the fish is conscious or not. The observation of behaviour can be carried out by assessment of self-initiated behaviour and induced responses.

The state of unconsciousness and insensitivity in fish can be assessed by measuring brain function in combination with the observation of behaviour. Brain function can be measured by recording the electroencephalogram (EEG) and evoked responses in the EEG.

Measurement of visual evoked responses in trout has been developed by Kestin *et al.*, (1991). A visual evoked response is the response in the brain to flashes of light directed towards the eyes. The absence of an average visual evoked response (VER) indicates that the fish must be unconscious and insensitive (Kestin *et al.* 1991).

Current slaughter of eels in the Netherlands and Germany

The slaughter process of eels (*Anguilla anguilla* L.) in the Netherlands starts with a so-called salt bath. Salt is put in a container with live eels but without water for desliming. Water is added after some time. The process of desliming lasts for about 20 min. Subsequently, the eels are washed and gutted to prepare them for brining and hot smoking. Verheijen and Flight (1997) concluded from observations of behaviour at the desliming process that the eels were probably not rendered instantaneously unconscious and insensitive. It is foreseen that legislation on the protection of eels at slaughter will come into force in the near future in the Netherlands. In April 1999 in Germany legislation on the protection of animals at slaughter came into force. This prohibits the traditionally used method in the fish industry, which consists of stunning/killing of eels in aqueous ammonia solutions or in salt. Since 1999 the only methods permitted for stunning/killing of eels are electrical stunning or a blow on the head.

   The present status of slaughter of eels in the Netherlands and the legislative requirements in Germany make clear that research is needed to improve slaughter procedures with respect to welfare and quality. In the present study various stunning/killing methods were investigated with respect to their effects on welfare. A method which may be humane, i.e. in accordance with the proposed general term of reference, was selected on the basis of the results. The selected method was then also studied with respect to quality. The Dutch commercial method was performed as the control method.

Material and methods

Fish

Prior to harvest the eels (*Anguilla anguilla* L.) were fasted in water of about 13°C for at least 7 days. The eels (weight 100-200 g) were in a silvering stage.

Stunning/killing methods

Chemical anaesthesia using AQUI-S^TM, BHA, BHT, propyl gallate and gallic acid
The eels (n=10) were placed in 50 l tap water at 18-19°C. Anaesthesia was performed by the addition of one of the following chemicals: AQUI-S^TM (Fish Transport System Ltd, Lower Hutt, New Zealand), BHT (butylated methyl phenol), BHA (butylated hydroxynanisole), propyl gallate and gallic acid to a final concentration of 0.003% (w/v). The eels were left for 45 min in the tank. AQUI-S^TM is accepted as a food-grade chemical for anaesthesia of fish in New Zealand. The active substance is eugenol. Clove oil consists of 80% eugenol (2-methoxy-4-propenylphenol) (Keene *et al.* 1998). BHT, BHA, propyl gallate and gallic acid are anti-oxidants, which are listed in an appendix of the Codex Alimentarius (CX-STAN 019-1981, 1999).

Mechanical stunning

Two methods of mechanical stunning were investigated: percussive stunning and the use of a pinbolt/hollow punch. Percussive stunning was carried out using a pneumatic stunning gun (Hewitt 1999). The head of the fish was hit at the point where the imaginary diagonals between the eyes and the opening of the opercula cross. The stunning gun was operated at 6 bar. During stunning the fish were restrained manually. The pneumatic stunning gun was also used for pin bolting (hollow punch). The head of the fish was hit at the point where the imaginary diagonals between the eyes and the opening of the opercula cross. The modification consisted of replacement of the piston by a hollow pipe 8 mm in diameter. When the gun is used, the hollow pipe protrudes about 5 cm.

Electrical stunning

Two methods of electrical stunning were investigated: head-only and whole-body. The electrical stunning equipment consisted of a variable transformer capable of delivering 50 Hz AC from 0 to 350 V. The output from the transformer was controlled by a timer and switching box. Two digital Fluke multimeters were used to measure voltage and stunning current.

For head-only stunning a pair of stunning tongs was fabricated so that the electrodes could be applied on either side of the head of the eel, when it was removed from the water and while restrained manually. This head-only application directs the current flow to the head and therefore maximises the flow of current through the brain. The electrodes consisted of a soft foam rubber material (2 cm in diameter) which could deform around the head, ensuring maximum electrical contact. The foam rubber was soaked in a saturated aqueous solution of NaCl prior to each application, in order to facilitate electrical conductivity. A stun duration of 1s was used.

For whole-body electrical stunning on a large scale the eels (20 kg) were placed in a box fitted with multiple stick electrodes and covered with tap water. For head-only stunning a short duration application (1 s) was used to determine whether stunning was immediate as required by the proposed legislation. For whole-body stunning a high current stun for 3 s was followed immediately by a long duration low current stun.

Commercial salt bath method

Live eels were placed in a dry tank and NaCl, or a combination of NaCl and aqueous Na_2CO_3, added for desliming. On average the duration of the desliming process was 20 min. The method is designated as the salt bath. After desliming the eels were gutted and stored in flake ice.

Desliming of eels after application of a research stunning/killing method

When the eels were stunned using a research method the fish could not be deslimed using the salt bath. Therefore, the eels were placed in a stainless steel tumbler containing a saturated solution of $Ca(OH)_2$ in water, the ratio of solution to eels being $3:100$ (w/w). The tumbler was run for 10 min at 10 rpm.

Assessment of welfare

Behavioural measurements

The behaviour of the fish prior to killing by gutting was observed. The observations can be grouped into two: self-initiated behaviour and stimuli responses. The descriptions of self-initiated behaviour and stimuli to responses were obtained by performing observations with conscious eels, eels anaesthetised for 10 min with 2 ml 2-phenoxyethanol per litre and decerebrated eels. From these observations, descriptions for self-initiated behaviours and responses to stimuli of conscious eels were obtained.

Self-initiated behaviour
Three self-initiated behaviours were recognised. First was the ability of the eel to maintain equilibrium such that, if the fish was turned upside down (if it was possible to catch it) it would right itself and adopt an S-shaped position when it was lying on the bottom of a tank. Second was that the eel would swim without any interference from the observer or with a slight disturbance (but not being touched) and was able to avoid a dip net. Lastly, the eel was breathing such that opercular movements pumped water over the gills.

Responses to stimuli
Four responses to stimuli were recognised. First was that the eel responded to a pinch on the tail by trying to escape and this response was more than a simple muscle contraction. Second, the eel responded to being pricked with a needle on the tail by trying to escape and this response was more than a simple muscle contraction. Thirdly, the eel responded to a 6 V electric shock applied to the mouth and this was more than a simple muscle contraction. Lastly, the vestibulo-ocular response was observed. The vestibulo-ocular response is the movement of the eyes compensating for changes in body postures in the longitudinal axis. The response is absent in the anaesthetised or decerebrated eel.

Interpretation of observed parameters
The eel was classified as conscious and sensitive when one or more of the following observations could be made: normal swimming, normal swimming but appearing to be sluggish, the fish was lying in the so-called S-shape on the bottom of the tank, disabled or moribund fish responded to catching and handling, responding to being pricked with a sharp needle, responding to stimulation with a 6 V shock. The eel was classified as unconscious and insensitive when the fish lay extended on its side. This body posture was accompanied by an absence of the vestibulo-ocular response. Breathing was absent or its rate and amplitude decreased when compared to a conscious eel.

Measurement of brain function
The brain function of eels before, during and after the application of a killing method was observed using a method adapted from Kestin *et al.* (1991). Eels weighing 700 to 850 g were anaesthetised with 2-phenoxyethanol. The skull was exposed and four holes were drilled through the cranium. Two silver recording electrodes were implanted on the left side of the brain, with the negative on the caudal margin of the optic lobe and the positive on the anterior margin of the optic lobe. Two silver ground electrodes were positioned at arbitrary points on the contralateral hemisphere, G1 centrally on the optic lobe and G2 (used to suppress alternating current (AC) artefact (mainly 50 Hz mains noise) just anterior to the rostral margin of the optic lobe. After at least four hours of recovery the EEG was recorded, together with the evoked response from photic stimulation. On line averaged evoked responses using 250 ms sweep duration were displayed on a digital oscilloscope so that recording could

continue until after the VERs were lost. The time to loss of VERs for each fish was determined (Kestin *et al.* 1991). The VERs of each fish were recorded prior to the application of the killing method and for ten minutes after each method was applied. During electrical stunning the wires from the fish were unplugged from the monitoring apparatus to prevent damage to this. At least five fish were used for assessment of a stunning/killing method.

Evaluation of stunning/killing with respect to quality

After stunning/killing the eels were gutted and stored in flake ice and samples taken at regular time intervals for assessment of post-mortem changes during storage. Separate eels were used for the evaluation of welfare and quality. Post-mortem changes in pH were assessed in homogenised fillets, as described by Oehlenschläger (1991). For each measurement five fish were analysed individually. The onset and resolution of *rigor mortis* in ten fish was assessed as described by Sørensen *et al.* (1997) and the deflection index (DI), calculated as described by Bito *et al.* (1983). Changes in the impedance of ten gutted eels during storage in flake ice were measured at regular intervals with the RT freshmeter (Jason & Richards 1975; Oehlenschläger & Nesvadba 1997). The readings were taken by drawing the electrodes across the fish from the dorsal to the abdominal side.

Fish were assessed using the Quality Index Method (QIM) as described in Bremner (1985). The scheme which was used by the panellists (n = 5) for assessment of whole eel (n = 10) is shown in Table 21.1.

Table 21.1 QIM scheme for deslimed gutted eels.

Quality parameter	Demerit points			
	0	1	2	3
Skin appearance				
dorsal	shining colour, very bright, clear contrast	less contrast	dull	dull, gritty, discoloration
abdominal and lower jaw	shining, white greyish, slime absent	slightly yellowish	yellowish, slight shrinkage, slimy	gritty, discoloration
Muscle				
Appearance at incision	glassy, brilliant	slightly dull	greyish, waxy	yellowish, slimy
Texture at dorsal side	firm, elastic	slightly soft	soft, not elastic	
Flesh odour	fresh water fish	neutral, milky	musty	spoilt

Sensory analysis of the fillet

At regular time intervals five fishes were filleted and the edible part was mixed and divided into five portions. The taste of the cooked fillet was assessed by a sensory panel, resulting in an average taste quality score of the panellists ($n = 5$) from 2 (bad) to 9 (excellent quality). The scheme which was used is shown in Table 21.2.

Table 21.2   Sensory scheme for cooked eel filet.

Overall quality	Taste description	Score
Good	creamy, fresh	8–9
	creamy slightly earthy	7–8
	earthy, milky	6–7
Acceptable	musty	5–6
	musty, bitter	4–5
Not acceptable	spoilt	lower than 4

Meaning of scores: 9 = excellent; 8 = very good; 7 = good; 6 = satisfactory; 5 = mediate; 4 = acceptable; 3 = rejectable; 2 = bad.

Data processing

Single factor ANOVA was used for processing the results obtained from analysis of gutted eel by using the QIM. The results obtained by using the other methods of analysis were treated prior to applying the two factor ANOVA. The treatment consisted of neglecting small differences in sampling times between similar samples from different batches of eels stunned/killed by different methods.

Results and discussion

Welfare

Chemical anaesthesia

AQUI-S™ and BHA could be used in principle for chemical anaesthesia (see Table 21.3). None of these substances rendered the eels instantaneously unconscious and insensitive as they showed co-ordinated swimming for at least 5 min. However, after 45 min the eels were not able to maintain their normal position in the water (the S-shape on the bottom of the tank) or to show co-ordinated swimming. The eels were lying extended on the bottom of the tank, the vestibulo-ocular response was lost and the frequency and amplitude of the opercular movements were reduced. These observations may indicate that the eels can be rendered unconscious and insensitive by the use of AQUI-S™ or BHA.

The eels recovered completely (they showed co-ordinated swimming or the ability to adopt an S-shape on the bottom of the tank) when they were placed in fresh aerated tap water. It is obvious that these chemicals have to be used in combination with a killing

Table 21.3 Effect of chemical anaesthesia on eels (each figure shows the number of animals showing the response).

Observations ($n = 10$)	AQUI-S^{TM}	BHA	BHT	Gallic acid	Propyl gallate
After 5 minutes					
swimming	10	10	10	10	10
no balance*	0	0	0	0	0
breathing	10	10	10	10	10
After 45 minutes					
swimming	0	0	10	10	10
no balance*	10	10	0	0	0
breathing	5–6	7–8	10	10	10
conditions	18°C	18°C	19°C	19°C	19°C
in water	83% O_2	77% O_2	70% O_2	56% O_2	56% O_2
	pH 7.27	pH 7.37	pH 7.38	pH 7.15	pH 7.23

* no balance = not able to maintain a normal position in the water.

method for humane slaughter, as in the next step of the slaughter process the eels have to be deslimed, using an aqueous salt solution. Kuhlmann *et al.* (2000) observed that BHA might stun eels quicker than AQUI-S^{TM}. BHT, gallic acid and propyl gallate did not stun the eels as they still showed co-ordinated swimming after 45 min (Table 21.3). Moreover, the use of BHT might be aversive as the eels tried to escape from the tank.

BHA, BHT, gallic acid and propyl gallate were selected by Kuhlmann *et al.* (2000), as they are chemical analogues of eugenol which is the active substance in AQUI-S^{TM}. Moreover, the chemical substances are antioxidants, which may be added to food. Chemical anaesthesia by using BHA did not result in loss of VERs in eels (Table 21.4). However, the observed behaviour may indicate that unconsciousness and insensitivity was induced.

Table 21.4 VER measurements in eels stunned using various methods.

Stunning method	Animals implanted	Successful recordings	Comments	VERs recovered (s)
BHA	15	14	No loss of VERs in any animal	
Head-only stun, 1.3 A, 225 V	5	5	VERs lost and then recovered	85 ± 4

Electrical stunning
Head-only electrical stunning (1.3 A, 225 V a.c.) resulted in loss of VERs in eels. All eels recovered (i.e. VERs could be detected) after 85 ± 4 s (Table 21.4). During stunning the eels

were in a tonic phase, as a substantial muscle contraction was visible. After stunning, the eels were placed in water when a clonic phase was observed. During the latter phase the eels showed vigorous movements without the ability to maintain a normal position in the water or co-ordinated swimming.

Observations of the immediate effects of whole-body stunning were difficult because of the methods used. But whole-body electrical stunning (36 A, 330 V a.c.) seemed to induce a clonic phase in six out of ten eels for approximately 60 s. The four other eels were not able to maintain a normal position in the water, and breathing and the vestibulo-ocular response were absent. These results suggest that unconsciousness and insensitivity were induced instantaneously. When whole-body electrical stunning was applied using 36 A, 330 V a.c., and was immediately followed by an application of 8.3 A, 45 V a.c., for 15 min to attempt to kill them, it was observed that 80% of the eels did not recover in 90 min. The remaining 20% of the eels might have recovered and would be classified as conscious and sensitive. The results show that electrical stunning is reversible but that most eels can be stunned and killed if the current is maintained for a prolonged period after the application.

Mechanical stunning

Percussive stunning did not render the eels unconscious and insensitive, as shown in Fig. 21.1. At t = 0 min, 33 ± 13%, and after 20 min 60 ± 20%, of the eels were still conscious. It was not possible to use higher pressures than 6 bar in the pneumatic stunning gun as this resulted in crushing the skull. Damage to the skull is not acceptable to the industry. There was a significant difference between percussive stunning and the salt bath. Statistical analysis also showed that there was interaction between the stunning/killing method and the period of time which elapsed. This was expected, as for both methods the number of conscious fish increased as time elapsed.

VERs could be recorded in one eel only and this showed that for 10 min during treatment with the salt bath (the control method) VERs were not lost. Observation of behaviour revealed that 15% of the eels showed co-ordinated swimming when they were placed in fresh water after the salt bath. This further suggests that the eels were not rendered unconscious and insensitive instantaneously by the salt bath.

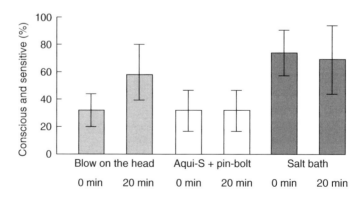

Fig. 21.1 Effects of percussive stunning, AQUI-S^{TM} in combination with the hollow punch and salt bath on indices of sensitivity in eels.

Stunning using the hollow punch may render the eels unconscious and insensitive, as no self-initiated behaviour, responses to stimuli or breathing could be detected (Fig. 21.1). The approach which was chosen in the latter trial (chemical anaesthesia with AQUI-S^{TM} followed by mechanical stunning), may be suitable for commercial slaughter as some chemicals (i.e. gases) are allowed in the EU (EU Council Directive 1993). Therefore, the eels which were stunned and killed by using AQUI-S^{TM} in combination with the hollow punch, were evaluated with respect to quality. The results were compared with those obtained with eels which were slaughtered by the commercial method.

Quality

pH, rigor mortis *and impedance*
Stunning/killing by using AQUI-S^{TM} in combination with the hollow punch resulted in a significantly higher initial pH of the fish flesh, compared to eels which were killed by the commercial method (Fig. 21.2). These results may indicate that the commercial method caused more stress in eels, compared to AQUI-S^{TM} in combination with the hollow punch. However, the pH values of the fish stunned/killed by the two methods differed after resolution of rigor. This is contrary to the observations made by Marx *et al.* (1997). In the case of the commercial method a decrease in pH followed by an increase, which is the typical pattern for changes in the course of pH post-mortem, was not observed (Fig. 21.2). It is possible that this was caused by a rapid decrease in pH during the application of the method but no measurements were made during this period.

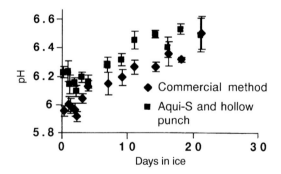

Fig. 21.2 Effects of slaughter by AQUI-S^{TM} in combination with the hollow punch and commercial slaughter on the pH in eels.

The course of *rigor mortis* in eels, which were slaughtered by the commercial method and AQUI-S^{TM} in combination with the hollow punch, is shown in Fig. 21.3. No significant differences could be detected in the course of the deflection index for both methods.

The course of changes in the impedance of the eels during storage in flake ice is shown in Fig. 21.4. There was a significant difference between impedance for the two slaughter

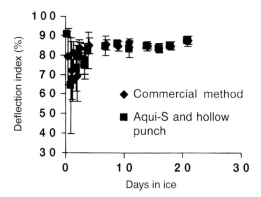

Fig. 21.3   Effects of slaughter by AQUI-S™ in combination with the hollow punch and commercial slaughter on *rigor mortis* in eels.

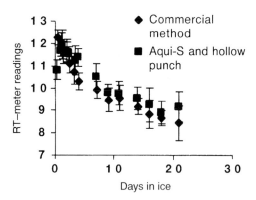

Fig. 21.4   Effects of slaughter by AQUI-S™ in combination with the hollow punch and commercial slaughter on the impedance of eels.

methods, probably because the application of the commercial method results in substantially more movements of the eels, leading to more damage to the muscle tissue, than the use of AQUI-S™ in combination with the hollow punch.

Quality Index Method
The eels which were slaughtered by the commercial method and AQUI-S™ in combination with the hollow punch were evaluated by using QIM. The results are shown in Fig. 21.5. A significant difference between the methods was found with the overall QIM score for the commercial method being higher than that for AQUI-S™ in combination with the hollow punch. The overall sensory quality of eels which were slaughtered by the latter method was higher compared to the commercial method.

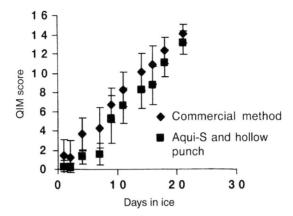

Fig. 21.5 Effects of slaughter by AQUI-S^{TM} in combination with the hollow punch and commercial slaughter on QIM scores of eels.

Taste of fillets
The taste of the fillets obtained from the eels slaughtered by the two methods was assessed (see Fig. 21.6). The overall taste was scored significantly higher for eels slaughtered by using AQUI-S^{TM} in combination with the hollow punch, compared to the commercial method. The fillets obtained by using the first method were rejected after 21 days of storage, whereas the fillets obtained by using the commercial method were rejected after 18 days. This effect was revealed by a significant interaction between the slaughter method used and the storage period in ice. The panel members observed that the use of AQUI-S^{TM} resulted in the presence of an odour and taste which was caused by the absorption of this substance by the fish flesh.

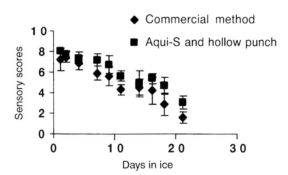

Fig. 21.6 Effects of slaughter by AQUI-S^{TM} in combination with the hollow punch and commercial slaughter on the taste of eel fillets.

Conclusion

In conclusion it can be stated that a more humane slaughter of eels than the current commercial method may lead to improvement of the flesh quality.

Acknowledgements

The authors would like to thank the EU for funding (FAIR CT97-3127) this study.

K.K. Brünner, J.W.M. Gouda, J.W. van der Heul, A. Kamstra, C. Lammers and A. Schelvis from the Netherlands Institute for Fisheries Research, IJmuiden, The Netherlands, and H.-J. Knaack and W. Westrup from Bundesforschungsanstalt für Fischerei, Institut für Biochemie und Technologie, are thanked for their skilful assistance, as are S.B. Wotton and J.L. McKinstry, Department of Clinical Veterinary Science, University of Bristol, UK, for their work on the brain measurements in eel and E. Lambooij from the Institute for Animal Science and Health (ID-Lelystad), Lelystad, the Netherlands, for critically reviewing the manuscript.

The authors also thank P. van Banning (Netherlands Institute for Fisheries Research, IJmuiden, the Netherlands) for making arrangements with the Dutch Animal Experiments Commission, which resulted in permission to perform the experiments with live eels, and the Dutch eel farmer and processor Royaal BV, Helmond, for providing facilities, skilful assistance and fruitful discussions.

References

Azam, K., Mackie, I.M. & Smith, J. (1989) The effect of slaughter methods on the quality of rainbow trout (*Salmo gairdneri*) during storage on ice. *International Journal of Food Science and Technology*, 24, 69–79.

Bito, M., Yamada, K., Mikumo, Y. & Amano, K. (1983) Studies on rigor mortis in fish–I. Difference in mode of rigor among some varieties of fish by modified Cutting's method. *Bulletin Tokai Regional Fisheries Research Laboratory*, 109, 89–96.

Blackmore, D.K. & Delany, M.W. (1988) Slaughter of stock. Publ. No. 118. Vet. Con. Ed. Massey University, Palmerston North, New Zealand.

Bremner, H.A. (1985) A convenient easy to use system for estimating the quality of chilled seafood. *Fish Processing Bulletin*, 7, 59–70.

Broom, D.M. (1988) The scientific assessment of animal welfare. *Applied Animal Behaviour Science*, 20, 5–19.

Broom, D.M. & Johnson, K.G. (1993) *Stress and Animal Welfare*. Chapman and Hall, London.

CX-STAN 019-1981 (1999) *Codex standard for edible fats and oils not covered by individual standards*. Codex Alimentarius Commission, FAO, Rome, Italy.

Erikson, U. (1997) *Muscle quality of Atlantic salmon* (Salmo salar) *as affected by handling stress*. Doctoral thesis, Norwegian University of Science and Technology, Trondheim, Norway.

EU Council Directive (1993) Council Directive 93/119/EC on the protection of animals at the time of slaughter and killing. *Official Journal of the European Communities*, 22 December 1993.

FAWC (1996) *Report on the welfare of farmed fish*. PB 2765, MAFF, UK.

Hewitt, L. (1999) A novel stunning system for the slaughter of poultry. In: *Poultry Meat Science* (eds R.I. Richardson & G.C. Mead). CABI Publishing, Wallingford, Oxon, UK.

Iwamoto, M., Yamanaka, H., Watabe, S. & Hashimoto, K. (1990) Studies on prolongation of the pre rigor period of fish. IV. Comparison of *rigor mortis* progress between wild and cultured plaice. *Bulletin of the Japanese Society of Scientific Fisheries*, **56**, 101–104.

Jason, A.C. & Richards, J.C.S. (1975) The development of an electronic fish freshness meter. *Journal of Physics (E) Scientific Instrumentation*, **8**, 826–30.

Kals, J., Kamstra, A. & Van de Vis, J.W. (1995) Effects of slaughter and processing on quality of African catfish (*Clarias gariepinus*) (in Dutch). *Aquacultuur*, **4**, 7–15.

Keene, J.L., Noakes, D.L.G., Moccia, R.D. & Soto C.G. (1998) The efficacy of clove oil as an anaesthetic for rainbow trout, *Oncorhynchus mykiss* (Walbaum). *Aquaculture Research*, **29**, 89–101.

Kestin, S.C. (1994) *Pain and stress in fish*. Royal Society for the Prevention of Cruelty to Animals, Causeway, Horsham, West Sussex, UK.

Kestin, S.C., Wotton, S.B. & Gregory, N.G. (1991) Effect of slaughter by removal from water on visual evoked activity in the brain and reflex movement of rainbow trout (*Oncorhynchys mykiss*). *The Veterinary Record*, **128**, 443–6.

Kuhlmann, H., Münkner, W., Van de Vis, J.W., Oehlenschläger, J. & Koch, M. (2000) Untersuchungen zur anästhesierenden Wirkung von AQUI-S und chemisch verwandten Verbindungen beim Aal (*Anguilla anguilla*). *Archiv für Lebensmittelhygiene* (submitted).

Lowe, T.E., Ryder, J.M., Carragher, J.F. & Wells, R.M.G. (1993) Flesh quality in snapper (*Pagus auratus*) affected by capture stress. *Journal of Food Science*, **58**, 770–74.

Marx, H., Brunner, B., Weinzierl, W., Hoffmann, R. & Stolle, A. (1997) Methods of stunning freshwater fish: impact on meat quality and aspects of animal welfare. *Zeitschrift für Lebensmittel-Untersuchung und -Forschung A*, **204**, 282–6.

Oehlenschläger, J. (1991) pH-Werte in Filets von fangfrischen Seefischen aus der Nordsee. *Informationen für die Fischwirtschaft*, **38** (3), 109–13.

Oehlenschläger, J. & Nesvadba, P. (1997) Methods for freshness measurement based on electrical properties of fish tissue. In: *Methods to determine the freshness of fish in research industry* (eds G. Olafsdóttir, J.B. Luten, P. Dalgaard, M. Careche, V. Verrez-Bagnis, E. Martinsdóttir & K. Heia). Proceedings of the final meeting of the Concerted Action 'Evaluation of Fish Freshness' AIRCT94 2283, Nantes Conference, November 12–14, 1997, pp 363–8. International Institute of Refrigeration, Paris, France.

Proctor, M.R.M., Ryan, J.A. & McLoughlin, J.V. (1992a) The effects of stunning and slaughter methods on changes in skeletal muscle and quality of farmed fish. Presented at *EC FAR International Conference on Upgrading and Utilization of Fishery Products*, 12–14 May 1992, Noordwijkerhout, The Netherlands.

Proctor, M.R.M., Dorgan, M. & McLoughlin, J.V. (1992b) The concentrations of adenosine triphosphate, glucose-6-phosphate, lactate and glycogen in skeletal muscle of marine and fresh water fish species anaesthetised with MS 222. *Proceedings of the Royal Irish Academy*, **92b**, 45–51.

Schreck, C.B., Olla, B.L. & Davis, M.W. (1997) Behavioural responses to stress. In: *Fish Stress and Health in Aquaculture* (eds G.K. Iwama, A.D. Pickering, J.P. Sumpter & C.B. Schreck) pp. 145–70. Cambridge University Press, Cambridge, UK.

Sigholt, T., Erikson, U., Rustad, T., Johansen, S., Nordvedt, T. & Seland, A. (1997) Handling stress and storage temperature affect meat quality of farm-raised Atlantic salmon (*Salmo salar*). *Journal of Food Science*, **62**, 898–905.

Sørensen, N.K., Brataas, R. Nyvold, T.E. & Lauritzen, K. (1997) Influence of early processing (pre-rigor) on fish quality. In: *Seafood from Producer to Consumer, Integrated Approach to Quality* (eds J.B. Luten, T. Børresen & J. Oehlenschläger), pp. 253–63. Elsevier Science BV, Amsterdam, The Netherlands.

Templeton, J. (1996) Seafood research. A New Zealand perspective. At *Making the Most of the Catch Seafood Symposium*, Brisbane, July 1996.

Thomas, P.M., Pankhurst, N.W. & Bremner, H.A. (1999) The effect of stress and exercise on post-mortem biochemistry of Atlantic salmon and rainbow trout. *Journal of Fish Biology*, 54, 1177—96.

Verheijen, F.J. & Flight, W.F.G. (1997) Decapitation and brining: experimental tests show that after these commercial methods for slaughtering of eel, *Anguilla anguilla* L., death is not instantaneous. *Aquaculture Research*, 28, 361–6.

Wendelaar Bonga, S. (1997) The stress response in fish. *Physiological Reviews*, 77 (3), 592–625.

Wiepkema, P.R. (1997) The emotional vertebrate. In: *Animal Consciousness and Animal Ethics* (eds M. Dol, S. Kasanmoentalib, S. Lijmbach, E. Rivas & R. van den Bosch), pp. 93–102. Van Gorcum and Comp BV, Assen, The Netherlands.

Chapter 22

Primary Processing (Evisceration and Filleting)

A.M. Bencze Rørå, T. Mørkøre and O. Einen

Institute of Aquaculture Research (AKVAFORSK), N-1432Ås, Norway

Introduction

'Primary processing' is the term used to describe the first steps of converting fish into a food item that is attractive to the consumers. Primary processing has a number of functions. Its main function is to enhance hygiene and in this way extend the shelf-life of the final product. Further, primary processing decreases the mass of the raw material which increases the cost-effectiveness of transportation of the semi-processed or final products. Another factor is the desire of producers to add product value and increase product differentiation. These are becoming increasingly important to secure current markets and to penetrate new ones. Primary processing also allows the separation of edible parts from inedible ones, securing optimal quality of the parts that are going to be used and minimising waste. Proper waste handling has become a major, and costly, problem in several countries. The choice of method and degree of processing depends on a number of factors. These include the kind and size of fish material, the technology which is available, economic aspects, consumer demands and market trends.

The principal processes of primary processing are the same for wild and for farmed fish. However, dealing with fish from aquaculture provides certain advantages over traditional fisheries. The logistics are often simpler, with the fish being grown close to the harvest facilities. This means that the processing of the fish can begin almost at the farm-site. Control of production also gives the processors better forehand knowledge of the amount, size distribution and quality of the fish that are to be processed. Aquaculture also allows a better control over a number of factors which affect the stress level of the fish, including handling, crowding in the nets and the use of anaesthetics. This gives the processor a greater control and the ability to at least predict and maybe influence the development of *rigor mortis*. Several of the factors which influence yield and the total quality of the primary products can be influenced prior to primary processing. These factors include breeding (Iwamoto *et al.* 1990c; Dunham 1995; Gjedrem 1997), feed and feeding (Sigurgisladottir *et al.* 1994; Hillestad *et al.* 1998), farming conditions (Einen *et al.* 1999), harvest handling (Herborg & Villadsen 1975; Dassow 1976; Botta *et al.* 1987; Berg *et al.* 1997) and the option of harvesting before maturation. Aquaculture offers the potential of reducing intrinsic variation before slaughter and also offers opportunities to satisfy consumer preferences for such matters as the colour (Gormley 1992; Christiansen *et al.* 1995; Nickell & Bromage 1998) and fat content (Hillestad & Johnsen 1994; Einen *et al.* 1998) of salmon.

This chapter reviews some important factors related to the quality and output of the evisceration and filleting processes, with emphasis on results from salmonid farming.

Hygiene

Microbial contamination of fish depends mainly on the microbial flora in the environment, the temperature of the environment, the method of catching, and handling conditions at harvest. The bacterial density in the open sea is low, a few colony-forming units (cfu) per cm^3, while coastal regions and coastal sediments may be heavily polluted with bacterial densities of up to 10^6 cfu per cm^3 (Hobbs 1987). Nematodes are rarely found in farmed fish while their wild counterparts can be heavily infested. The skin of fish farmed in, and the skin of wild fish captured in, clean, cold, surface waters usually has bacterial densities ranging from one to ten bacteria per cm^2. The bacterial load of the gills is usually ten to a hundred times higher, and that of the intestines can be as much as 10^9 times higher, depending on the feeding status of the fish. The muscle tissue of fresh healthy fish is sterile (Liston 1980). The skin microflora of fresh fish taken from cold waters is composed predominantly of Gram-negative bacteria, mainly *Psychrobacter, Acinetobacter, Alteromonas, Pseudomonas, Flavobacterium* and *Vibrio* (Hobbs 1987). The bacterial load increases during ice storage, mainly as a result of a very rapid growth of *Alteromonas putrefaciens* and *Pseudomonas* spp. (Barile *et al.* 1985), both of which are well adapted to refrigerated conditions.

Handling, storage and chilling of the fish changes the number, distribution and composition of the microflora. There is usually an initial lag period which lasts until the resolution of *rigor mortis*. Following this, the rate of penetration of bacteria into the muscles depends on the barrier properties of the skin and whether skin cuts and abrasions are present. These facilitate bacterial migration. In whole, uneviscerated fish, the muscles are also invaded, during the period of autolysis, by bacteria from the alimentary tract. Moreover, exposed layers of fillets are especially vulnerable to bacterial penetration.

Chilling and *rigor mortis*

Most species lose quality between harvest and consumption, but damage, rate of degradation and loss of shelf-life can be minimised by controlling the temperature. The length of time from harvest to proper chilling of the fish depends on the core temperature of the fish at harvest. Therefore it takes potentially longer to cool fish reared at high water temperatures. Temperature fluctuations during transport and processing often occur, typically with an increase during filleting and a decrease after packaging. The temperature can fluctuate at reloading locations during transport. Further temperature fluctuations may occur during retail display and sale, before the product finally reaches the domestic refrigerator. These temperature fluctuations reduce the quality and the product spoils and loses shelf-life faster. It has therefore been recommended that fish are cooled to the temperature of melting ice (0°C) as soon as possible, and that they are held at that temperature during handling, transport and processing (FAO 1973). Chilling should, in general, be rapid, but some reports have indicated that ageing some tropical species a few hours before chilling prevents cold-shock reactions and a subsequent loss in yield of the fillets, without causing a significant reduction in shelf-life (Curran *et al.* 1986).

During *rigor mortis* the red muscle can contract by up to 52% of its original length, while the white muscle can contract by about 15%. This may be due to the higher contraction efficiency of red muscle, or to inhibition of rigor shortening by P_i (Wilkie 1986) and Ca^{2+} (Takahashi *et al.* 1995), both of which reach higher concentrations in white muscle than in red muscle. It seems that rigor sets in faster in small, active fish than in larger fish and flat fish (Huss 1983).

The effect of temperature during harvest and storage on the onset and duration of *rigor mortis* depends on the acclimatisation temperature of the fish prior to death. For some species, fish which have been acclimatised to a cold temperature enter rigor more rapidly, while for other species the opposite is true. For example, stunning and killing trout by hypothermia gave the fastest onset of *rigor mortis* for rainbow trout (Azam *et al.* 1990), and the decline of ATP levels was faster at 0–3°C than it was at 10–15°C in several other fish species (Iwamoto *et al.* 1985, 1987, 1990a,b). On the other hand, rigor onset takes place more slowly in carp acclimatised to cold waters (0–10°C) than in those acclimatised to warmer waters (20°C) (Hwang *et al.* 1991). Ushio *et al.* (1991) found calcium ion-pyroantimonate deposits in the sarcoplasmic reticulum immediately after spiking and in fish stored in the pre-rigor state at 10°C, but not at the onset of rigor in fish stored at 0°C. Muscle ATP degraded faster at 0°C than at 10°C. Rapid calcium ion release from the sarcoplasmic reticulum at 0°C seemed to stimulate the rapid onset of *rigor mortis*, as in the 'cold-shortening' phenomenon. In general, enzyme activity declines with decreasing temperature, within the 'normal' range in which the enzyme can perform. Lower temperatures therefore decrease the activity of both myosin ATPase and Ca^{2+}-ATPase (Iwamoto *et al.* 1988) in cold water acclimatised fish, resulting in decreased contraction. In warm water adapted fish, on the other hand, the muscle spontaneously responds to the cold by increasing the frequency of contractions in order to produce heat. Thus, it is not surprising that fish acclimatised to warmer waters show a faster onset of *rigor mortis* at low temperatures than fish already acclimatised at those temperatures. This emphasises the importance of controlling and adapting the temperature regime to species, size and harvesting conditions in order to achieve a *rigor mortis* development suitable for further packaging, transportation and processing.

Evisceration

Most farmed species are eviscerated before they are sold to the consumer. However, some small species which are sold in portion size (such as sea bream and sea bass) are kept whole as a sign that the fish is fresh. The rate of spoilage of fillets and of whole fish held on ice depends on the species (Fernandez-Salguero & Mackie 1987) and on the state of feeding. Actively feeding fish deteriorate more rapidly when held uneviscerated (Meyer *et al.* 1986), and most farmed fish species are starved for a short period prior to harvest. This reduces the amount and the activity of digestive enzymes, and a major objective of subsequent evisceration is to remove the source of digestive enzymes. Shewan (1961) has pointed out that the main advantage of gutting is to prevent autolytic spoilage rather than bacterial contamination. In non-starved fish, the proteolytic activity of viscera will be able to cause a violent autolysis after death, which may give strong off-flavours, particularly in the belly area, or even cause belly-burst. On the other hand, the belly area and cut surfaces are exposed to air during gutting, which makes them more susceptible to oxidation and discoloration.

Another advantage of evisceration is that it may allow more rapid chilling. Price *et al.* (1991) showed that eviscerated albacore, *Thunnus alalunga*, cooled faster than round or bled albacore.

The flesh of healthy live fish is sterile, as the immune system of the fish prevents bacteria from growing in it. When the fish dies, the immune system collapses and bacteria proliferate more freely. On the skin surface, bacteria mainly colonise the scale pockets. During storage, they invade the flesh by moving between the muscle fibres. Another important purpose of evisceration is to prevent the invasion of parasites from the intestines into the muscle. In some countries, compulsory evisceration of certain species has been introduced to minimise the risk of nematodes and parasites invading the edible parts of the muscle (Chory 1988).

A number of authors recommend that fish be bled, as well as eviscerated, as soon as possible after death (Townley & Lanier 1981; Costakes *et al.* 1982; Strom & Lien 1984; Valdimarsson *et al.* 1984) while others have found that delayed gutting does not have any negative effects (Ravesi *et al.* 1985; Meyer *et al.* 1986; Moser 1986). These conflicting reports suggests that there is no universal bleeding and gutting procedure for fish. Thus, many factors should be considered before deciding whether or not gutting is advantageous, including the age of the fish, the species, amount of lipid, farming area and harvesting method.

In the fish industry of today, deheading and evisceration are largely mechanised (although hand processing does occur). A great variety of machines are available, which differ in such features as design, output capacity, the size and species range of the processed material, the way in which the operation is performed and the theoretical yield. There are special machines for scaling, gutting, deheading, combined deheading and gutting, nobbing, filleting, skinning, cutting and meat separation.

The head is rarely used for human consumption and can be removed in most cases. The head of the fish accounts for a large proportion of its weight, and deheading is thus desirable in order to reduce the mass of the material. The principal requirement is that deheading should cause the smallest possible loss of muscular tissue. Evisceration (or gutting) is usually done by a clean cut between the posterior end of the gut and close to the head. The gutting cut is placed in the most fat-rich part of the belly for two reasons. This method both avoids damaging the muscle parts and makes bacterial contamination of the muscle from the surroundings of the fish more difficult. Some machines eviscerate the fish without cutting the belly open by sucking the viscera out through the mouth of the fish and rinsing and cleaning the belly part through the mouth. This method of evisceration avoids cutting the fish but makes it difficult to control the quality of the gutting operation. The flesh in the belly part of the fish is vulnerable to bacterial contamination after evisceration and it is therefore very important that the membranes in this part, which protect the muscle, are kept intact.

Slaughter yield

The increasing cost of producing fish means that it is important to recover as much valuable flesh as possible, and this has encouraged greater attention to improving the yield of edible

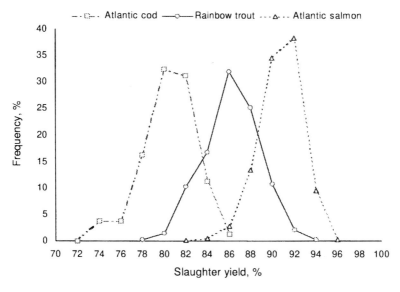

Fig. 22.1 Slaughter yield, calculated as gutted weight/ungutted weight*100, for farmed Atlantic cod (BW = 1.5–2 kg, *n* = 80), rainbow trout (BW = 1.5–5 kg, *n* = 325) and Atlantic salmon (BW = 2–5 kg, *n* = 800).

portions. The slaughter yield differs between species, but there are significant differences in slaughter yield also within one species (Fig. 22.1). Excess energy is deposited differently in different species. Excess energy is mainly stored in the liver of farmed or heavily fed cod, and the liver can comprise up to 15% of the body weight. In this species, the muscle fat is stable at about 1%, irrespective of feeding status. In rainbow trout and salmon, on the other hand, excess energy is deposited to a large extent as fat in the muscle and surrounding the intestines. It is possible to alter the fat content of both the viscera and muscle by altering the feed and the feeding regime (Einen *et al.* 1998, 1999). The production strategy prior to sea-transfer of Atlantic salmon, however, does not seem to have any impact on slaughter yield although the deposition pattern in the fillet differs (Mørkøre & Solbakken 1999). A recent study found slaughter yields of 80% for farmed cod and 88% for wild cod. The difference was mainly due to enlargement of the liver of the farmed fish (Mørkøre, unpub. data).

Filleting

Many species are filleted to satisfy consumer and market demands. Filleting is also important for logistics, economics, the addition of value along the marketing chain, and for the separation of edible parts from inedible ones. Filleting can be performed either by machine or by hand. A skilled worker can achieve a very high yield by hand-filleting, but hand-filleting is labour-intensive, hard and risky work. Therefore, most companies which fillet large quantities of fish use machines. The basic operations of machine filleting are: cutting along the upper and lower appendices on the spine, cutting over the ribs and cutting along the vertebrae. The ideal filleting machine has a high capacity, produces even surface cuts, is simple and convenient to operate, and is easy and cheap to maintain. Filleting machines are composed of three basic parts:

(1) The internal handling system (linear or rotary holders handling the fish), grippers, saddle-shaped holders, conveyer belts or chains with holding bite or needles
(2) The control-adjustment system
(3) The filleting tool unit, typically disk knives.

Most fish are symmetrically built, and so each operation is performed by a pair of symmetrical knives. Modern filleting machines are fast, they are often adapted to specific species and they are easy to adjust for fish size, thereby achieving higher yields. These newer machines are very flexible, but shapes and sizes outside of the range for which the machine are designed will result in reduced yield and might also damage both the machines and the fillets.

During filleting, it is normal to remove just the bones and fins (and the head if that has not been removed in a previous operation) from the flesh. Different standards of trimming are used, ranging from just removing the backbone to full removal of all visible fat, pinbones and the skin. Some customers require that all intermuscular bones are removed, to produce what is known as a 'v-cut' fillet. In this case, a higher proportion of meat is removed together with the appendices, which can reduce the yield by as much as 25%.

It is generally accepted that for larger farmed species it is best to process the fish once rigor has resolved, usually 2-4 days after harvest. However, a study of filleting gutted cod (*Gadus morhua*) 1, 4 or 7 days post-mortem showed that fillets produced seven days post-mortem had the longest shelf life (Shaw *et al.* 1984). The problem with early industrial filleting is that it is difficult to know when the onset of rigor occurs, resulting in a lot of unwanted handling while the fish are in rigor. As mentioned earlier, farming of fish makes it easier to control the metabolic rate at harvest, the transportation and harvesting stress, and to chill the fish more rapidly, all of which give a greater extent of control over rigor development. Rested harvesting and efficient chilling delay the onset of rigor, and it is possible that filleting could be carried out during this period. Pre-rigor filleting gives very fresh processed fish with no or little fillet gaping (Andersen *et al.* 1994), but with a change in shape (Sørensen *et al.* 1997). The fresh fillets can be sent to the market a few hours after harvest. The flesh is tight after pre-rigor filleting, and it is difficult to remove the pinbones of the fillet. A pre-rigor fillet will have a different general quality (texture and colour), different cooking properties (Sørensen *et al.* 1997) and different salt uptake and distribution (Fennema 1990) than a post-rigor fillet. It is therefore essential for further use and processing of the fillets to know whether the fillets have been produced pre- or post-rigor.

Fillet gaping

Fillet gaping is a serious problem for the filleting industry. Fillets with gaping are difficult to process and unattractive to the consumer. The causes of gaping are not well understood, but the following factors decrease the amount and the severity of gaping: harvesting fish in a rested state (Jerrett *et al.*1996), rapid cooling, including pre-mortal chilling (Skjervold *et* *al.* 1996), minimal harvest handling (Sigholt *et al.* 1997), optimal feeding strategy (Love 1988; Einen *et al.* 1999), season (Lavety *et al.* 1988) and rapid expedition to the market (Andersen *et al.* 1994). Gentle primary processing is also an important factor in avoiding fillet gaping.

Fig. 22.2 shows the frequency distribution of gaping scores of right and left side fillets of Atlantic salmon graded just after the filleting machine. (Gaping is scored on a scale from 0–5, where 0 = no gaping and 5 = severe gaping.) The left side fillets all had gaping scores between 0 and 2, with a mean score of 1.1. No left side fillets showed severe gaping. The mean gaping score of the right side fillets, on the other hand, was 2.4, and several had severe gaping. This illustrates the considerable impact that treatment, in this case the sharpness of the knives, can have on an important quality trait such as gaping.

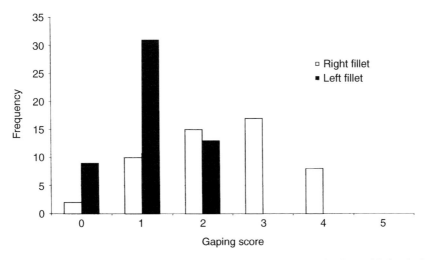

Fig. 22.2   Fillet gaping scores (0 = no gaping, 5 = severe gaping) of right and left side fillets of farmed Atlantic salmon evaluated immediately after machine filleting.

Fillet yield

Profit margins in the fish processing industry are often small and even small differences in yield have a considerable impact on the economy of fish processing companies. Figure 22.3 shows the filleting and trimming losses at six companies processing commercial Atlantic salmon of 3–4 kg weight, trimming according to the Norwegian B-standard. The filleting loss for these companies was from 16.8% to 23.8%, and the trimming loss from 10.7% and 12.4%. The total processing loss lay between 31% and 35%. This large difference in yield is a serious matter for the individual companies and shows that there is potential for improvement.

Fillet yield depends on the species and on the structural anatomy. Fish with large heads and frames relative to their musculature give a lower yield than those with smaller heads and

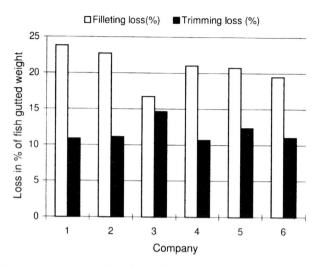

Fig. 22.3 Filleting and trimming loss (%) of farmed Atlantic salmon of 3–4 kg processed in different companies.

frames. The filleting yield for farmed species ranges from under 40% to over 70% depending on species, size and shape. Machine processing of well-chilled fish often gives higher yields than the processing of warm fish (Connell 1990). Hoffman *et al.* (1992) found that farmed male catfish yielded 46.7% fillets while wild males yielded 44.2%. The yield for farmed females was 38.9% and for wild females 44%. Farmed females had a higher gonadal mass which strongly reduced the dressout percentage. A study of tilapia (*Oreochromis niloticus*) and channel catfish (*Ictalurus punctatus*) reared on the same diet, size approximately 600 g, showed that processing yield (total fish weight – head, skin and viscera) was 51.0% for tilapia, and 60.6% for channel catfish. Fillet yield was also lower for tilapia (25.4%) than catfish (30.9%) (Clement & Lovell 1994).

The effects of weight and shape on the trimmed fillet yield of Atlantic salmon are shown in Fig. 22.4. The yield is a function of both size and shape and increases up to a condition factor of 1.5, showing that large and wide fish give the best yield. The normal variation in condition factor for Atlantic salmon is from 0.7 to 1.9, for large rainbow trout from 0.9 to 2.7 and for farmed cod from 0.7 to 1.5, which means that selection and grading before processing can improve the yield. The effects of fat content on filleting and trimming loss differ between species depending on how the fat is distributed. Generally, we have seen very little effect of fat content on filleting loss, with a slight tendency for increased trimming loss for higher fat contents (Rørå *et al.* 1998). Fillet yield can be affected by the farming and feeding conditions before harvesting farmed fish. Einen *et al.* (1999) studied Atlantic salmon, reducing the ration levels for the last 110 days prior to harvest. They found that the fillet yield was reduced if the ration was reduced below 50%. Similarly, a study in which Atlantic salmon were fasted for up to 86 days prior to harvest showed a significant reduction in the fillet yield, starting after the fish were starved for 30 days (Einen *et al.* 1998).

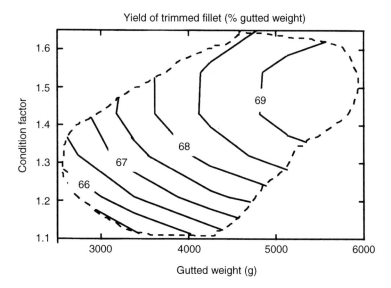

Fig. 22.4 Fillet yield (%) of trimmed fillet of farmed Atlantic salmon in relation to gutted weight and condition factor.

Conclusions

Farmed fish are processed by similar methods to those used for wild fish. The aims of primary processing are to increase the shelf-life of the fish, to provide a fresh product for sale and to enable it to be presented in a manner that is attractive to the buyer/consumer. Aquaculture makes it possible to alter the size and shape distribution, chemical composition, texture and pigmentation of the farmed fish prior to harvest. It is therefore possible to produce a fish which is specially designed to satisfy certain criteria which give optimal yield and quality from the primary processing. Many of the techniques used for the primary processing of farmed fish are based on experience gained from wild fish. Knowledge on the optimal chilling procedure and the best time to start primary processing is for most species scarce. Thus, future scientific work is needed to enable the special advantages of farmed fish to be investigated and exploited.

References

Andersen, U.B., Stromsnes, A.N., Steinsholt, K. & Thomassen, M.S. (1994) Fillet gaping in farmed Atlantic salmon (*Salmo salar*). *Norwegian Journal of Agricultural Sciences*, **8**, 165–79.
Azam, K., Strachan, N.J.C., Mackie, I.M., Smith, J. & Nesvadba, P. (1990) Effect of slaughter method on the progress of rigor in rainbow trout (*Salmo gairdneri*) as measured by an image processing system. *International Journal of Food Science and Technology*, **25**, 477–82.

Barile, L.E., Estrada, M.H., Milla, A.D., Reilly, A. & Villadsen, A. (1985) Spoilage patterns of mackerel (*Rastrelliger faudhni* Matsui). 2. Mesophilic and psychrophilic spoilage. *ASEAN Food Journal*, **1**, 121–6.

Berg, T., Erikson, U. & Nordtvedt, T.S. (1997) Rigor mortis assessment of Atlantic salmon (*Salmo salar*) and effects of stress. *Journal of Food Science*, **62**, 439–46.

Botta, J.R., Kennedy, K. & Squires, B.E. (1987) Effect of method of catching and time of season on the composition of Atlantic cod (*Gadus morhua*). *Journal of Food Science*, **52**, 922–4.

Chory, H. (1988) Regulations regarding the wholesomeness of fish and shellfish. *Bundesgesetzblatt*, part I, 1570–1 (in German).

Christiansen, R., Struksnaes, G., Estermann, R. & Torrissen, O-J. (1995) Assessment of colour in Atlantic salmon, *Salmo salar* L. *Aquaculture Research*, **25**, 311–21.

Clement, S. & Lovell, R.T. (1994) Comparison of processing yield and nutrient composition of cultured Nile tilapia (*Oreochromis niloticus*) and channel catfish (*Ictalurus punctatus*). *Aquaculture*, **119**, 299–310.

Connell, J.J. (1990) *Control of Fish Quality*. Fishing News Books, Oxford, UK.

Costakes, J., Connors, E. & Paquette, G. (1982) *Quality at Sea-Recommendations for On-board Quality Improvement Procedures*. New Bedford Seafood Producers Association and New England Fisheries Development Foundation, New Bedford, USA.

Curran, C.A., Poulter, R.G., Brueton, A. & Jones, N.R. (1986) Effects of handling treatment on fillet yields and quality of tropical fish. *Journal of Food Technology*, **21**, 301–10.

Dassow, J.A. (1976) Handling fresh fish. In: *Industrial Fishery Technology* (ed. M.E. Stansby) pp. 45–64. Robert E. Krieger, New York.

Dunham, R.A. (1995) The contribution of genetically improved aquatic organisms to global food security. *International Conference On Sustainable Contributions of Fisheries to Food Security*, Kyoto, Japan, 4–9 December, KC/FI/95/TECH/6, 111pp.

Einen, O., Waagan, B. & Thomassen, M.S. (1998) Starvation prior to slaughter in Atlantic salmon (*Salmo salar*). I. Effects on weight loss, body shape, slaughter- and fillet-yield, proximate and fatty acid composition. *Aquaculture*, **166**, 85–104.

Einen, O., Mørkøre, T., Rørå, A.M.B. & Thomassen, M.S. (1999) Feed ration prior to slaughter – a potential tool for managing product quality of Atlantic salmon (*Salmo salar*). *Aquaculture*, **178**, 149–69

FAO (1973) *Code of Practice for Fresh Fish*. FAO Fisheries Circular C318, Food and Agriculture Organization, Rome.

Fennema, O.R. (1990) Comparative water holding properties of various muscle foods. *Journal of Muscle Food*, **1**, 363–81.

Fernandez-Saluero, J. & Mackie, I.M. (1987) Comparative rates of spoilage of fillets and whole fish during storage of haddock (*Melanogrammus aeglefinus*) and herring (*Clupea harengus*) as determined by formation of non-volatile and volatile amines. *International Journal of Food Science and Technology*, **22**, 385–90.

Gjedrem, T. (1997) Flesh quality improvement in fish through breeding. *Aquaculture International*, **5**, 197–206.

Gormley, T.R. (1992) A note on consumer preference of smoked salmon colour. *Irish Journal of Agriculture and Food Research*, **31**, 199–202.

Herborg, L. & Villadsen, A. (1975) Bacterial infection/invasion of fish flesh. *Journal of Food Technology*, **10**, 507–13.

Hillestad, M. & Johnsen, F. (1994) High-energy/low-protein diets for Atlantic salmon: effects on growth, nutrient retention and slaughter quality. *Aquaculture*, 124, 109–16.

Hillestad, M., Johnsen, F., Austreng, E. & Åasgård, T. (1998) Long-term effects of dietary fat level and feeding rate on growth, feed utilization and carcass quality of Atlantic salmon. *Aquaculture Nutrition*, 4, 89–97.

Hobbs, G. (1987) Microbiology of fish. In: *Essays in Agricultural and Food Microbiology*, (eds J.R. Morris & G.L. Pettifer). John Wiley and Sons, London.

Hoffman, L.C., Casey, N.H. & Prinsloo, J.F. (1992) Carcass yield and fillet chemical composition of wild and farmed African sharptooth catfish, *Clarias gariepinus. Production, Environment and Quality*, Special Publication, no 18, pp. 421–32. European Aquaculture Society, Ghent, Belgium.

Huss, H.H. (1983) *Fersk Fisk, Kvalitet og Holdbarhed* (in Danish). Fiskeriministeriets Forsøgslaboratorium, Bygn. 221. DTH, Lyngby, Denmark.

Hwang, G-C., Ushio, H., Watabe, S., Iwamoto, M. & Hashimoto, K. (1991) The effect of thermal acclimation on *rigor mortis* progress of carp stored at different temperatures. *Nippon Suisan Gakkaiski*, 57, 541–8.

Iwamoto, M., Ioka, H. & Yamanaka, H. (1985) Relation between *rigor mortis* of sea bream and storage temperatures. *Bulletin of Japanese Society of Scientific Fisheries*, 51, 443–6.

Iwamoto, M., Yamaka, H., Watabe, S. & Hashimoto, K. (1987) Effects of storage temperatures on *rigor mortis* and ATP degeneration in plaice *Paralichthys olivaceus* muscle. *Journal of Food Science*, 52, 1514–17.

Iwamoto, M., Yamaka, K., Abe, H., Ushio, H., Watabe, S. & Hashimoto, K. (1988) ATP and creatine phosphate breakdown in spiked plaice muscle during storage and activities of some enzymes involved. *Journal of Food Science*, 53, 1662–5.

Iwamoto, M., Yamaka, K., Abe, H., Watabe, S. & Hashimoto, K. (1990a) *Rigor mortis* progress and its temperature-dependency in several marine fishes. *Nippon Suisan Gakkaishi*, 56, 93–9.

Iwamoto, M., Yamaka, K., Abe, H., Watabe, S. & Hashimoto, K. (1990b) Comparison of rigor mortis progress between wild and cultured plaices. *Nippon Suisan Gakkaishi*, 56, 101–104.

Iwamoto, R.N., Myers, J.M. & Hershberger, W.K. (1990c) Heritability and genetic correlations for flesh coloration in pen-reared coho salmon. *Aquaculture*, 86, 181–90.

Jerrett, A.R., Stevens, J. & Holland, A.J. (1996) Tensile properties of white muscle in rested and exhausted chinook salmon (*Oncorhynchus tshawytscha*). *Journal of Food Science*, 61, 527–32.

Lavety, J., Afolabi, O.A. & Love, R.M. (1988) The connective tissue of fish. IX. Gaping in farmed species. *International Journal of Food Science and Technology*, 23, 23–30.

Liston, J. (1980) Microbiology in fisheries science. In: *Advances in Fish Science and Technology* (ed. J. Connell). Fishing News Books, Oxford, England.

Love, R.M. (1988) *The Food Fishes – Their Intrinsic Variation and Practical Implications*. Farrand Press, London.

Meyer, B., Samuels, R. & Flick, G. (1986) *A seafood quality program for the mid-Atlantic region*, Part II. A report submitted to the Mid-Atlantic Fisheries Development Foundation. Virgina Polytechnic Institute and State University, Sea Grant, Blackburg.

Moser, M.D. (1986) *Maine Groundfish Association vessel quality handling project*. A report submitted to the New England Fisheries Development Foundation.

Mørkøre, T. & Solbakken, R. (1999) Product quality in Atlantic salmon in relation to production strategy. *Fish Farming International*, 2, 23.

Nickell, D.C. & Bromage, N.R. (1998) The effect of dietary lipid level on variation of flesh pigmentation in rainbow trout (*Oncorhynchus mykiss*). *Aquaculture*, 161, 237–51.

Price, R.J., Melvin, E.F. & Bell, J.W. (1991) Post-mortem changes in chilled round, bled and dressed albacore. *Journal of Food Science*, 56, 318–21.

Ravesi, E.M., Licciardello, J.J., Tuhkunen, B.E. & Lundstrom, R.C. (1985) The effects of handling or processing treatments on storage characteristics of fresh spiny dogfish, *Squalus acanthias*. *Marine Fisheries Review*, 47, 48–52.

Rørå, A.M.B., Kvåle, A., Mørkøre, T., Rørvik, K-A., Steien, S.H. & Thomassen, M.S. (1998) Process yield, colour and sensory quality of smoked Atlantic salmon (*Salmo salar*) in relation to raw material characteristics. *Food Research International*, 31, 601–609.

Shaw, S.J., Bligh, E.G. & Woyewoda, A.D. (1984) Effects of delayed filleting on quality of cod flesh. *Journal of Food Science*, 49, 979–80.

Shewan, J.M. (1961) The microbiology of sea-water fish. In: *Fish as Food* (ed. G. Borgstrom), Vol. 1, pp. 487–560. Academic Press, New York.

Sigholt, T., Erikson, U., Rustad, T., Johansen, S., Nordtvedt, T.S. & Seland, A. (1997) Handling stress and storage temperature affects meat quality of farm-raised Atlantic salmon (*Salmo salar*). *Journal of Food Science*, 64, 898–905.

Sigurgisladottir, S., Parrish, C.C., Lall, S.P. & Ackman, R.G. (1994) Effect of feeding natural tocopherols and astaxanthin on Atlantic salmon (*Salmo salar*) fillet quality. *Food Research International*, 27, 23–32.

Skjervold, P.O., Fjaera, S.O. & Christoffersen, K. (1996) Pre-mortal chilling of farmed salmon (*Salmo salar*). In: *Refrigeration and Aquaculture*, pp. 167–73. International Institute of Refrigeration, Paris.

Strom, T. & Lien, K. (1984) Fish handling on board Norwegian fishing vessels. In: *Fifty Years of Fisheries Research in Iceland*. Proceedings of a Jubilee Seminar (ed. A. Moller), p.15.

Sørensen, N.K., Brataas, R., Nyvold, T.E. & Lauritsen, K. (1997) Influence of early processing (pre-rigor) on fish quality. In: *Seafood from Producer to Consumer, Integrated Approach to Quality* (eds J.B. Luten, T.Børressen, J. Oehlenschläger) pp. 253–63. Elsevier, Amsterdam.

Takahashi, K., Hattori, A. & Kuroyanagi, H. (1995) Relationship between the translocation of paratropomyosin and the restoration of rigor-shortened sarcomeres during post-mortem ageing of meat. A molecular mechanism of meat tenderization. *Meat Science*, 40, 413–23.

Townley, R.R. & Lanier, T.C. (1981) Effects of early evisceration on the keeping quality of Atlantic croaker (*Micropogon undulatus*) and grey trout (*Cynoscion regalis*) as determined by subjective and objective methodology. *Journal of Food Science*, 46, 863–7.

Ushio, H., Watabe, S., Iwamoto, M. & Hashimoto, K. (1991) Ultrastructural evidence for temperature-dependent $Ca2+$ release from fish sarcoplasmic reticulum during *rigor mortis*. *Food Structure*, 10, 267–75.

Valdimarsson, G., Matthiasson, A. & Stefansson, G. (1984) The effect of onboard bleeding and gutting on the quality of fresh, quick frozen, and salted products. In: *Fifty Years of Fisheries Research in Iceland*, p. 61. Proceedings of a Jubilee Seminar (ed. A. Moller).

Wilkie, D.R. (1986) Muscular fatigue: effects of hydrogen ions and inorganic phosphate. *Federation Proceedings*, 45, 2921–3.

Chapter 23

Lipid Oxidation in Fatty Fish During Processing and Storage

I. Undeland,

SIK – The Swedish Institute for Food and Biotechnology, P.O. Box 54 01, S-402 29 Göteborg, Sweden

Introduction

The lipids of fish are rich in polyunsaturated n-3 fatty acids, a feature that is on the one hand nutritionally important and on the other sensorially critical. For many years, it has been thought that, for example, eicosapentaenoic acid, EPA (C20 : 5, n-3) and docosahexaenoic acid, DHA (C22 : 6, n-3), have positive effects on various cardiovascular diseases (Dyerberg et $al.$ 1978). However, their high degree of unsaturation and close proximity to strong pro-oxidative systems predispose them to oxidation, which converts them into compounds that negatively affect the quality attributes of the fish. Most well known is the rancid flavour, but changes in colour, texture and nutritional value also develop as consequences of lipid oxidation. To what extent these attributes will affect the spoilage of the fish is closely related to the tissue type considered and the type of handling the fish is subjected to. Following a short introduction into the lipid oxidation process and its reactants, some of the results reported in these two fields will be reviewed.

Lipid oxidation in fish – mechanisms and reactants

The oxidation process

The lipid oxidation process is an autocatalytic chain reaction that takes place through several intermediate steps: initiation, propagation, chain branching and termination (Hultin 1992). Initiation in fish can be brought about by enzymes, light or, as shown in Fig. 23.1, an initiator (X·). Formation of alkyl radicals (L·) and peroxy radicals (LOO·) propagates the oxidation reaction. The first relatively stable products are the tasteless and odourless hydroperoxides (LOOH), or primary oxidation products. In the chain branching step, these products are broken down into radicals and secondary products such as aldehydes and ketones. The secondary products often have very low odour thresholds and are responsible for fishy, trainy, oily or rancid flavours of stored fish (Milo & Grosch 1993). Some of these

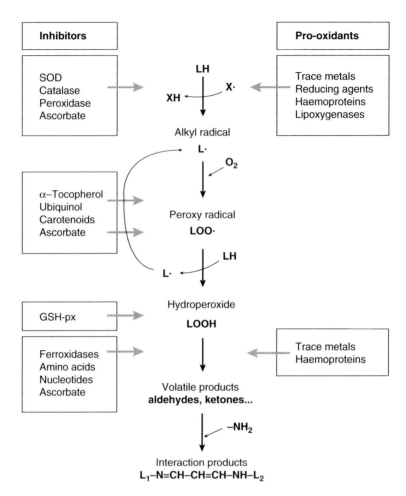

Fig. 23.1 Schematic illustration of the lipid autoxidation process including some of the native inhibitors and pro-oxidants found in fish tissues. LH = fatty acid, X' = initiator, L' = alkyl radical, LOO' = peroxy radical, LOOH = hydroperoxide, SOD = Superoxide dismutase, GSH-px = Glutathione peroxidase.

products can react further with compounds containing free amino groups, yielding tertiary products affecting the colour and texture of the fish (Aubourg *et al.* 1995, 1997). In parallel with these three steps, termination reactions, i.e. the interaction between two radicals with the formation of non-radical species, are continuously brought about.

Compounds participating in the oxidation process

Lipids
The fish lipids can be roughly divided into neutral lipids (NL) and polar lipids (PL), the former being the primary constituents of the fat deposits and the latter being responsible for

the function and structure of the cell membranes. It is generally believed that the PL are more prone to oxidation than the NL (Halpin 1984) due to their higher degree of unsaturation and their larger contact surface with the aqueous phase of the cell (Hultin 1994).

From both NL and PL, free fatty acids (FFA) can be hydrolysed post-mortem by the action of lipases and phospholipases, respectively (Shewfelt 1981). Although FFA oxidise more rapidly than esterified fatty acids when lipids are stored in bulk (Hultin 1992), the situation in fish muscle is more complex. In this case it has been found that NL hydrolysis increases the oxidation whereas PL hydrolysis inhibits it (Shewfelt 1981).

Oxygen

After harvest of the fish, the three principal oxygen sources are (Flink & Goodhart 1978):

(1) Oxygen present in situ
(2) Oxygen transported throughout the intact muscle
(3) Oxygen incorporated through processing.

Although ground state oxygen ($^3O_2$) is crucial in the propagation of lipid oxidation, it cannot initiate oxidation due to the rule of spin conservation (Harris & Tall 1989). However, mechanisms in the fish tissues can convert $^3O_2$ into species with initiating abilities, e.g. the hydroperoxyl radical, $HO_2\cdot$, the hydroxyl radicals, $\cdot OH$, and singlet oxygen $^1O_2$ (Hultin 1992).

Pro-oxidants

Most well known among the pro-oxidants in fish are the transition metals such as iron and copper, both impacting on several steps of the oxidation chain (Fig. 23.1). These metals are generally most active in their reduced state, while reducing compounds such as $O_2\cdot^-$, ascorbate and enzyme systems associated with both mitochondria and microsomes are also critical for the onset of oxidation (Hultin 1994). A large part of the iron in fish muscle is bound in haemoproteins such as myoglobin, haemoglobin and cytochromes. Both haemoglobin and myoglobin have catalytic properties due to their ability to break down hydroperoxides and their suggested activation by H_2O_2 into a highly reactive porphyrin cationic radical, $P^+\text{-}Fe^{4+} = O$ (Kanner *et al.* 1987). Lipoxygenases and cyclooxygenases are examples of pro-oxidative enzymes that can insert $^3O_2$ into an unsaturated fatty acid (Harris & Tall 1989).

Inhibitors

According to Fig. 23.1, native lipid oxidation inhibitors in the fish can interfere either with the catalysts (preventive inhibitors) or directly with the radicals (true antioxidants). Most preventive inhibitors act by removing active oxygen species, e.g. superoxide dismutase (SOD), catalase, peroxidase (Hultin 1992) and ascorbate (Petillo *et al.* 1998), or by oxidising or chelating transition metals. Oxidation can be executed e.g. by ferroxidases and chelation, by for example nucleotides, amino acids, free fatty acids and ascorbate (Hultin 1994).

The most well-known true antioxidant in fish is α-tocopherol, which is located in the lipid interior of the membranes (Brannan & Erickson 1996). Other lipid-soluble antioxidants are ubiquinol (Petillo *et al.* 1998) and the carotenoid pigments (Christophersen *et al.* 1992), the latter being of particular importance in salmonids. Among aqueous antioxidants in fish tissue are ascorbate, important for example in α-tocopherol regeneration, and glutathione peroxidase (GSH-px) (Hultin 1994).

The function and relevance of most pro- and antioxidants depend both on their own concentration and on the presence of other compounds in their close surroundings. As a result, many of the compounds mentioned here may both accelerate and inhibit oxidation (Decker & Hultin 1992), which is important to keep in mind when interpreting the findings discussed below.

Tissue-related differences in the development of oxidation

The nature, microenvironment, and thus the oxidative stability of the lipids, vary widely between different parts of the fish. To minimise oxidative changes during handling and processing, removal of critical tissues is therefore important, and a lot of work has been carried out to identify such tissues.

Among various dorsal and ventral sections from whole block frozen mackerel, Icekson *et al.* (1998) found that the dorsal and tail sections increased the most in secondary oxidation products (measured as thiobarbituric reactive substances, TBARS) during storage at − 18°C. The instability of the tail section was attributed to a higher proportion of dark muscle tissue. Tichivangana and Morrisey (1982) focused on the increase in TBARS values in dark and light muscle from head, mid and tail regions of a range of different fish species after cooking and refrigerated storage. The mid-regions were generally the most oxidised, and in all species, the dark muscle had higher TBARS values than the light one. In mullet, Lee and Toledo (1977) found the dark muscle, together with the tissue along the visceral cavity and the bone marrow, to be the tissues most rapidly increasing in TBARS during storage at 3°C. A range of studies have focused on differences in the stability between dark muscle, light muscle and skin. Consistently, the white muscle has been reported as the most stable (Ke *et al.* 1977; Yamaguchi *et al.* 1984), which is in accordance with its low level of pro-oxidants.

In relation to dark muscle and skin, there is less agreement; Visvanathan Nair *et al.* (1976) and Toyomizu and Hanoka (1980) reported the skin to be the least stable of the two, whereas the inverse was found by Yamaguchi *et al.* (1984). Compositional factors reported as critical in the skin are the high content of lipids (Yamaguchi *et al.* 1984), trace metals (Ke & Ackman 1976) and the presence of lipoxygenases (Harris & Tall 1989). In the dark muscle, the abundance of haemoproteins, low molecular weight (LMW) metals, microsomal enzymes fat and PL are among critical factors discussed (Tichivangana & Morrisey 1982; Hultin 1992, 1994). Our own work with light muscle, dark muscle and skin from herring (Undeland *et al.* 1998a) revealed that the contradictions in the literature largely originate in the storage conditions used, i.e. the degree of tissue disintegration applied, and in the type of oxidation measure used. When keeping differences in the access to oxygen and adjacent tissues to a minimum by storing the three tissues separately, hydroperoxides were formed in the order: dark muscle > skin > light muscle, whereas when stored as an intact fillet, the corresponding order became: skin > dark muscle > light muscle (Fig. 23.2). These data revealed that the dark muscle has the most unfavourable composition with respect to lipid

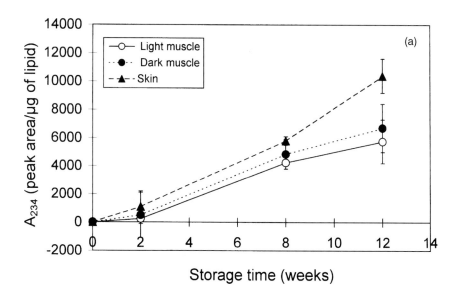

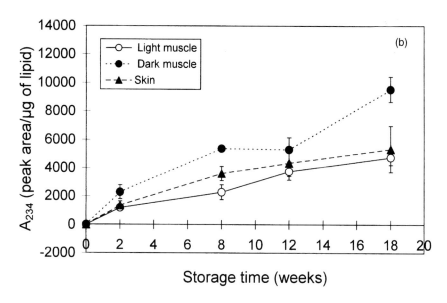

Fig. 23.2 Increase in primary and secondary oxidation products (measured as total absorbance at 234 nm, A_{234}) in herring ($\bigcirc$) light muscle ($\bullet$) dark muscle and ($\blacktriangle$) skin when stored (a) as an intact fillet versus (b) separated from each other at $-18°C$. (Adapted from Undeland *et al.* 1998a, published with permission from *Journal of the American Oil Chemists Society*.) Error bars indicate the span between maximum and minimum values.

oxidation, but also that it is fairly well protected by the skin and light muscle within the intact fillet. Similar findings were reported by Yamaguchi *et al.* (1984) when storing sardine light muscle, dark muscle and skin at −5°C, either within the round fish, or as separate minces. In mackerel, however, the skin was found to oxidise fastest under both these storage conditions (Toyomizu & Hanoka 1980).

With respect to the influence of the oxidation measure used, the order we obtained for hydroperoxide development in the separately-stored herring tissues i.e. dark muscle > skin > light muscle was changed into light muscle > dark muscle > skin when tertiary products were followed. Compositional analyses revealed that the high level of iron, copper, and aqueous pro-oxidative activity in the dark muscle gave rise to its rapid hydroperoxide formation, whereas in the white muscle, the rapid tertiary product formation originated in the high proportion of PL in the lipids. To evaluate further how intrinsic factors and oxygen diffusion through the fish tissue influence the development of oxidation, we have also focused on the formation of oxidation products in various horizontal layers of herring fillets: the under-skin layer, middle part and inner part (Undeland *et al.* 1998b).

In both skinned and skin-on fillets, as well as under both ice and frozen storage, the under-skin layer oxidised most rapidly, which was primarily related to its high degree of dark muscle. However, in fillets stored skinned, the rate of oxidation in the under-skin layer was three times faster than in skin-on fillets, and in the middle part up to twice as high, again pointing to the protective properties of the skin. Oxygen diffusion through the muscle was also a limiting factor. Although analyses of aqueous pro-oxidative activity ranked the three layers as under-skin > middle part > inner part, the middle part still oxidised slower than the inner part during storage, illustrating its reduced oxygen access. In accordance with this, Freeman and Hearnsberger (1994) found less development of TBARS and total volatiles (Fig. 23.3) in the inner 3 mm of skinned catfish fillets than in surface samples from the skin side and visceral side. The present results fit well into the observations of Lawrie (1974), that oxygen diffuses some 1–4 mm into muscle tissues, depending on the mitochondrial content and degree of tissue disruption.

Lipid oxidation as affected by storage and processing

The lipids, pro- and antioxidants constitute a system which is highly dynamic and changes drastically from harvest to consumption. In order to develop antioxidative systems that are efficient throughout the entire processing chain, knowledge about these changes is essential.

Ice storage

Ice storage of fish induces an array of changes in the tissue affecting oxidation, e.g. a decrease in reducing capacity (Hultin 1994), increase in free iron (Decker & Hultin 1992), activation of haemoproteins (Kanner *et al.* 1987) and membrane disintegration (Petillo *et al.* 1998). However, as bacterial growth is not totally inhibited by ice storage, the importance of lipid oxidation in reducing the quality of ice-stored fish is limited. According to Ackman and Ratnayake (1992), rancidity is only crucial as a quality index in frozen fish, whereas above

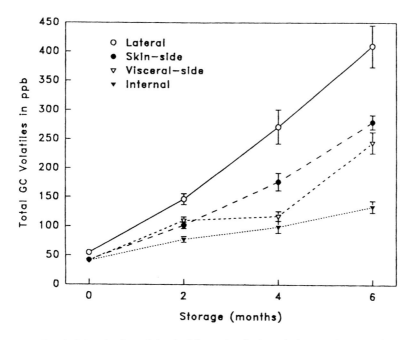

Fig. 23.3 Total GC volatiles of the (○) lateral side (i.e. dark muscle), (●) skin side, (▽) visceral side and (▼) internal from channel catfish fillets after frozen storage (Freeman & Hearnsberger 1994, published with permission from *Journal of Food Science*). Error bars indicate standard errors.

0°C, bacterial spoilage is of major importance. This statement has been confirmed by a number of studies (Smith *et al*. 1980; Harris & Tall 1989; Aubourg *et al*. 1995).

However, in the literature, large losses of antioxidants have also been reported during ice storage of fish (Petillo *et al*. 1998), as well as substantial increases in both primary (Sen & Bhandary 1978), secondary (Aubourg *et al*. 1997) and tertiary (Aubourg *et al*. 1995, 1997) lipid oxidation products. A possible link between bacterial and oxidative spoilage has also been discussed, with the bacteria and/or their products blamed for both inhibiting (Castell & MacLean 1964) and enhancing (Vercellotti *et al*. 1992) oxidation in fish. However, in the study of Lee and Toledo (1977), as well as in our work with ice stored herring (Undeland *et al*. 1999a), no such link was established. In the latter, a significant increase in oxidation products was seen after 2–3 days, which was 4–5 days earlier than the total bacterial count had increased significantly. The former also showed a higher correlation with sensory changes, i.e. in 'rancid', 'sharp', 'old', 'fresh' and 'shellfish' odours, which was in accordance with the work of Mendenhall (1972).

Accompanying the oxidation product formation, large losses of antioxidants were also noted, as well as a 75% decrease in total aqueous pro-oxidative activity. As regards the antioxidants, the order of relative loss was α-tocopherol > ascorbic acid > glutathione peroxidase activity (Fig. 23.4), which differed from the order found in channel catfish fillets (Brannan & Erickson 1996) and mackerel (Petillo *et al*. 1998) (Fig. 23.5), where the aqueous

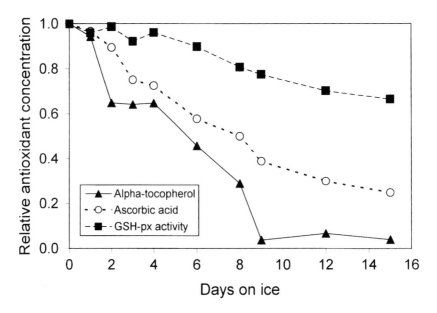

Fig. 23.4   Loss of (▲) α-tocopherol, (○) ascorbic acid and (■) glutathione peroxidase (GSH-px) activity in fillets of herring during storage on ice. (Adapted from *Undeland et al.* 1999, published with permission from *Journal of Agricultural and Food Chemistry*). Results are expressed as the relative fraction of the level found at time 0.

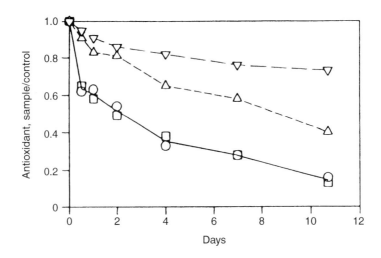

Fig. 23.5   Loss of (○) ascorbic acid, (□) total glutathione, (△) α-tocopherol and (▽) ubiquinone-10 from mackerel light muscle during storage on ice (Petillo *et al.* 1998, published with permission from *Journal of Agricultural and Food Chemistry*). Results are expressed as the relative fraction of the control sample (stored at −62°C) at a given time.

antioxidants were consumed first. The large pro-oxidant loss observed could indicate that oxidation triggered by iron or by different types of enzymes is of highest importance in the early stages of storage, whereas later on autoxidation becomes the dominating reaction (Lee *et al.* 1996).

Frozen storage – influence of pre-freezing storage

The effects of freezing on lipid oxidation are opposing. So, the temperature decrease suppresses the lipid oxidation reaction rate, while for example freezing-induced tissue dehydration, concentration of catalytic solutes and mechanical disruption of membranes increase it (Fennema 1985). To keep the initial freshness of fish during prolonged frozen storage is therefore a challenge, and there are many factors which need to be rigorously controlled. Among those studied are the conditions and length of pre-freezing storage (Kolakowska 1981), the degree of tissue disruption (Smith *et al.* 1980; Kolakowska 1981), oxygen accessibility (Christophersen *et al.* 1992; Undeland *et al.* 1998a), access to UV light (Christophersen *et al.* 1992;) freezing methods (Jhaveri & Constantinides 1981), storage temperature (Ke *et al.* 1977) and thawing conditions (Lee & Toledo 1977). The general conclusions have been that pre-freezing storage in refrigerated seawater (RSW) followed by minimised tissue disruption, removal of oxygen and light, fast freezing, low freezer storage temperatures and fast thawing reduces rancidity and gives the longest shelf-life. However, the effect of the duration of pre-freezing storage on oxidation is one of the topics still not fully understood.

Although most studies have pointed to a faster increase in oxidation products when pre-freezing storage is extended (Hiltz *et al.* 1976; Ke & Ackman 1976; Kolakowska 1981), some interesting findings have indicated that processing and storage of very fresh fish increase the susceptibility towards oxidation during subsequent frozen storage (Kolakowska 1981; Richards *et al.* 1998). Extensive blood contamination during the filleting as well as antioxidative effects of compounds formed during proteolysis, lipid-protein interactions, and lipid hydrolysis have been cited as possible explanations. In our own study (Undeland & Lingnert 1999), the results showed that herring fillets held on ice for more than three days greatly increased their formation rate of oxidation products at $-18°C$. Particularly, the period 3–6 days on ice appeared to be critical, which was ascribed to an extensive rise in radicals and pre-formed hydroperoxides, the reaching of critical antioxidant levels (Brannan & Erickson 1996), lipolysis (Shewfelt 1981) and decompartmentation (Hultin 1992). Partial least square regression analysis (PLS) of the data showed that the ice storage period in general had a greater effect than frozen storage on changes in oxidation products and antioxidants. As an example, one day on ice gave the same increase in peroxide value (PV) as did three weeks at $-18°C$.

Filleting and mincing

According to the above, the oxidation rate during storage of fish largely depends on access to oxygen (Flink & Goodhart 1978; Undeland *et al.* 1998a,b). A lot of data therefore shows how lipid oxidation in fish is affected, for example by filleting, skinning and mincing operations which also cause tissue decompartmentation (Hultin 1994), trace metals contamination (Lee & Toledo 1977) and increased lipid molecule mobility etc. From comparisons between round and filleted fish, it has been shown that fillets are the least stable with respect to lipid oxidation (Mendenhall 1972). This instability is further exacerbated by skinning the fillets prior to storage (Undeland *et al*. 1998b). Richards *et al*. (1998) reported that filleting under water can improve the storage stability of fish fillets prepared from very fresh fish, since blood is removed and the oxygen access diminished. Adding antioxidants to the cutting water further increased the stability. Stabilisation of fish fillets can also be achieved by deep skinning, i.e. the removal of the skin together with 3–4 mm of the under-skin tissue (Hultin 1994). In normally skinned fillets, the lipids in this tissue very quickly become oxidized (Undeland *et al*. 1998a,b).

The general view is that mincing fish is a risky operation from a lipid oxidation point of view. This has been confirmed from comparisons of mince with skin-on fillets (Undeland & Lingnert 1995) as well as with whole fish (Kolakowska 1981). However, from comparisons with skinned fillets, which is the most accurate comparison considering that the skin tissue is not present in the mince, only one study was found showing a lower storage stability of mince (Hiltz *et al*. 1976), whereas several, e.g. Undeland *et al*. (1998c) and Brannan and Erickson (1996) revealed the opposite. In our own study (Fig. 23.6), where the surface and interior of the mince cubes studied were 5 and 20–25 times more stable than the fillets, respectively, the positive effect of mincing was difficult to explain. The only compositional

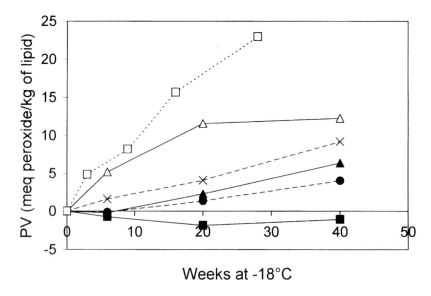

Fig. 23.6   Increase in hydroperoxides (measured as peroxide value, PV) in (▲) the surface of crude herring mince samples, (■) the centre of crude herring mince samples, (△) the surface of washed herring mince samples, (×) the surface of herring mince pre-cooked at 100°C for 38 min, (●) the surface of herring mince pre-cooked at 55°C for 38 min and (□) skinned herring fillets during storage at −18°C. (Adapted from Undeland 1998 and Undeland *et al*. 1998c, published with permission from *Journal of Agricultural and Food Chemistry*.)

changes we noted during mincing were slight decreases in total fat, C20 : 1 and C20 : 5, as well as a slight rise in C22 : 6 in the fat. It has been suggested, however, that mincing activates antioxidative enzymes, e.g. aldehyde dehydrogenase (Brannan & Erickson 1996) or phospholipase (Shewfelt & Hultin 1983). An increased contact surface between the lipids and antioxidant is also possible.

Washing

Washing of fish mince lowers the levels of fat, pigments and pro-oxidants, which, from an oxidation point of view, is generally regarded as positive. However, both increased (Ekstrand *et al.* 1993; Undeland *et al.* 1998c) and decreased (Lee & Toledo 1977) oxidation rates have been observed in washed mince, which is probably due to the washing procedure being more or less extensive. The lowered stability of washed mince obtained during studies in our lab (Undeland *et al.* 1998c) (Fig. 23.6) was primarily ascribed to an extensive wash out of antioxidants along with the pro-oxidants. The levels of selenium and GSH-px activity decreased by 33% and 38%, respectively, and the washing water residue showed significant antioxidative properties. Additional explanations may be the relatively higher levels of PL (Halpin 1984) and FFA (Hultin 1992) we found in the washed mince lipids. Kelleher *et al.* (1992) reported that washing fish mince with an antioxidant solution efficiently increased the storage stability since it compensated for washing-induced antioxidant losses. The best effect was obtained when the antioxidants had already been introduced at the grinding step, and were thus present at the moment of cellular breakage.

Pre-cooking

From a lipid oxidation point of view, the shelf-life of frozen and chilled 'ready-to-eat/ready-for-the-microwave oven' is highly affected by the pre-cooking step. Previous studies have illustrated how heating of muscle tissues can activate haemoproteins (Eriksson *et al.* 1971), inhibit antioxidative enzymes (Lee *et al.* 1996), inhibit catalytic enzymes (Wang *et al.* 1991) and form antioxidants (Fujita *et al.* 1994). It is therefore not surprising that pre-cooking of fish has been reported to both increase (Fujita *et al.* 1994) and decrease (Wang *et al.* 1991) the subsequent storage stability.

As part of our work with herring and lipid oxidation, the effect of pre-cooking herring mince at 55°C and 100°C was evaluated (Undeland *et al.* 1998c). As shown in Fig. 23.6, samples treated at 55°C developed primary and secondary oxidation products significantly more slowly than the raw mince samples during subsequent storage at $-18°C$. In contrast, 100°C increased the development rate of most oxidation products. A slightly pro-oxidative effect was also found by prolonging the cooking times. The stabilising effect of mild heating was believed to arise by inactivation of catalytic enzymes such as lipoxygenases and microsomal enzymes in the mince, without affecting the haemoproteins. At 100°C, on the other hand, catalytic changes in the haemoprotein structures (Eriksson *et al.* 1971) together with antioxidant inactivation (Lee *et al.* 1996) and general thermal acceleration of non-enzymatic reaction rates (Fennema 1985), most likely became the deciding factors.

Conclusions

To increase the storage stability of fatty fish, research has indicated that during ice storage, attention should not only be paid to bacterial growth, but also to the development of lipid oxidation. With respect to the latter, there appear to be two critical pre-processing ice storage periods that need to be controlled: one between harvest and gutting/filleting and the other between gutting/filleting and further processing/freezing. The former should be long enough to let the blood coagulate, whereas the latter should not exceed two days to minimise pro-oxidative changes and avoid the development of rancid odour. To protect the sensitive tissues right under the skin, i.e. the subcutaneous fat layer and the dark muscle, the skin should be kept on for as long as possible during storage of the fish, or should be removed together with underlying tissues at an early stage. Minced fish should be frozen and stored in large blocks to decrease the surface to volume ratio.

For both fillets and mince, removal of oxygen and light are essential factors to prolong the storage stability. If washing is to be undertaken, this should be done as quickly after filleting/mincing as possible, and, to compensate for losses of native inhibitors, preferably with antioxidants added to the water. Pre-cooking of fish should be carried out using the mildest possible conditions to avoid catalytic side effects. Possibly, a short heat-treatment/pre-cooking step inactivating catalytic enzymes in the sensitive surface tissue could be used as an antioxidative treatment early in the processing chain to increase the shelf life/oxidative stability of farmed fish.

References

Ackman, R.G. & Ratnayake, W.M.N. (1992) Non-enzymatic oxidation of seafood lipids. In: *Advances in Seafood Biochemistry. Composition and Quality* (eds G.J. Flick, & R.E. Martin), pp.245–67. Technomic Publishing Co, Inc. Lancaster, Basel.

Aubourg, S., Gallardo, J.M., Medina, I. & Pérez-Martin, R. (1995) Fluorescent compound formation in sardine muscle during refrigeration and frozen storage. *Proceedings of European Food Chemistry VIII*, Vienna, 3, 579–83.

Aubourg, S., Sotelo, C.G. & Gallardo, J.M. (1997) Quality assessment of sardines during storage by measurement of fluorescent compounds. *Journal of Food Science*, 62, 295–8.

Brannan, R.G. & Erickson, M.C. (1996) Quantification of antioxidants in channel catfish during frozen storage. *Journal of Agricultural and Food Chemistry*, 44, 1361–6.

Castell, C.H. & MacLean, L. (1964) Rancidity in lean fish muscle. III. The inhibiting effect of bacterial activity. *Journal of the Fisheries Research Board of Canada*, 21, 1371–7.

Christophersen, A-G., Bertelsen, G., Andersen, H.J., Knuthsen, P. & Skipsted, L.H. (1992) Storage life of frozen salmonoids. Effect of light and packaging conditions on carotenoid oxidation and lipid oxidation. *Zeitschrift für Lebensmittel-Untersuchung und Forschung*, 194, 115–19.

Decker, E.A. & Hultin, H.O. (1992) Lipid oxidation in muscle foods via redox iron. In: *Lipid Oxidation in Foods*, ACS Symposium Series 500 (ed. A.J. St. Angelo), pp.33–54. American Chemical Society, Washington DC.

Dyerberg, L., Bang, H.O., Stofferesen, E., Minacada, S. & Vane, J. (1978) Eicosapentaenoic acid and prevention of thrombosis and atherosclerosis. *Lancet*, 2, 117–19.

Ekstrand, B., Gangby, I., Janson, R-M., Pettersson, A. & Åkesson, G. (1993) *Lipid stability related to development of herring mince products*. 23rd annual WEFTA meeting, Göteborg, Sweden.

Eriksson, C.E., Olsson, P.A. & Svensson, S.G. (1971) Denaturated hemoproteins as catalysts in lipid oxidation. *Journal of the American Oil Chemists Society*, 48, 442–7.

Fennema, O.R. (1985) Water and ice. In: *Food Chemistry* (ed. O.R. Fennema), pp.23–68. Marcel Dekker, Inc. New York.

Flink, J.M. & Goodhart, M. (1978) Transport of oxygen through ice and frozen minced fish. *Journal of Food Processing and Preservation*, 2, 229–48.

Freeman, D.W. & Hearnsberger, J.O. (1994) Rancidity in selected sites of frozen catfish fillets. *Journal of Food Science*, 59, 60–63, 84.

Fujita, Y., Oshima, T. & Koizumi, C. (1994) Increase in the oxidative stability of sardine lipids through heat treatment. *Fisheries Science*, 60, 289–94.

Halpin, B.E. (1984) *On the possible role of the sarcoplasmatic reticulum in lipid oxidation in fish*. MS thesis, University of Massachusetts, Amherst, USA.

Harris, P. & Tall, J. (1989) Rancidity in fish. In: *Rancidity in Foods* (ed. J.C. Allen & R.J. Hamilton), pp.256–72. Blackie Academic and Professional, London.

Hiltz, D.F., Smith Lall, B., Lemon, D.W. & Dyer, W.J. (1976) Deteriorative changes during frozen storage in fillets and minced flesh of silver hake (*Merluccius bilinearis*) processed from round fish held in ice and refrigerated sea water. *Journal of the Fisheries Research Board of Canada*, 33, 2560–67.

Hultin, H.O. (1992) Lipid oxidation in fish muscle. In: *Advances in Seafood Biochemistry. Composition and Quality* (ed. G.J. Flick, J.R. & R.E. Martin), pp.99–122. Technomic Publishing Co., Inc. Lancaster, Basel.

Hultin, H.O. (1994) Oxidation of lipids in seafoods. In: *Seafoods Chemistry, Processing; Technology and Quality* (ed. F. Shahidi & J.R. Botta), pp.49–74. Blackie Academic and Professional, London.

Icekson, I., Drabkin, V., Aizendorf, S. & Gelman, A. (1998) Lipid oxidation levels in different parts of the mackerel, *Scomber scombrus*. *Journal of Aquatic Food Product Technology*, 7, 17–29.

Jhaveri, S.N. & Constantinides, S.M. (1981) Chemical composition and shelf life study of grayfish (*Squalus acanthias*). *Journal of Food Science*, 47, 188–92.

Kanner, J., German, B. & Kinsella, J.E. (1987) Initiation of lipid peroxidation in biological systems. *CRC Critical Reviews in Food Science and Nutrition*, 25, 317–64.

Ke, P.J., & Ackman, R.G. (1976) Metal-catalyzed oxidation in mackerel skin and meat lipids. *Journal of the American Oil Chemists Society*, 53, 636–40.

Ke, P.J., Ackman, R.G., Linke, B.A & Nash, D.M. (1977) Differential lipid oxidation in various parts of mackerel. *Journal of Food Technology*, 12, 34–47.

Kelleher, S.D., Silva, L.A., Hultin, H.O. & Wilhelm, K.A. (1992) Inhibition of lipid oxidation during processing of washed, minced Atlantic mackerel. *Journal of Food Science*, 57, 1103–108.

Kolakowska, A. (1981) The rancidity of frozen Baltic herring prepared from raw material with different initial freshness. *Refrigeration Science and Technology*, 4, 341–8.

Lawrie, R.A. (1974) The eating quality of meat. In: *Meat Science*, 2nd edn (ed. R.D. Lawrie), pp.289–92. Pergamon Press, Oxford.

Lee, C.M. & Toledo, R.T. (1977) Degradation of fish muscle during mechanical deboning and storage with emphasis on lipid oxidation. *Journal of Food Science,* 42, 1646–9.

Lee, S.K., Mei, L. & Decker, E.A. (1996) Lipid oxidation in cooked turkey as affected by added antioxidant enzymes. *Journal of Food Science,* 61, 726–8, 795.

Mendenhall, V.T. (1972) Oxidative rancidity in raw fish fillets harvested from the Gulf of Mexico. *Journal of Food Science,* 37, 547–50.

Milo, C. & Grosch, W. (1993) Changes in the odorants of boiled trout (*Salmo fario*) as affected by the storage of the raw material. *Journal of Agricultural and Food Chemistry,* 41, 2076–81.

Petillo, D., Hultin, H.O., Krzynowek, J. & Autio, W.R. (1998) Kinetics of antioxidant loss in mackerel light and dark muscle. *Journal of Agricultural and Food Chemistry,* 46, 4128–37.

Richards, M.M., Kelleher, S.D. & Hultin, H.O. (1998) Effect of washing with and without antioxidants on quality retention of mackerel fillets during refrigerated and frozen storage. *Journal of Agricultural and Food Chemistry,* 46, 4363–71.

Sen, D.P. & Bhandary, C.S. (1978) Lipid oxidation in raw and cooked oil sardine (*Sardinella longiceps*) fish during refrigerated storage. *Lebensmittel Wissenschaft Technologie,* 2, 124–7.

Shewfelt, R.L. (1981) Fish muscle lipolysis – a review. *Journal of Food Biochemistry,* 5, 79–100.

Shewfelt, R.L. & Hultin, H.O. (1983) Inhibition of enzymic and non-enzymic lipid peroxidation of flounder muscle sarcoplasmicreticulum by pretreatment with phospholipase A2. *Biochimica et Biophysica Acta,* 751, 432–8.

Smith, J.G.M., Hardy, R. McDonald, I. & Templeton, J. (1980) The storage of herring (*Clupea harengus*) in ice, refrigerated sea water and at ambient temperature. Chemical and sensory assessment. *Journal of the Science of Food Agriculture,* 31, 375–85.

Tichivangana, J.Z., and Morrisey, P.A. (1982) Lipid oxidation in cooked fish muscle. *Irish Journal of Food Science and Technology,* 6, 157–63.

Toyomizu M. & Hanoka, K. (1980) Lipid oxidation of the minced ordinary muscle of fish during storage at $-5°C$ and susceptibility to lipid oxidation. *Bulletin of the Japanese Society of Scientific Fisheries,* 46, 1007–10.

Undeland, I. & Lingnert, H. (1995) Measurement of oxidative changes in fatty fish using UV- and chemiluminescence detection. *Proceedings from the 18th Nordic lipid symposium,* pp.139–53. Reykjavik, Iceland, June 18–21.

Undeland, I., Ekstrand, B. & Lingnert, H. (1998a) Lipid oxidation in herring (*Clupea harengus*) light muscle, dark muscle and skin, stored separately or as intact fillets. *Journal of the American Oil Chemists Society,* 75, 581–90.

Undeland, I., Stading, M. & Lingnert, H. (1998b) Influence of skinning on lipid oxidation in different horizontal layers of herring (*Clupea harengus*) during frozen storage. *Journal of the Science of Food and Agriculture,* 78, 441–50.

Undeland, I., Ekstrand, B. & Lingnert, H. (1998c) Lipid oxidation in minced herring (*Clupea harengus*) during frozen storage. Effect of washing and pre-cooking. *Journal of Agricultural and Food Chemistry,* 46, 2319–28.

Undeland, I. (1998) *Lipid oxidation in fillets of herring* (Clupea harengus) *during processing and storage.* PhD thesis, Chalmers University of Technology, Göteborg, Sweden.

Undeland, I., Hall, G. & Lingnert, H. (1999) Lipid oxidation in fillets of herring (*Clupea harengus*) during ice storage. *Journal of Agricultural and Food Chemistry,* 47, 524–32.

Undeland, I. & Lingnert, H. (1999) Lipid oxidation in fillets of herring (*Clupea harengus*) during frozen storage. Influence of pre-freezing storage. *Journal of Agricultural and Food Chemistry*, 47, 2075–81.

Vercellotti, J.R., St Angelo, A.J. & Spanier, A.M. (1992) Lipid oxidation in foods. In: *Lipid Oxidation in Foods*, ACS Symposium Series 500 (ed. A.J. St. Angelo) pp.1–11. American Chemical Society, Washington DC.

Viswanathan Nair, P.G., Gopakumar, K. & Rajendranathan Nair, M. (1976) Lipid hydrolysis in mackerel (*Rastrelliger kanagurta*) during frozen storage. *Fish Technology*, 13, 111–14.

Wang, Y-J., Miller, L.A. & Addis, P.B. (1991) Effect of heat inactivation of lipoxygenase on lipid oxidation in lake herring (*Coregonus artredii*). *Journal of the American Oil Chemists Society*, 68, 752–7.

Yamaguchi, K., Nakamura, T. & Toyomizu, M. (1984). Preferential lipid oxidation in the skin of round fish. *Bulletin of the Japanese Society of Scientific Fisheries*, 50, 5869–74.

Chapter 24

Modified Atmosphere Packaging of Chilled Fish and Seafood Products

B.P.F. Day

Campden and Chorleywood Food Research Association, Chipping Campden, Gloucestershire, GL55 6LD, UK

Introduction

Fish and seafood products are highly perishable and spoilage is principally the result of microbial and oxidative mechanisms. Microbial activity causes a breakdown of fish protein with resulting release of undesirable fishy odours. Oxidative rancidity of unsaturated fats in oily fish also results in other additional offensive odours and flavours. These spoilage mechanisms can be delayed by lowering the temperature, either by refrigeration or by freezing. Microbial spoilage is prevented by freezing fish, but other deteriorative changes take place, although at a much slower rate. The marketing of chilled fish as opposed to frozen fish is favoured by the major UK retailers since chilled foods are perceived as being fresher and of higher quality than their frozen counterparts. In combination with good refrigeration and handling conditions, modified atmosphere packaging (MAP) is an increasingly popular preservation technique for significantly improving the product image, reducing the wastage and extending the quality shelf-life of chilled fish and seafood products.

This chapter provides some brief background information on the MAP of fish and seafood products along with limited market data and a discussion of the safety and quality assurance aspects of such products. A few selected new developments are also briefly highlighted.

Background and market information

The preservation action of carbon dioxide (CO_2) on flesh foods has been known for over a century. In the 1930s, Coyne demonstrated that fillets or whole fish at ice temperature could be kept twice as long if stored in an atmosphere containing a minimum of 25% CO_2 (Coyne 1932). Many other researchers also demonstrated that CO_2 atmospheres prolonged the shelf-life of whole fish and shellfish but commercialisation of MAP for retail packs of fish and seafood products did not materialise until the late 1970s (Davis 1998). As a result of improvements to packaging materials and machinery, MAP systems for fish and other perishable food products became firmly established during the early 1980s in retail supermarkets, particularly in the UK and France (Day 1993).

Over the last decade, the development and commercialisation of MAP for fresh chilled food products has been most rapid in the UK and France which hold about 40% and 35% of the European MAP market, respectively (Day 1993). Although reliable market figures for the total European MAP market are notoriously difficult to compile, it was estimated that 2000 million MA packages were produced in 1990 in contrast to only 650 million units in 1986. Of these MA packages, it was estimated that 250 million MA fish packs were produced in 1990 in contrast to only 50 million MA fish packs in 1986 (Day 1993). By comparison, a recently published market research report estimated that 2595 million MA packages (of which 9% were MA packs of fish and shellfish) were sold in the UK alone in 1997, with significant growth in demand to 3370 million units predicted for 2002 (MSI 1998)

Although the retail MAP of chilled fish and seafood products is now well established in Europe, it has hardly made an impact in the US market because of longer shelf-life requirements along with lingering food safety concerns (Rose 1992). For the past ten years, the US National Maritime Fisheries Service has maintained strict control over the certification and hence commercialisation of chilled vacuum and MA packed fish. To date, only a handful of companies have applied for certification and they must provide documented evidence that the fish is packed no later than five days after capture; storage, transportation and display are at a temperature below $3°C$; the packs must be clearly labelled with storage conditions; and the product is to be sold within 10 days of packing (Rose 1992). Because of the demographics of the US market and the less sophisticated chilled distribution and national retail network compared with the UK and parts of Europe, it is not surprising that so few US companies are prepared or able to meet the requirements laid down above.

MAP gas mixtures

The gas mixtures used in the MAP of fish and seafood products depend on the specific species and pack format, but for almost all products these will be some combination of CO_2, oxygen (O_2) and nitrogen (N_2). Typically these gases are supplied to food manufacturers by industrial gas companies in the form of cylinder, cryogenic, N_2 dispensing or air separation systems.

For MAP applications, gases supplied in cylinders are the most common option. Cylinder gases can be either a single gas or a predetermined mixture depending on the requirements of the food manufacturer customer. Where high production levels make the use of pre-mixed gases uneconomic, then single gas cylinders and on-site mixing offer better flexibility. However, for situations where a limited number of products are MA packed and/or production rates are relatively low, then pre-mixed cylinder gases offer a good supply option, especially since the gas mixture composition is assured by the gas supplier.

As stated previously, MAP is a very effective technique for delaying microbial spoilage and oxidative rancidity in chilled fish and seafood products. MAP is particularly effective at extending the shelf-life of white fish products (e.g. cod, haddock, plaice, whiting, coley, shark, hake, halibut, skate, Dover and lemon sole). For retail MA packs of chilled white fish, crustaceans (e.g. prawns, shrimps, crab and lobster) and shellfish (e.g. mussels and cockles) products, gas mixtures containing 35–45% CO_2, 25–35% O_2 and 25–35% N_2 are recommended (SFIA 1985; Day 1992; Air Products PLC 1995). Gas mixtures containing 35–45% CO_2 and 55–65% N_2 are recommended for retail MA packs of chilled oily or smoked fish products (e.g. herring, mackerel, salmon, trout, sardines, whitebait and Greenland halibut). For bulk MAP of all types of fish and seafood products, gas mixtures of 40–60% CO_2 and 40–60% N_2 are recommended. Tables 24.1 and 24.2 list information on the MAP of raw, high fat, oily fish products (Day & Wiktorowicz 1999). More detailed information can be accessed via the Air Products Plc website at http://www.airproducts.com/freshline.

The inclusion of CO_2 is necessary for inhibiting common aerobic spoilage bacteria. However, for retail packs of fish and other seafood products, too high a proportion of CO_2 in the gas mixture can induce pack collapse, excessive drip, and in cold-eating seafood products such as crab, an acidic, sherbet-like flavour. O_2 is necessary to prevent growth of *Clostridium botulinum* type E, and to reduce drip, colour changes and bleaching in white fish, crustacean and shellfish MA packs (SFIA 1985; Day 1992; Air Products PLC 1995). However, O_2 is preferentially excluded from oily fish retail MA packs and bulk MA packs so as to inhibit oxidative rancidity.

It is interesting to note that other gases have been used experimentally or on a limited commercial basis to extend the shelf-life of fish and seafood products. For example, carbon monoxide has been shown to be very effective at maintaining the red stripe of salmon and is currently used commercially in Norway. However, its use in the UK and other EU countries is not permitted due to safety concerns (Day 1992).

MAP machinery, materials and pack format

There are numerous MAP techniques used for extending the shelf-life of fish and seafood products (Stammen *et al.* 1990; Day 1992; Air Products PLC 1995; Day & Wiktorowicz 1999). The most popular MA retail pack consists of a clear thermoformed semi-rigid tray which is hermetically sealed to a clear flexible barrier lidding film on a thermoform-fill-seal machine. The semi-rigid tray is typically constructed of polyethylene/PVC and is ribbed on the bottom to allow passage of gas to the underside of the product. The lidding film is most commonly constructed of polyester/polyethylene with a sandwich layer of PVDC to provide gas barrier properties. A gas/product ratio of 3 : 1 is recommended (SFIA 1985). A variation of this pack format combines the advantages of both vacuum skin packaging (VSP) and MAP. This is done by producing a standard VSP pack, but with a permeable top web skin. After this process, the product enters a second sealing station where gas is introduced and a third web is used to place a barrier lidding film on the pack (Day 1993).

Although the semi-rigid tray and lidding film pack format is undoubtedly practical and hygienic, certain retailers and consumers perceive it as giving the product an over manu-factured look compared with the traditional expanded polystyrene tray and cling film overwrap format. For this reason, some retailers have decided to revert back to traditional retail packaging of their white fish range. In order to benefit from the extended shelf-life potential of MAP during distribution, the master pack technique can be utilised. This consists of placing several traditionally packed products into a large gas barrier bag and flushing with the desired gas mixture using a snorkel-type machine. Extended shelf-life is

Table 24.1   A summary of information relating to MAP of raw high-fat oily fish (food items, gas mixes, labelling and storage requirements, spoilage mechanisms and food poisoning hazards).

Food items	Bluefish
	Carp
	Eel
	Greenland halibut
	Herring
	Mackerel
	Pilchards
	Rock salmon
	Salmon
	Sardines
	Sprats
	Swordfish
	Trout
	Tuna
	Whitebait
Recommended gas mix	
Retail	40% CO_2, 60% N_2
Bulk	70% CO_2, 30% N_2
Labelling requirements**	'packaged in a protective atmosphere'
Storage temperature	
Legal maximum*	8°C
Recommended	−1°C to +2°C
Achievable shelf-life	
In air	2–3 days
In MAP	4–6 days
Principal spoilage mechanisms	Oxidative rancidity
	Microbial, e.g.
	Pseudomonas species
	Acinetobacter/Moraxella species
	Lactobacillus species
	Flavobacterium species
	Alteromonas species
	Streptococcus species
	Shewanella species
Possible food poisoning hazards	*Clostridium botulinum* (non-proteolytic E, B and F)
	Vibrio parahaemolyticus

* Food Safety (Temperature Control) Regulation 1995. There is some flexibility to vary this legal maximum when scientifically justified.
** As required by Regulation 33 of the Food Labelling Regulations 1996.

Table 24.2 A summary of information relating to MAP of raw high-fat oily fish (machines, types of package and materials).

Typical MAP machines	
Retail	TFFS – Thermoform-fill-seal
	CBDF – Cryovac BDF system
	PTLF – Preformed tray and lidding film
	TWTFFS – Three web thermoform-fill-seal
Bulk	VC – Vacuum chamber
	ST – Snorkel-type
Typical types of package	
Retail	Tray and lidding film
	Cryovac BDF system
	Three web pack
Bulk	Bag-in-box
	Master pack
Examples of typical MAP materials	
Retail	Tray
	UPVC/LDPE
	HDPE
	EPS/EVOH/LDPE
	Lidding film
	PET/PVDC/LDPE
	PA/PVDC/LDPE
	PC/EVOH/EVA
	Cryovac BDF shrink film
Bulk	PA-LDPE
	PA/EVOH/LDPE

Abbreviations of typical MAP materials: EPS, Expanded polystyrene; EVA, Ethylene vinyl acetate; EVOH, Ethylene vinyl alcohol; HDPE, High density polyethylene; LDPE, Low density polyethylene; PA, Polyamide (nylon); PC, Polycarbonate; PET, Polyester; UPVC, Unplasticised polyvinyl chloride.

achieved throughout distribution until the bag is opened prior to retail display. Alternatively, some retailers are using coloured barrier foam trays and lidding film for their MA packed fish and seafood product ranges (Air Products PLC 1995).

Quality assurance and safety aspects

The MAP of chilled fish and seafood products has been marketed in the UK for many years and during this period these MA packs have provided the consumer with high quality, safe and fresh products. However, it is important for all personnel involved in the manufacture and handling of MA packed chilled fish and seafood products to be vigilant about possible food safety hazards. It is imperative that the food safety of such products is not compromised by complacency and poor manufacturing or handling practices (SFIA 1985; Day 1992).

Some concern has been expressed regarding the vacuum or MAP of chilled fish and seafood products because of the possibility of growth and toxin production of anaerobic bacteria such as *Clostridium botulinum* type E. This issue has been extensively reviewed and recommendations for assuring the safety of chilled perishable foods with particular emphasis on the risks of botulism, are now available (Betts 1996). Amongst the many recommendations is one that states that chilled perishable food packed under reduced O_2 levels with an assigned shelf-life of more than 10 days should contain one or more controlling factors in addition to chill temperatures to prevent growth and toxin production by *Clostridium botulinum*. Such controlling factors include heat treatment, acidity, salt level and water activity (Betts 1996).

Regarding the MAP of chilled fish and seafood products in the UK, the key issue is the ten day shelf-life which is never exceeded commercially. Hence, the risk of botulism is negligible and this is borne out by the fact that there has not been a single incident of botulism associated with such products in the UK (Betts 1996). Nevertheless, it is still important for all personnel to be vigilant and practise good manufacturing and handling procedures. Fish and seafood of the highest quality should be used and maintenance of recommended chilled temperatures and good hygiene is essential (SFIA 1985; Day 1992; Air Products PLC 1995). In addition, quality assurance procedures such as microbiological testing, seal integrity testing, gas analysis and temperature monitoring should be routinely performed to assure the safety and extended shelf-life of MA packed products (Day 1992; Air Products PLC 1995). The continuing commercial success of MA packed fish and seafood products in the UK is reliant on maintaining such high standards.

Conclusion

Fish and seafood products are highly perishable and the shelf-life of these products is limited by microbial and oxidative spoilage. So long as good manufacturing and handling conditions are followed, MAP is a very effective preservation technique for improving the image, reducing wastage and extending the quality shelf-life of a wide range of chilled fish and seafood products.

References

Air Products PLC (1995) *The Freshline® Guide to Modified Atmosphere Packaging (MAP)*, pp.1–66. Air Products PLC, Basingstoke, Hampshire, UK.

Betts, G.D. (1996) *A code of practice for the manufacture of vacuum and modified atmosphere packaged chilled foods*. Guideline No. 11, pp.1–114. Campden and Chorleywood Food Research Association, Chipping Campden, Gloucestershire, UK.

Coyne, F.P. (1932) The effect of carbon dioxide on bacterial growth with special reference to the preservation of fish. *Journal of the Society of Chemical Industry*, April, 119–21.

Davis, H.K. (1998) Fish. In: *Principles and applications of modified atmosphere packaging of foods*, 2nd edn (ed. B.A. Blakistone), pp.194–239. Blackie Academic and Professional, London, UK.

Day, B.P.F. (1992) *Guidelines for the good manufacturing and handling of modified atmosphere packed food products.* Technical Manual No. 34, pp.1–79. Campden and Chorleywood Food Research Association, Chipping Campden, Gloucestershire, UK.

Day, B.P.F. (1993) Recent developments in MAP. *European Food and Drink Review*, Summer edition, 87–95.

Day, B.P.F. & Wiktorowicz, R. (1999) MAP goes on-line. *Food Manufacture*, 74(6), 40–41.

MSI (1998) *Data report: Modified atmosphere packaging: UK*, pp.1–75. Marketing Strategies for Industry (UK) Limited, Chester, UK.

Rose, P. (1992) Why the US fish industry isn't hooked on MAP. *Packaging Week*, August 5th edn, 11.

SFIA (1985) *Guidelines for the handling of fish packed in a controlled atmosphere*. Sea Fish Industry Authority, Hull, UK.

Stammen, K., Gerder, D. & Caporaso, F. (1990) Modified atmosphere packaging of seafood. *CRC Food Science and Nutrition*, 29(5), 301–31.

Chapter 25

Rigor Measurements

U. Erikson

SINTEF Fisheries and Aquaculture, N-7465 Trondheim, Norway

Introduction

At the time of death, the muscle is relaxed and has a limp elastic texture. Eventually the whole fish gradually becomes hard and stiff. This phenomenon is called *rigor mortis* (rigor) and the onset occurs when the adenosine triphosphate (ATP) content is nearly depleted. This is because the detachment of the myosin and actin filaments are energetically dependent on ATP. The characteristic development of rigor starts in parts of the fish and progressively includes the whole fish. In salmonids the impression is that the rigor process starts in the neck region even in cases where the fish were subjected to excessive ante-mortem stress and where vigorous tail beating occurred with its anticipated ATP depletion. The rigor process continues with a gradual increase in rigor strength up to a maximal value. Subsequently, the rigor resolves until the muscle once again becomes limp, usually after 2–4 days (ice storage). However, the muscle is still not as elastic as before rigor onset. Tenderisation may be caused by a weakening of the Z-discs of the myofibrils (Hultin 1984; Seki & Tsuchiya 1991) or of myosin-actin junction (Yamanoue & Takahashi 1988), or by a degradation of connective tissue (Seki & Watanabe 1984; Montero & Borderias 1990; Ando *et al.* 1993).

If rapid ATP-depletion in the muscle occurs, either ante- or post-mortem, such as if the fish is subjected to handling stress during harvesting (Erikson *et al.* 1997) or if the fish is not effectively chilled after killing (Stroud 1968), the time to rigor onset is dramatically reduced (see Fig. 25.1). In such cases strong rigor tensions occur causing a disruption of the muscle cell-myocommata junctions. In turn, the flesh may gape and the economic and technological value of the fish could be reduced.

Particularly in some older literature, data showing onset and duration of rigor for different species are sometimes tabulated. It follows from the above that this type of information should at best be used as an indication only, unless pre-slaughter handling procedures have been described thoroughly (which is usually not the case). When instrumental methods are used to determine rigor, the location on the fish for the assessment should be taken into consideration. The ATP-degradation pattern along the fish muscle may not be uniform, consequently the rigor development at different locations may also be different in the early phase of rigor (Berg *et al.* 1997).

It has been pointed out that the mechanism regarding rigor onset is not fully understood since onset in unstressed fish apparently occurs at a higher ATP-content than in stressed fish

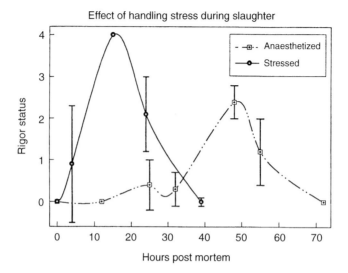

Fig. 25.1   Rigor status of anaesthetised and stressed Atlantic salmon (*Salmo salar*) during ice storage. Mean ± SD (*n* = 12). Experimental data: Commercially farmed fish (4.4 ± 1.0 kg) fasted for 15 days. The fish were acclimated to a seawater temperature of 15°C under laboratory conditions. The fish were anaesthetised (Benzocaine 68 mg/L) in the holding tank or chased to exhaustion. Fish in both groups were killed by a blow to the head. The fish were immediately gutted, but not bled prior to chilling. (Unpubl. obs. of U. Erikson, A. Sverdrup, J.E. Steen and T. Rosten.)

(Korhonen *et al.* 1990; Mochizuki & Sato 1994). It could be hypothesised that this may be due to a possible non-uniform distribution of the high-energy phosphates along the fish. Excision of muscle samples for analysis reflects the macroscopic average content in the sample which may not necessessarily be identical to the levels at a cellular level. Furthermore, such a non-uniform rigor development may be the cause of the lower rigor tensions commonly observed in unstressed fish (Berg *et al.* 1997). This hypothesis may be extended to explain the strong rigor tensions at high storage temperatures, where the common factor with stressed fish is a rapid ATP-depletion more or less simultaneously over the *whole* fish length.

   Electricity is commonly used for stunning or tenderisation of mammal muscles. In contrast, fish muscle is inherently tender due to its lower content of connective tissue. This also means that the mechanical strength of the fish muscle is considerably lower. The use of electricity as a killing method may cause a rapid depletion of ATP, initiating a rapid and strong rigor. Furthermore, it has been reported that the fish in such cases developed tonic spasms and did not bleed well (DC-current) or suffered induced convulsions (AC-current) (Boggess *et al.* 1973).

   If the fish are in rigor shortly after slaughter, this should be an incentive to improve fish handling routines during harvesting by reducing ante-mortem handling stress. Once rigor has started, further handling and processing should ideally be avoided until the fish are in the post-rigor state. However, this view should be weighed against the loss of freshness and costs of storage. If fish in rigor are processed, the filleting yield will be poor and the handling involved may cause gaping (Huss 1995). Furthermore, fish in rigor should not be forcibly straightened from a bent position during filleting or packing as this may damage the fish and cause gaping (Lavety 1984).

Pre-rigor frozen whole fish and fillets can give good products if they are carefully thawed at a low temperature to give rigor time to pass while the muscle is still frozen (Huss 1995). On the other hand, several fish species frozen in rigor show a sharp increase in gaping compared with pre-rigor frozen fish (Love *et al*. 1969).

In addition to their use in research, rigor assessments used in the industry can provide valuable information for improving harvesting and processing routines. Due to the close relationship with post-mortem degradation of glycogen, ATP and related compounds, the longer the time to rigor onset the fresher the fish. In addition, it has been shown that if the pre-rigor period is prolonged, the flesh texture may be improved (Jerrett & Holland 1998; Jerrett *et al*. 1998). Thus, if fish can be presented in the market still in rigor, rigor may be considered an inherent indicator of quality.

Factors affecting rigor development

Handling stress, physical condition and killing method

As already mentioned, handling stress during slaughter depletes high-energy phosphates and glycogen stores, initiating a rapid rigor onset as shown in several species (Izquierdo-Pulido *et al*. 1992; Lowe *et al*. 1993; Mochizuki & Sato 1994; Berg *et al*. 1997; Sigholt *et al*. 1997). Also, the maximal muscle tensions in stressed fish during rigor have been shown to be considerably higher than in unstressed fish (Korhonen *et al*. 1990; Nakayama *et al*. 1992) affecting flesh texture. Depending on species, glycogen in the muscle may be used as fuel during the initial phase of fasting. In such species, heavy glycogen depletion during fasting prior to slaughter may result in a shorter pre-rigor period. Nakayama *et al*. (1996) demonstrated that spinal cord destruction in addition to stabbing the medulla oblongata delayed rigor development the most. The peak value of rigor tension was small, resulting in high muscle breaking strength due to the extremely gradual development of rigor tension resulting in less weakening of structure. Conversely, when fish are killed using electroshock, a shorter rigor period will result compared with fish killed by a blow to the head, or notably with asphyxiated fish (Amlacher 1961).

Species

The speed of the rigor progression differs among species (Bito *et al*. 1983) and on whether a given species is cultivated or not. For instance, rigor started earlier in cultured specimens of red sea bream (*Pagrus major*) than in their wild counterparts both at 0°C and 10°C (Iwamoto & Yamanaka 1986). Similarly, wild plaice (*Paralichtys olivaceus*) showed slower rigor progress than cultured at 10°C, but both wild and cultured fish showed accelerated rigor when stored at 0°C (Iwamoto *et al*. 1990). These differences might of course reflect differences in pre-slaughter handling.

ATPase activity

Fish adapt to the habitat temperature by altering ATPase activity. When the habitat temperature is lowered, the activity of the enzyme is increased (Tsuchimoto *et al.* 1988). For example, the myofibrillar ATPase activity from cold acclimated (1°C) goldfish (*Carassius auratus*) muscle was higher than in the warm acclimated muscle (26°C) (Johnston *et al.* 1975) and similarly carp reared at 10°C had higher ATPase activity than those reared at 26°C (Misima *et al.* 1993). But, once a given fish is chilled after slaughter, the myofibrillar Mg^{2+}-ATPase activity decreases with decreasing storage temperature and the ATP-depletion rate is slowed down increasing the pre-rigor period. Tsuchimoto *et al.* (1998) reported that a ranking of myofibrillar Mg^{2+}-ATPase activity among fish species, as well as between cultured and wild fish of the same species, was not consistent with the different muscle contraction rates. This was explained by suggesting that the main factor was the different Ca^{2+}-concentrations around the myofibrils. Thus, ATP-depletion rates do not necessarily depend on enzyme activity per se, but can also be the result of an increased content of Ca^{2+}-ions surrounding the contractile proteins due to a leakage of the sarcoplasmic reticulum at lower temperatures.

Storage temperature

Storage temperature has a major effect on rigor development (Bito *et al.* 1983; Iwamoto *et al.* 1987, 1990; Hwang *et al.* 1991; Mochizuki & Sato 1994; Lee *et al.* 1998). In cold water species, rigor development is commonly delayed as the temperature is gradually reduced towards 0°C. Storage at higher temperatures not only results in a more rapid rigor development, but it also means that a very strong rigor is obtained. The resulting high tensions in muscle may, as already mentioned, cause fillet gaping due to weakening of the connective tissue (Burt *et al.* 1970; Huss 1995). However, ice storage may not always extend the pre-rigor period as has been shown for tropical fish or fish acclimated to high water temperatures. For example, in plaice (*Paralichthys olivaceus*) spiked in the brain, rigor started after 6 h at 0°C and after 15 h at 10°C (Iwamoto *et al.* 1988). Also, the depletion of ATP was slower between 5°C and 15°C than at 0°C (Iwamoto *et al.* 1987) due to poor sarcoplasmatic reticulum uptake of Ca^{2+}-ions, activating Mg^{2+}-ATPase (Watabe *et al.* 1989). Similarily, cold-acclimated (10°C) carp (*Cyprinus carpio*) had a slower rigor progress with related biochemical changes than their warm-acclimated (30°C) counterparts during storage both at 0°C and 10°C. At 20°C though, the effect was opposite which was explained by the higher myofibrillar Mg^{2+}-ATPase activity of the cold-acclimated fish (Hwang *et al.* 1991).

Thaw rigor

When fish is frozen pre-rigor, rigor may proceed slowly during frozen storage or, depending on storage time and storage temperature, it can occur during thawing (thaw rigor). The phenomenon is probably more likely to occur in cases when good routines for freezing (rapid freezing) and slaughter (rested harvesting) are applied, where ATP and glycogen are preserved to some extent. Thaw rigor is characterized by rapid and intense muscle contractions compared with rigor in unfrozen muscle causing high drip losses and a harder flesh texture.

Ma *et al.* (1992) showed that the strong contractions during thawing were due to a rupture of the sarcoplasmic reticulum membranes during freezing and thawing causing Ca^{2+}-release activating the myofibrillar Mg^{2+}-ATPase. Cold storage for several weeks is a way to prevent thaw rigor as the flesh has time to pass through rigor in the frozen state. This procedure had no detrimental effects on whole fish or fillet quality (Stroud 1968). Moreover, Bito (1986) suggested that thaw rigor is likely to be depressed by slow thawing in the range of $-2\,°C$ to $0\,°C$.

Handling during rigor assessments

As well as maintaining a constant storage temperature, minimal and careful handling of the fish during rigor assessment is a prerequisite to obtaining accurate and comparable results. Berg *et al.* (1997) showed that when the fish were repeatedly taken out of storage boxes for rigor evaluation using different methods, this clearly shortened both the rigor duration and intensity compared with fish that were left in the boxes and evaluated solely by using a gentle sensory method. Also, excision of muscle samples may affect rigor development. Ando *et al.* (1996) demonstrated that the rigor development was accelerated when samples (1 g) were cut from the dorsal muscle.

Rigor-related phenomena

In some tropical species cold-shock stiffening may occur within a few minutes after killing and chilling the fish quickly in ice (Curran *et al.* 1986a). It was concluded that cold shock stiffening was different from rigor stiffening. The implications of the cold shock reaction were reduced filleting and processing yields, as well as a high drip loss and occurrence of gaping (Curran *et al.* 1986b).

Johansen *et al.* (1996) observed an almost immediate strong rigor-like stiffening when Atlantic salmon (*Salmo salar*) (14 to 24 % total fat content in muscle) acclimated to $5\,°C$ were chilled in seawater at $-1.5\,°C$ after slaughter. However, no such stiffening occurred at $-0.5\,°C$. The stiffness seemed to be related to the outer layer of the body (including skin) and the phenomenon was not due to either freezing, cold shock or early rigor. It was speculated that the phenomenon was related to fatty acid hardening (fat depots and red muscle). However, no detrimental effects on the flesh quality were found after three and eight days of ice storage. To distinguish this phenomenon from true rigor, simply place the fish in ambient atmosphere and the stiffness will disappear after a few minutes. Possibly, a similar effect might have been observed by Tomlinson *et al.* (1964) who reported that when steelhead trout (*Onchorhyncus mykiss*) was chilled in refrigerated seawater ($0\,°C$), the fish became stiff without any detectable loss of ATP.

Methods

Rigor assessments can be carried out either on whole fish or on excised muscle samples. In contrast to mammals, fish flesh lacks convenient attachment sites (such as tendons) for instrumental measurements of excised muscle. This, as well as the anatomical complexity and the relatively fragile nature of fish muscle, may make such methods less quantitative in fish. The sampling or measuring locations for the assessments should be taken into consideration as the anatomy (muscle fibre orientation) of the fish (Jerrett & Holland 1998), as well as the rigor development (Berg *et al.* 1997), may be different along the fish.

Rigor index (RI)

Rigor index (RI) is by far the most commonly used method for rigor assessment, almost to the point that it has become the standard method. RI is a modification of Cutting's method from 1939 and was introduced by Bito *et al.* (1983). The sag of the tail is measured when the front half of the fish's body is placed on a horizontal table (Fig. 25.2). The RI is calculated from the equation:

$$\text{Rigor index } (\%) = 100 \, (D - D_o) \, / \, D_o$$

where D_o and D represent the distance of the base of the caudal fin from the horizontal line of the table, as measured pre-rigor and at subsequent intervals during storage, respectively. Note that it is important that the fish are stored flat between measurements. During the depletion of the high-energy phosphates and glycogen, along with the corresponding reduction in the pH of the muscle, the RI gradually starts to increase and by the time the high-energy phosphates are almost fully depleted, full rigor is attained (RI = 100%) (Iwamoto *et al.* 1987; Watabe *et al.* 1991). Consequently, the RI method correlates well with the underlying biochemistry related to rigor. However, Chen and Kong (1997) reported that the RI (whole body) developed faster than the RI of the fibres calculated as the percentage of the muscle shortening. It was concluded that the whole body-based RI method was less sensitive than the RI of the fibres. Fig. 25.3 shows a typical RI curve for unstressed Atlantic salmon. After about 8 h post-mortem, the first signs of rigor onset were observed. Subsequently, the rigor strength gradually increased until maximum rigor was attained at about 30 h. Thereafter, the rigor strength diminshed until the fish were in the post rigor state at

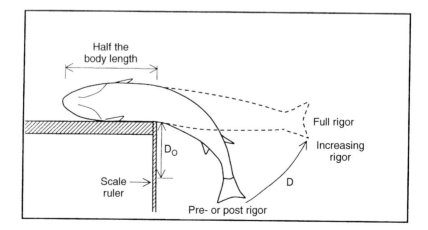

Fig. 25.2   The rigor index method.

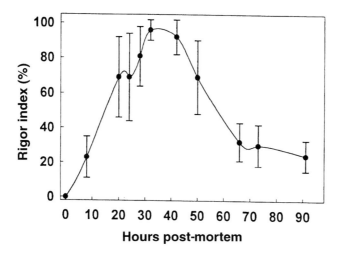

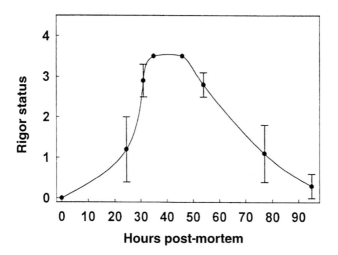

Fig. 25.3   Rigor index and rigor status (with the same individual fish) of unstressed Atlantic salmon (*Salmo salar*) during ice storage. Mean ± SD (*n* = 12). Experimental data: Commercially farmed fish (5.6 ± 1.1 kg) fasted for 15 days. The fish were acclimated to a seawater temperature of 15°C under laboratory conditions. The fish were netted individually and killed by a blow to the head within 5–10 s. The fish were immediately gutted, but not bled prior to chilling. (Unpubl. obs. of U. Erikson, A. Sverdrup, J.E. Steen and T. Rosten.)

about 70 h post-mortem. Typically (at least for Atlantic salmon), the curve levelled off at RI > 0 % (initial value).

Rigor status

This subjective method is based on a scale as described by Curran *et al.* (1986a,b) and was used by Johansen *et al.* (1996), Berg *et al.* (1997) and Sigholt *et al.* (1997). The scale description can be simplified to:

0   Pre- or post-rigor
1   Rigor onset (first sign of stiffness, for instance in neck or tail region)
2   Rigor (a larger area is clearly in rigor)
3   Whole fish rigor
4   Stronger rigor;
5   Very strong rigor (the fish is extremely stiff, rod-like)

The method is rapid and it is not necessary to take the fish out of boxes, thus minimising the effects of handling. The tail is gently lifted and the muscle along the fish length is gently touched to get a total impression of the stiffness. Using the descriptive scale above does require some training, but depending on skill, a simpler scale can be used, e.g. 0–3. Using the same individual fish, the method (scale: 0–4) correlated well with the RI method as shown in Fig. 25.3 except that the rigor onset was detected earlier using the RI method. Figure 25.1 shows the full potential of rested harvesting and the time available for pre-rigor processing. The curves represent the extremes of the influence of handling stress because the fish were either chased to exhaustion (the fish could easily be lifted by the tail out of the holding tank without opposition) or anaesthetised (no vigorous swimming activity). It should be mentioned that the rigor development of the stressed fish is typical of commercially slaughtered salmonids based on CO_2-anaesthesia. Notably, the duration of rigor per se was not much different between treatments. Rather, a displacement in time (about 30 h) was seen where the pre-rigor time made up the difference. Note that there is no standard deviation at the peak value (rigor status 4) of the stressed fish curve showing that all fish exhibited a strong rigor within a narrow period of time. Also, the peak value of the anaesthetised fish was lower (rigor status 2–3). As mentioned previously, this may be beneficial for the post-rigor flesh texture. Since the rigor status is not a continuous variable, it is not mathematically correct to present data as mean ± SD values as shown in the figure, but for the sake of simplicity and for easy comparison with the RI-method, it is used here.

Isometric tension

A muscle strip is excised and mounted with one end in a vice placed in a humidity chamber at constant temperature where the sample may be suspended in physiological saline. The upper end is attached to one arm of a balance. The muscle length is kept constant during measurements. The force applied to maintain the length constant is recorded. The tension development during isometric rigor contractions and the following decline are recorded using transducer systems or a balance. The curve shape (tension versus time) is affected by the different factors discussed previously (Buttkus *1963*; Burt *et al.* 1970; Korhonen *et al.* 1990; Nakayama *et al.* 1992, 1996; Jerrett & Holland 1998; Tsuchimoto *et al.* 1998).

Isotonic tension

The extensibility of the muscle can be measured by allowing the muscle to shorten under a constant loading and unloading cycle. The force applied to the muscle is kept constant while the change in length or elasticity due to rigor is recorded (Schmidt *et al.* 1968).

Typically, an excised muscle strip kept in a chilled physiological saline is attached to a fixed bottom clamp, while the upper clamp is connected to a beam via a steel thread. The rigor contractions (length versus time) can be measured using a linear variable differential transformer connected to a recorder. It has been shown that the curve shape is clearly affected by the ante-mortem stress. Also, the rigor development using the method correlates well with the results from isometric tension measurements (Nakayama *et al.* 1997).

Low frequency vibration

A non-destructive low frequency vibration (LFV) method was devised based on an electrical signal source, a vibrator and two sensors. During measurement the fish was placed on a thin plastic film held by a frame. The sensor touched the plastic film with the fish on top. Sensors and source were controlled by a PC and all data were logged to file. A low-frequency (1–200 Hz) vibrational force was applied and the mechanical impedance of the muscle was measured. The frequency response profiles differed from fish to fish as well as on different locations along single fish. The only obvious trend was that each rigor state always represented a change in profile at a given location. It was concluded that correct classification of different rigor states based on a visual assessment of the profiles would be practically impossible.

Artificial Neural Networks (ANN) are a useful technique in classification of data. Data from LFV and a rigorometer (see below) were used as input data to describe different rigor states defined by a subjective method (rigor status). The concept of ANN is based on the recognition of data compared to data from a training phase. When using LFV data as input, a higher number of correct classifications were obtained compared with using rigorometer data as input. Moreover, the changes in mechanical impedance combined with ANN coincided well with the subjective evaluation of rigor. With further improvement of the method using LFV + ANN it was considered well suited for automatic rigor classification in a commercial slaughter line (Berg *et al.* 1997).

Rigorometer

An instrument was based on a design by Amlacher (1961). A circular-ended piston was pressed vertically into the fish just above the lateral line (three locations) by applying a force on a spring. The force needed to press down the piston until a limiting plate touched the skin was measured by a sensor and displayed. The rigor development was followed using the average value from three locations on the same fish. Although the method could distinguish between unstressed and stressed fish, it was clear that, compared to fish assessed by another method (rigor status), the rigor duration was considerably shortened as a result of using the rigorometer. It was concluded that repeated measurements at particular positions resulted in softer texture and consequently too low readings. Due to small differences in force, the instrument was not considered suitable for rigor assessments using random fish from a batch. However, the method may be used to follow an approximate rigor course on single fish (Berg *et al.* 1997).

Reduction in body length or length of excised muscle samples

Due to muscle contractions during rigor, the body length will be reduced. Johansen *et al.* (1996) demonstrated that this feature can be used (at least on large fish) for rigor assessments. The body length was reduced by maximally 2.5% and a good correlation with stiffness score (rigor status) was found.

Alternatively, the contraction (%) of excised white muscle samples can be measured. The sample is typically placed on a glass plate in a chamber with a water-saturated atmosphere at constant temperature. The length of the muscle is measured immediately after excision (pre-rigor) and during the following lapse of time (Lee *et al.* 1998; Tsuchimoto *et al.* 1998). Tsuchimoto *et al.* (1998) regarded the contractile percentage of the excised muscle strips as a simple and useful method as several samples can be measured simultaneously and no particular measuring apparatus is required. The method distinguished between species and between cultured and wild fish of the same species. Moreover, the results agreed well with the corresponding changes in isometric tension. Jerrett *et al.* (1998) used time lapse video to study the rigor contraction development on excised white muscle strips as affected by fatigue and temperature. A scale ruler was immersed in the same focal plane as the muscle strips in a bath with marine Ringers solution. The changes in strip length were monitored directly from a monitor.

Indirect methods

Due to the close relationship between rigor onset and the depletion of muscle ATP (with a concomitant increase in IMP) and glycogen (with increase of lactate and H^+), measurements of the patterns of change in metabolites may be used as indicators of rigor development. As the changes are more or less complete in the early rigor phase, these methods cannot be used to monitor the resolution of rigor. The ATP-related metabolites are usually analysed using the rather laborious method of making extracts followed by HPLC-analysis. Alternatively, the ATP : IMP ratio can be analysed using a simple method proposed by Khan and Frey (1971). The absorbance at 258 and 250 nm of muscle perchloric acid extracts is determined. The method has been shown to correlate well with corresponding measurements of pH, RI and rigor tension (Korhonen *et al.* 1990; Nakayama *et al.* 1996, 1997).

In chinook salmon (*Oncorhynchus tshawytscha*) Jerrett and Holland (1998) showed that rigor contraction onset coincided with a muscle pH of 6.6, a value that was not strongly affected by exercise. Furthermore, the contractions ended when the muscle pH levelled off at pH 6.2–6.3. Similarly, Nakayama *et al.* (1992) reported that the muscle pH in carp was about pH 6.55 at the onset of the tension generation irrespective of whether the fish were unstressed or stressed. In sea bream, the time when pH reached the ultimate value (pH 6.2–

6.5, depending on killing method) coincided with the time when full rigor was attained (Nakayama *et al.* 1996).

Other methods

Shear strength
Montero and Borderias (1990) employed shear strength measurements on trout (*Salmo irideus*) fillet slices using the method of Møller (1980–1981). During rigor they observed a slight but significant drop in shear strength due to tenderisation of connective tissue.

Degree of sag
The whole, intact fish is clamped by the tail in a vertical position. Depending on stiffness, the angle of sag could visually be divided into four stages (pre-rigor, onset, full rigor and post-rigor) (Korhonen *et al.* 1990).

Differential scanning calorimetry
Park and Lanier (1988) used differential scanning calorimetry (DSC) to study the post-mortem thermal behaviour of excised dorsal muscle. A large exothermic peak attributed to ATP hydrolysis was detected in the DSC thermogram (cal/g versus temperature). Chen and Kong (1997) reported that DSC can be used as a rapid method for assessing rigor development and the corresponding formation of the acto-myosin complex. The method correlated well with depletion of ATP and RI of the fibres.

AC-impedance
As the pH drops post-mortem, there is a close relationship with the reduction of electrical resistance and with a higher permeability to free ions in the muscle tissue (Amlacher 1961). AC-impedance has been used to study the effect of stress on rigor development with platinum needle electrodes inserted directly into the muscle. The electrodes were kept in the same position during the experiments. A low sinusoidal potential (amplitude 10 mV) was applied at 25 different frequencies (1 Hz to 65 kHz) using a frequency response analyser. From the transient changes, represented as impedance plots (Nyquist diagrams), it was deduced that the capacitive nature of intact cell membranes was evident in pre-rigor muscle. During rigor, the capacitive contribution to the overall impedance gradually diminished whereas in post-rigor muscle, only minor changes occurred as the plot indicated that mainly simple resistive conditions prevailed. Physically, the changes were attributed to degradation of membranes and consequently to increased permeability (lower resistance). These changes occurred faster in stressed fish and the changes correlated well with depletion of ATP. It was concluded that the method may be used to monitor stress and rigor-related changes during the early post-mortem period (U. Erikson and B. Johansen, unpubl. obs.).

Conclusions

Ideally, rigor measurements should be performed non-invasively. Depending on the method used, the location along the muscle for the assessment should be taken into consideration.

Although several approaches for objectively measuring rigor in fish have been published, no commercial instrument is currently available for standardised procedures. Several suggested methods require complex equipment and may be time consuming in use. Simple manual methods such as rigor index or rigor status can be considered adequate for many purposes as they clearly differentiate between different pre-slaughter treatments and different post-mortem storage conditions. However, in cases when rigor is studied at cellular levels, methods based on excision of muscle samples must be utilised.

References

Amlacher, E. (1961) Rigor mortis in fish. In: *Fish as Food* (ed. G. Borgstrom), Vol. 1, pp.385–409. Academic Press, New York.

Ando, M., Toyohara, H., Shimizu, Y. & Sakaguchi, M. (1993) Postmortem tenderization of fish muscle due to weakening of pericellular connective tissue. *Nippon Suisan Gakkaishi*, 59, 1073–76.

Ando, M., Banno, A., Haitani, M., Hirai, H., Nakagawa, T. & Makinodan, Y. (1996) Influence on post-mortem rigor of fish body and muscular ATP consumption by the destruction of spinal cord of several fishes. *Fisheries Science*, 62, 796–9.

Berg, T., Erikson, U. & Nordtvedt, T.S. (1997) *Rigor mortis* assessment of Atlantic salmon (*Salmo salar*) and effects of stress. *Journal of Food Science*, 62, 439–46.

Bito, M. (1986) The influence of freshness of the fish, freezing temperature, thawing rate and thawing temperature on thaw rigor. *Bulletin of Tokai Regional Fisheries Research Laboratory*, 119, 25–31.

Bito, M., Yamada, K., Mikumo, Y. & Amano, K. (1983) Studies on rigor mortis of fish – I. Difference in the mode of *rigor mortis* among some varieties of fish by modified Cutting's method. *Bulletin of Tokai Regional Fisheries Research Laboratory*, 109, 89–96.

Boggess Jr., T.S., Heaton, E.K., Shewfelt, A.L. & Parvin, D.W. (1973) Techniques for stunning channel catfish and their effects on product quality. *Journal of Food Science*, 38, 1190–93.

Burt, J.R., Jones, N.R., McGill, A.S. & Stroud, G.D. (1970) Rigor tensions and gaping in cod muscle. *Journal of Food Technology*, 5, 339–51.

Buttkus, H. (1963) Red and white muscle of fish in relation to *rigor mortis*. *Journal of the Fisheries Research Board of Canada*, 20, 45–58.

Chen, T.-Y. & Kong, M-S. (1997) Combined use of traditional and DSC methods in monitoring rigor mortis development of iced chub mackerel. *Journal of the Chinese Agricultural Chemical Society*, 35, 333–41.

Curran, C.A., Poulter, R.G., Brueton, A. & Jones, N.S.D. (1986a) Cold shock reactions in iced tropical fish. *Journal of Food Technology*, 21, 289–99.

Curran, C.A., Poulter, R.G., Brueton, A., Jones, N.R. & Jones, N.S.D. (1986b) Effect of handling treatment on fillet yields and quality of tropical fish. *Journal of Food Technology*, 21, 301–10.

Erikson, U., Beyer, A.R. & Sigholt, T. (1997) Muscle high-energy phosphates and stress affect K-values during ice storage of Atlantic salmon (*Salmo salar*). *Journal of Food Science*, 62, 43–7.

Hultin, H.O. (1984) Postmortem biochemistry of meat and fish. *Journal of Chemical Education*, **61**, 289–98.

Huss, H.H. (ed.) (1995) Quality and quality changes in fresh fish. *FAO Fisheries Technical Paper 348*, FAO, Rome.

Hwang, G-C., Ushio, H., Watabe, S., Iwamoto, M. & Hashimoto, K. (1991) The effect of thermal acclimation on *rigor mortis* progress of carp stored at different temperatures. *Nippon Suisan Gakkaishi*, **57**, 541–8.

Iwamoto, M. & Yamanaka, H. (1986) Remarkable differences in *rigor mortis* between wild and cultured specimens of red sea bream *Pagrus major*. *Bulletin of the Japanese Society of Scientific Fisheries*, **52**, 275–9.

Iwamoto, M., Yamanaka, H., Watabe, S. & Hashimoto, K. (1987) Effect of storage temperature on *rigor mortis* and ATP degradation in plaice *Paralichthys olivaceus* muscle. *Journal of Food Science*, **52**, 1514–17.

Iwamoto, M., Yamanaka, H., Abe, H., Ushio, H., Watabe, S. & Hashimoto, K. (1988) ATP and creatine phosphate breakdown in spiked plaice muscle during storage, and activities of some enzymes involved. *Journal of Food Science*, **53**, 1662–5.

Iwamoto, M., Yamanaka, H., Watabe, S. & Hashimoto, K. (1990) Comparison of *rigor mortis* progress between wild and cultured plaices. *Nippon Suisan Gakkaishi*, **56**, 101–104.

Izquierdo-Pulido, M., Hatae, K. & Haard, N.F. (1992) Nucleotide catabolism and changes in texture indices during ice storage of cultured sturgeon, *Acipenser transmontanus*. *Journal of Food Biochemistry*, **16**, 173–92.

Jerrett, A.R. & Holland, A.J. (1998) Rigor tension development in excised rested, partially exercised and exhausted chinook salmon white muscle. *Journal of Food Science*, **63**, 48–52.

Jerrett, A.R., Holland, A.J. & Cleaver, S.E. (1998) Rigor contractions in rested and partially exercised chinook salmon white muscle as affected by temperature. *Journal of Food Science*, **63**, 53–6.

Johansen, S., Rustad, T., Erikson, U. & Nordtvedt, T.S. (1996) Effect of pre-packing temperatures on Atlantic salmon quality. *Proceedings from Refrigeration and Aquaculture*, 20–22 March, Bordeaux, France, pp.189–97.

Johnston, I.A., Davison, W. & Goldspink, G. (1975) Adaptions in Mg^{2+}-activated myofibrillar ATPase activity induced by temperature acclimation. *FEBS Letters*, **50**, 293–5.

Khan, A.W. & Frey, A.R. (1971) A simple method for following *rigor mortis* development in beef and poultry meat. *Canadian Institute of Food Science and Technology Journal*, **4**, 139–41.

Korhonen, R.W., Lanier, T.C & Giesbrecht, F. (1990) An evaluation of simple methods for following rigor development in fish. *Journal of Food Science*, **55**, 346–8, 368.

Lavety, J. (1984) *Gaping in farmed salmon and trout*. Torry Advisory Note no. 90. Torry Research Station, Aberdeen.

Lee, K.H., Tsuchimoto, M. Onishi, T., Wu, Z., Jabarsyah, A., Misima, T. & Tachibana, K. (1998) Differences in progress of *rigor mortis* between cultured red sea bream and cultured Japanese flounder. *Fisheries Science*, **64**, 309–13.

Love, R.M., Lavéty, J. & Steel, P.J. (1969) The connective tissues of fish. II. Gaping in commercial species of frozen fish in relation to *rigor mortis*. *Journal of Food Technology*, **4**, 45–9.

Lowe, T.E., Ryder, J.M., Carragher, J.F. & Wells, R.M.G. (1993) Flesh quality of snapper, *Pagrus auratus*, affected by capture stress. *Journal of Food Science*, **58**, 771–3, 796.

Ma, L.B., Yamanaka, H., Ushio, H. & Watabe, S. (1992) Studies on the mechanism of thaw rigor in carp. *Nippon Suisan Gakkaishi*, **58**, 1535–40.

Misima, T., Mukai, H., Wu, Z., Tachibana, K. & Tsuchimoto, M. (1993) Resting metabolism and myofibrillar Mg^{2+}-ATPase activity of carp acclimated to different temperatures. *Nippon Suisan Gakkaishi*, **59**, 1213–18.

Mochizuki, S. & Sato, A. (1994) Effects of various killing procedures and storage temperatures on post-mortem changes in the muscle of horse mackerel. *Nippon Suisan Gakkaishi*, **60**, 125–30.

Montero, P. & Borderias, J. (1990) Effect of *rigor mortis* and ageing on collagen in trout (*Salmo irideus*) muscle. *Journal of the Science of Food and Agriculture*, **52**, 141–6.

Møller, A.J. (1980–81) Analysis of Warner-Bratzler shear pattern with regard to myofibrillar and connective tissue components of tenderness. *Meat Science*, **5**, 247–60.

Nakayama, T., Liu, D-J. & Ooi, A. (1992) Tension change of stressed and unstressed carp muscles in isometric rigor contraction and resolution. *Nippon Suisan Gakkaishi*, **58**, 1517–22.

Nakayama, T., Toyada T. & Ooi, A. (1996) Delay in *rigor mortis* of sea-bream by spinal cord destruction. *Fisheries Science*, **62**, 478–82.

Nakayama, T., Matsuhisa, M., Yamaura, M., Sumiyoshiyama, T. & Ooi, A. (1997) Delayed example in *rigor mortis* of spinal cord destroyed plaice detected by measurements of isotonic contraction and isometric tension. *Fisheries Science*, **63**, 830–34.

Park, J.W. & Lanier, T.C. (1988) Calorimetric changes during development of *rigor mortis*. *Journal of Food Science*, **53**, 1312–14, 1372.

Schmidt, G.R., Cassens, R.G. & Briskey, E.J. (1968) Development of an isotonic-isometric rigorometer. *Journal of Food Science*, **33**, 239–41.

Seki, N. & Tsuchiya, N. (1991) Extensive changes during storage in carp myofibrillar proteins in relation to fragmentation. *Nippon Suisan Gakkaishi*, **57**, 927–33.

Seki, N. & Watanabe, T. (1984) Connectin content and its postmortem changes in fish muscle. *Journal of Biochemistry*, **95**, 1161–7.

Sigholt, T., Erikson, U., Rustad, T., Johansen, S., Nordtvedt, T.S. & Seland, A. (1997) Handling stress and storage temperature affect meat quality of farm-raised Atlantic salmon (*Salmo salar*). *Journal of Food Science*, **62**, 898–905.

Stroud, G.D. (1968) *Rigor in fish – the effect on quality*. Torry Advisory Note no. 36. Torry Research Station, Aberdeen.

Tomlinson, N., Geiger, S.E. & Kay, W.W. (1964) Apparent onset of *rigor mortis* in steelhead trout (*Salmo gairdneri*) in the absence of loss of adenosine triphosphate from the ordinary muscle. *Journal of the Fisheries Research Board of Canada*, **21**, 857–9.

Tsuchimoto, M., Misima, T., Utsugi, T., Kitajima, S., Yada, S. Senta, T. & Yasuda, M. (1988) Resolution characteristics of ATP related compounds in fishes from several waters and the effect of habitat temperatures on the characters. *Nippon Suisan Gakkaishi*, **54**, 683–9.

Tsuchimoto, M., Yamaga, T., Lee, K.H., Wu, Z., Misima, T. & Tachibana, K. (1998) The influence of Ca^{2+} concentration around myofibrillar Mg^{2+}-ATPase on the speed and pattern of *rigor mortis* in fish species or cultured and wild fish. *Fisheries Science*, **64**, 148–54.

Watabe, S., Ushio, H., Iwamoto, M., Yamanaka, H. & Hashimoto, K. (1989) Temperature-dependency of *rigor mortis* of fish muscle: Myofibrillar Mg^{2+}-ATPase activity and Ca^{2+} uptake by sarcoplasmic reticulum. *Journal of Food Science*, **54**, 1107–10, 1115.

Watabe, S., Kamal, M. & Hashimoto, K. (1991) Postmortem changes in ATP, creatine phosphate, and lactate in sardine muscle. *Journal of Food Science*, **56**, 151–3, 171.

Yamanoue, M. & Takahashi, K. (1988) Effect of paratropomyosin on the increase in sarcomer length of rigor-shortened skeletal muscles. *Journal of Biochemistry*, **103**, 843–7.

Chapter 26
Measurement of Fish Flesh Colour

D.H.F. Robb

Department of Clinical Veterinary Science, University of Bristol, Langford, Bristol, BS40 5DU, UK

Introduction

Fish flesh is a widely marketed product and it is rigorously assessed by consumers prior to buying in order to determine whether it meets their demands of quality. A prime factor in their visual assessment of the product is its colour. Colour is judged to be an important indicator of the freshness of the product and the state of the animal. For instance, a grey coloured cod may be spoiled, while a pale coloured salmon may have been diseased. Up to 40% of the consumer's decision to buy a product may be determined by the colour of that product (Rasekh *et al.* 1970).

Flesh colour is especially important for some species of fish. Salmonid fish flesh is pigmented by carotenoid pigments giving it the characteristic red colour. Tuna flesh is pigmented by myoglobin giving it a red to brown colour. These colours are likely to vary according to a variety of factors. In salmonids, genetic background, stage of sexual maturation and weight have significant effects (Torrissen & Naevdal 1988). The colour of southern bluefin tuna flesh can be affected by the degree of pre-slaughter stress which the fish are placed under (Alistair Smart, pers. comm.). Similar results have been observed for Atlantic salmon (*Salmo salar*), showing that increasing the degree of muscle activity at slaughter affects the colour of the flesh (Robb & Warriss 1997; Frost 1999).

Because of the variations in flesh colour discussed above, fish farmers and producers need to have information about the colour of their products and about the colours demanded by the consumers. Consumers' requirements can be identified from consumer trials, where a range of coloured products are presented to consumers and their preferences recorded. From this the preferences of different markets can be identified. To assess colour, producers require techniques for routine monitoring. Ideally the techniques should be very robust, rapid, portable and cheap. They should also be repeatable by different operators, so that comparable measurements are made. However, the demands of the producers can be very different from those of scientists researching fish flesh quality. The research may involve direct colour assessment in order to be relevant to the demands of the producers. Alternatively, it may require determination of the pigment content of the flesh for example, in fish such as salmonids.

This chapter will introduce a series of methodologies commonly used for assessing the colour of fish flesh. The advantages and disadvantages of each method will be highlighted and the potential uses suggested.

Subjective rating

The simplest way of assessing colour is to ask an assessor to score the colour of the fish subjectively. To carry this out efficiently the assessor needs information on the colour required. This can be gained from personal experience after a period of exposure to the range of colours normally found (Rasekh *et al*. 1970; Little *et al*. 1979; Chen *et al*. 1984; Waagbø *et al*. 1993), or by training (Foss *et al*. 1984, 1987; Storebakken *et al*. 1987).

Commercially, the assessor will look at each fish and score them on a scale ranging from strongly disliked to strongly liked. This score can then be used to grade the fish into quality bands. For research purposes this can be refined and reference standards developed. Subjective rating is a fairly robust technique for commercial application. It can be very relevant to the consumers' requirements, providing that the assessor is correctly trained. It is also important that the assessors are screened for their ability to discriminate. For colour testing this screening should include tests for colour blindness and also the Munsell tristimulus test. This determines the ability of the assessor to discriminate between different colours and shades of those colours. This method is very rapid and easy to carry out. The results can be use to grade individual fish or batches. However, being subjective it is open to variations between assessors. For research programmes this requires the use of multiple assessors and care in analysis of the results. This limits its use to commercial assessment or one-off assessments within a carefully designed scientific trial.

Care should be taken with subjective ratings as the intensity of the colour increases. Above certain pigmentation levels the human eye cannot determine differences between pigment concentrations (Foss *et al*. 1984, 1987). Such saturation levels could affect the outcome of certain trials, for instance those investigating the effect of dietary pigment level on flesh pigment level, where the pigment concentration may reach levels beyond which the eye cannot discriminate.

Colour card rating

The next step, to reduce the variation between assessors, is to introduce a series of standards. These can be used to score individual fish. Standardised colour surfaces were used by Bolton *et al*. (1967), but Skrede *et al*. (1990) developed a series of cards specifically designed for salmonids. Colour card scales can be made by assessing the full range of colours found in fish. From this range it is then possible to pick a series of representative colours. These can then be assigned scores. Once the standard colour series is determined it can be used to compare individual fish. Lighting of the fish and colour cards is especially important. The use of different lights and angles of lighting will affect the scores. For a single situation, for instance within a trial, this may not be a problem. However, for the routine monitoring of flesh in industry, or carrying out repeated measures over time, the same lighting conditions should be used. Standard lighting conditions are recommended by the producers of commercially available cards.

Colour cards are very rapid to use and give a more reproducible result than subjective rating alone. Little training is required to use them, but screening of operators for colour blindness and the ability to differentiate between colours is essential. The results can be used to grade individual fish or batches, depending on the demands of the consumer. Christiansen *et al.* (1995) found the Roche colour card to be a very good indicator of the carotenoid concentration of the flesh of Atlantic salmon, especially at the higher concentrations. These authors found a good linear relationship between the colour card number and the concentration of astaxanthin in the flesh. Colour card measurements can be used for scientific trials, but care is required as there is still a subjective element in this method. Therefore several assessors are required for each fish in order to allow for bias.

Tristimulus meter

Colour can be measured objectively with meters. Colours can be split into different amounts of the 'real primaries': pure red, green and blue light. Any colour can be produced by combining these pure colours. Using this, measurements of the 'imaginary primaries' X, Y and Z, which are tristimulus values, can be used to plot any colour as a point in a sphere. Several such spheres, known as colour spaces, exist: the $CIEL^*a^*b^*$ and CIELUV spaces. All use the X, Y, and Z values to plot the colours (Warriss 1996). $CIEL^*a^*b^*$ has the advantage over CIELUV in that it is closer to the visual perception of humans: differences between colours as judged by humans are approximately the same as perceived by the system (Warriss 1996). The L^* value describes the 'lightness', the a^* the 'red-greenness' and b^* the 'yellow-blueness' (Fig. 26.1).

The colour of an object can be measured using a tristimulus meter. This directs a pulse of light at the surface of the object and measures the reflected light in terms of the 'imaginary

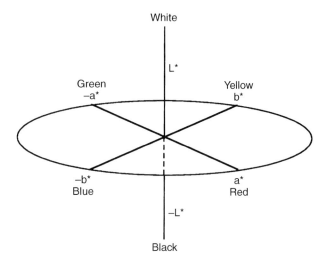

Fig. 26.1   The $CIEL^*a^*b^*$ (1976) colour sphere.

primaries'. These can then be used to calculate the CIEL*a*b* coordinates. The readings from the tristimulus meter are used to describe the colour of the subject. The L* value describes the lightness of the material. The other measurements are obtained from the polar coordinates a* and b*. These are the angle of hue and chroma. The angle of hue describes what is commonly known as the colour. It is calculated as:

Hue = arctan (b*/a*) for a* > 0 and b* > 0
Hue = 180° + arctan (b*/a*) for a* < 0
Hue = 360° − arctan (b*/a*) for a* > 0 and b* < 0

The colour of a sample (•) is dependent on the angle θ as shown in Fig. 26.2. The chroma defines the saturation of the colour. This means the purity or lack of dullness of the colour. It is calculated as:

Chroma = sqrt (a*2 + b*2)

The length of the line C in the figure describes the chroma.

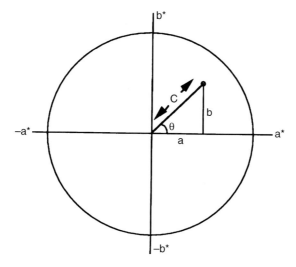

Fig. 26.2 The calculation of the angle of hue (θ) and the chroma (C) from the measurement a and b.

The measurements of a* and b* alone are often quoted in scientific papers. While it is possible to use these measurements, the determination of the angle of hue and the chroma gives far more information on material such as flesh.

Chromameters have been widely used in salmonid research to assess the colour of the flesh. However, often the data have been poorly used and the results confused. The main example of this poor use of data is the use of just the a* or just the b* values. These results are then discussed in terms of the colour of the flesh. However, as is shown from the description of the colour space above, using just the a* value does not give full information about the colour.

It has been found that with increasing astaxanthin concentration in salmonid muscle L*

values and the angle of hue decrease (Choubert 1982; Skrede & Storebakken 1986; Christiansen *et al*. 1995; Nickell & Bromage 1998a), while the chroma increases (Christiansen *et al*. 1995). Nickell & Bromage (1998a) found that the chroma and the angle of hue were significantly, but non-linearly, correlated with flesh astaxanthin concentrations in rainbow trout. This supported the findings of Christiansen *et al*. (1995) who found significant non-linear relationships between astaxanthin concentration in the flesh and chroma and the angle of hue.

The non-linear relationship between the angle of hue or the chroma values and the pigment concentrations observed above is a direct result of the colour space system responding in the same way as human perception. As the pigment concentration increases, the human eye is less able to perceive differences between samples. Similarly, with the hue and chroma readings there is less difference between the measurements at the higher pigment concentrations.

Christiansen *et al*. (1995) concluded that the tristimulus meter may be a good tool for identifying poorly pigmented fillets. However, they observed that the colour of the fillet varies over the whole surface and the meter did not allow for this if just one reading was taken. This problem is the same for all measurement systems. It may be possible in the future to scan the surface of the whole fillet to discriminate between good and bad fillets. However, at present spot readings are all that can be used.

Spectrophotometry

The colour of some fish flesh is determined by the level of pigments. The best known example of this is the flesh of salmonids. These fish have a characteristic red flesh because of carotenoid pigments bound to the muscle (Henmi *et al*. 1989). The carotenoid pigments are derived from the diet and there have been many studies to determine how diet affects pigment deposition. To understand this, methods of determining pigment concentration have been developed.

Probably the simplest way of accurately determining the pigment concentration in a sample of flesh is to use spectrophotometry. The pigment is extracted from the flesh sample in a suitable solvent and the absorbance of light at a specified frequency determined in a spectrophotometer. The technique is sensitive to the types of pigment and the solvents used (de Ritter & Purcell 1981). It is therefore essential that this is understood and stated when discussing results to ensure no confusion.

There are various techniques for carrying out spectrophotometry on pigments in fish flesh, worked up independently in different laboratories (for example Steven 1949; Torrissen & Naevdal 1984; Choubert 1985; Bjerkeng & Johnsen 1995). The results of the analysis should be accurate, but the method cannot easily separate different pigments of similar colours. An example of this in salmon may be astaxanthin and canthaxanthin, the two carotenoid pigments commonly added to salmonid diets. The procedure is good for determining the level of pigment in the flesh but it is not a substitute for a colour measurement, such as the three discussed above. The colour of a sample is affected by many factors, including the level of pigment. Therefore spectrophotometry is limited to a scientific analytical tool for certain types of trials.

High pressure liquid chromatography (HPLC)

HPLC is a very accurate technique for measuring pigment concentration. There are many different HPLC procedures appropriate for fish pigments (Torrissen 1986; Schiedt *et al.* 1988; Choubert & Blanc 1993; Nickell & Bromage 1998b) which reflect the different equipment available in different laboratories and the experiences of the researchers. However, the basic principles are the same for all methods.

A sample of flesh is weighed, blended and the pigment extracted using a suitable solvent. The solvent is evaporated off leaving the pigment dissolved in the flesh lipid. This is redissolved in a known volume of solvent and an aliquot injected into the HPLC instrument. The rate of travel through the HPLC column is dependent on the characteristics of the molecules in the sample. Thus different pigments elute at different times. The time of elution of the pigments gives information on their identity and the size of the peak their concentration.

The method is potentially very accurate. But it is complicated and a great deal of care is required to ensure this accuracy. As with spectrophotometry it is useful as a scientific tool for determining pigment levels; however it does not give information on the colour of the sample, although the level of pigment obviously affects this. The role of HPLC in the scientific analysis of carotenoid concentration determination is reviewed by Pfander *et al.* (1994).

Sampling point

Having decided which method is most suitable for the study, it is also important to decide on a sampling site. Several sites are described in the literature. Bell *et al.* (1998) used a sample cut from the fish according to the requirements of the product certification scheme for Scottish quality farmed salmon. This sample is taken as a 2–3 cm thick steak cut just anterior to the dorsal fin and can be used to assess the colour of the steaks or the pigment content of the white muscle. A second sample cut is defined by the Scottish salmon industry and is used to sample the colour of fillets. Taken from the area below the dorsal fin, between the anterior and posterior edges of this fin, the sample is then trimmed to remove the belly flaps and bone (Anon. 1995). The Norwegian industry standard uses a sample cut from the fish between the posterior edge of the dorsal fin and the vent (Norsk Standard 1994). This is for colour or pigment analysis. Robb (1998) used samples taken from the white muscle just behind the head of the fish, trimmed to remove the fat depots and any red muscle, for pigment analysis. Colour analysis, using a tristimulus meter, was carried out on the same section as in the Scottish quality farmed salmon scheme (Anon. 1995).

Figure 26.3 shows the variety of sampling sites chosen for sampling for colour and pigment concentration. Katikou (1998) carried out a detailed mapping of lipid concentrations in the fillet of Atlantic salmon. The lipid concentration of the different areas varies significantly and this will affect the results of colour analyses as lipid concentration appears to affect the visualisation of the colours (Little *et al.* 1979). However, the pigment concentration seems uniform across the fillet, except for a non-significant rise in the concentration in the tail region (Bell *et al.* 1998).

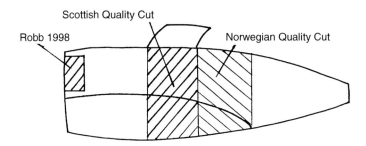

Fig. 26.3   Three sample points used for colour and pigment analyses.

The variation in sampling points has occasionally risen from the need to make other analyses on the same fillet. For instance, Robb (1998) required flesh from the more usual sample points for sensory analyses. However, there is a need to introduce a convention for sampling procedures. For colour sampling, due to the variation in colour along the fillet, it is necessary to take at least three measurements along the length of the fish (see Fig. 28.10b in Chapter 28). However, for pigmentation analyses there appears to be less need for an exact site following the results of Bell *et al.* (1998). Despite this, it would be preferable to stipulate a single position.

Conclusions

There are three methods for analysing colour and two for pigment analysis. Colour analysis is fairly simple, but requires some care, especially when dealing with subjective measurements. Tristimulus meters are valuable tools for measuring colour objectively, but the interpretation of the data must be carried out carefully in order to use the tool effectively. Pigment analysis, although not a substitute for colour analysis, is routinely carried out in salmonid research. There are three basic methods for this, although the exact techniques vary greatly between researchers. The benefits and limitations of the methods are summarised in Table 26.1. Finally, it is important to decide on and state the sampling points. This ensures uniformity for measurements and enables comparison of results. It would be preferable if the researchers and the industry could decide on set sample sites and keep to these in future research.

Table 26.1   The scope and limitations of the methods discussed.

Method	Ease of use	Relevance to consumer	Relevance to research
Subjective scoring	High	High	Medium/low
Colour card	High	High	Medium
Tristimulus	Medium/high	Medium/high	Medium/high
Spectrophotometry	Medium	Medium/low	Medium/high
HPLC	Medium/low	Medium/low	High

References

Anon (1995) *Operating manual for the Product Certification Schemes for Scottish Quality Farmed Salmon and Smoked Scottish Quality Salmon.* Scottish Quality Salmon Ltd., Inverness, Scotland.

Bell, J.G., McEvoy, J., Webster, J.L., McGhee, F., Millar, R.M. & Sargent, J.R. (1998) Flesh lipid and carotenoid composition of Scottish farmed Atlantic salmon (*Salmo salar*). *Journal of Agricultural and Food Chemistry*, 45, 119–27.

Bjerkeng, B. & Johnsen, G. (1995) Frozen quality of rainbow trout (*Oncorhynchus mykiss*) as affected by oxygen, illumination, and fillet pigment. *Journal of Food Science*, 60, 284–8.

Bolton, R.S., Mann, J.H. & Gushue, W. (1967) Use of standardised colour surfaces in the grading of canned salmon for colour. *Journal of the Fisheries Research Board of Canada*, 24, 1613–21.

Chen, H-M., Meyers, S.P., Hardy, R.W. & Biede, S.L. (1984) Colour stability of astaxanthin pigmented rainbow trout under various packaging conditions. *Journal of Food Science*, 49, 1337–40.

Choubert, G. (1982) Method for colour assessment of canthaxanthin pigmented rainbow trout (*Salmo gairdneri* Rich.). *Sciences des Aliments*, 2, 451–63.

Choubert, G. (1985) Effets des antioxydants (ethoxiquine et BHT) sur la stabilité de la canthaxanthine au cours de la granulation; conséquences sur la pigmentation de la truite arc-en-ciel. *Annals de Zootechnologie*, 34, 1–10.

Choubert, G., & Blanc, J-M. (1993) Muscle pigmentation changes during and after spawning in male and female rainbow trout, *Oncorhynchus mykiss*, fed dietary carotenoids. *Aquatic Living Resources*, 6, 163–8.

Christiansen, R., Struksnæs, G., Estermann, R., & Torrissen, O.J. (1995) Assessment of flesh colour in Atlantic salmon, *Salo salar* L. *Aquculture Research*, 26, 311–21.

Foss, P., Storebakken, T., Shiedt, K., Liaaen-Jensen, S., Austreng, E. & Streiff, K. (1984) Carotenoids in diets for salmonids. I. Pigmentation of rainbow trout with individual optical isomers of astaxanthin in comparison with canthaxanthin. *Aquaculture*, 41, 213–26.

Foss, P., Storebakken, T., Austreng, E. & Liaaen-Jensen, S. (1987) Carotenoids in diets for salmonids. V. Pigmentation of rainbow trout and sea trout with astaxanthin and astaxanthin dipalmitate in comparison with canthaxanthin. *Aquaculture*, 65, 293–305.

Frost, S. (1999) Welfare and Quality: What is the Relationship? Presentation at: *Innovations for Seafood '99*, 21–23 April 1999, Surfer's Paradise, Queensland, Australia.

Henmi, H., Hata, M. & Hata, M. (1989) Astaxanthin and/or canthaxanthin-actomyosin complex in salmon muscle. *Nippon Suisan Gakkaishi*, 55, 1583–9.

Katikou, P. (1998) *The distribution of lipids in the fillets of farmed Atlantic salmon* (Salmo salar): *determination by means of a rapid extraction method*. Dissertation for MSc Thesis, University of Bristol.

Little, A.C., Martinsen, C. & Sceurman, L. (1979) Colour assessment of experimentally pigmented rainbow trout. *Colour Research and Application*, 4, 92–5.

Nickell, D.C. & Bromage, N.R. (1998a) The effect of timing and duration of feeding astaxanthin on the development and variation of fillet colour and efficiency of pigmentation in rainbow trout (*Oncorhynchus mykiss*). *Aquaculture*, 169, 233–46.

Nickell, D.C. & Bromage, N.R. (1998b) The effect of dietary lipid level of the variation of flesh pigmentation in the rainbow trout (*Oncorhynchus mykiss*). *Aquaculture*, 161, 237–51.

Norsk Standard (1994) *NS 9401 Atlantik laks*. Referanse-prøveuttak for bedømmelse av kvalitet. Translation from Norwegian.

Pfander, H., Riesen, R. & Niggli, U. (1994) HPLC and SFC of carotenoids – scope and limitations. *Pure and Applied Chemistry*, 66, 947–54.

Rasekh J., Kramer, A., & Finch, R. (1970) Objective evaluation of canned tuna sensory quality. *Journal of Food Science*, 35, 417–23.

de Ritter, E. & Purcell, A.E. (1981) Carotenoid analytical methods. In: *Carotenoids as Colourants and Vitamin A Precursors*, (ed. J.C. Bauernfield), pp.815–923. Academic Press, London.

Robb, D.H.F. (1998) *Some factors affecting the flesh quality of salmonids: pigmentation, composition and eating quality*. PhD Thesis, University of Bristol.

Robb, D. & Warriss, P. (1997) How killing methods affect salmonid quality. *Fish Farmer*, 6, 48–9.

Schiedt, K., Mayer, H., Vecchi, M., Glinz, E. & Storebakken, T. (1988) Metabolism of carotenoids in salmonids. *Helvetica Chimica Acta*, 71, 881–6.

Skrede, G. & Storebakken, T. (1986) Characteristics of color in raw, baked and smoked wild and pen-reared Atlantic salmon. *Journal of Food Science*, 51, 804–808.

Skrede, G., Risvik, E., Huber, M., Enersen, G. & Blümlein, L. (1990) Developing a colour card for raw flesh of astaxanthin-fed salmon. *Journal of Food Science*, 55, 361–3.

Steven, D.M. (1949) Studies on animal carotenoids. II. Carotenoids in the reproductive cycle of the brown trout. *Journal of Experimental Biology*, 26, 295–303.

Storebakken, T., Foss, P., Schiedt, K., Austreng, E., Liaaen-Jensen, S. & Manz, U. (1987) Carotenoids in diets for salmonids. IV. Pigmentation of Atlantic salmon with astaxanthin, astaxanthin dipalmitate and canthaxanthin. *Aquaculture*, 65, 279–92.

Torrissen, O.J. (1986) Pigmentation of salmonids – a comparison of astaxanthin and canthaxanthin as pigment sources for rainbow trout. *Aquaculture*, 53, 271–8.

Torrissen, O.J. & Naevdal, G. (1984) Pigmentation of salmonids – genetic variation in carotenoid deposition in rainbow trout. *Aquaculture*, 38, 59–66.

Torrissen, O.J. & Naevdal, G. (1988) Pigmentation of salmonids – variation in flesh carotenoids of Atlantic salmon. *Aquaculture*, 68, 305–10.

Waagbø, R., Sandnes, K., Torrissen, O.J., Sandvin, A. & Lie Ø. (1993) Chemical and sensory evaluation of fillets from Atlantic salmon (*Salmo salar*) fed three levels of n-3 polyunsaturated fatty acids at two levels of vitamin E. *Food Chemistry*, 46, 361–6.

Warriss, P.D. (1996) Instrumental measurement of colour. In: *Meat Quality and Meat Packaging*, (eds S.A. Taylor, A. Raimundo, M. Severini & F.J.M. Smulders), pp.221–31. ECCEAMST, Utrecht, The Netherlands.

Chapter 27

Proximate Analysis of Fish with Special Emphasis on Fat

K.Fjellanger, A. Obach and G. Rosenlund

Nutreco Aquaculture Research Centre, Stavanger, Norway

Introduction

A good quality product meets customers' specifications and expectations. To produce good quality fish the farmer needs clearly defined quality targets and a quality plan. The fish have to be monitored at regular intervals and any deviations from the quality plan must be corrected.

From a quality point of view chemical analysis can be a tool to better describe and communicate aspects related to fish quality. An accurate description can only be achieved if the applied tests are related to internationally accepted reference methods. Furthermore, composition of fish will vary depending on the part of the fish that is analysed. This variation is, as shown below, mainly related to differences in fat content. Also, a wide range of methods are used to measure fat in farmed fish. As a result, communication problems regarding fish quality expressed as fat content are likely to occur.

Sampling methods

For practical reasons it is common to sample just a small part of the fish for analysis. Therefore, choice of the sampling method is the first step that can lead to differences in results related to fish composition. Some processors, such as the French smokeries Fécamp and Labeyrie, were amongst the first to define sub-samples for their analyses. In Norway, sampling has been standardised to the Norwegian Quality Cut (NS-9402), whereas in Japan it is more common to inspect fillet quality in the anterior part of the fish (Fig. 27.1).

The importance of standardising the sampling method is demonstrated by the differences in fat content found in different sub-samples of rainbow trout (Table 27.1). The reported values are fat levels measured in a trimmed fillet which was cut into the five parts shown in Fig. 27.1, and further split along the lateral line into ventral and dorsal sub-samples. The fat content was determined using a Soxtec instrument with dichloromethane as solvent (Foss Tecator AB 1993). There was a clear trend to increasing fat levels in the caudal-cranial direction, in agreement with previous reports (Hardy & King 1989; Einen *et al*. 1998). A similar, but less pronounced gradient was seen from the dorsal to the ventral part of the fish

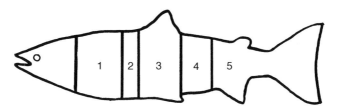

Fig. 27.1 Standard sub-samples of farmed salmon used for analysis. 1. Front part; 2. Labeyrie cut; 3. Fécamp cut; 4. Norwegian Quality Cut; 5. Tail part.

Table 27.1 Fat level (% in wet weight) in five cuts of trimmed rainbow trout fillets. Each cut was further divided in a dorsal and a ventral sub-sample. (Nutreco ARC, unpubl. data).

	Dorsal	Ventral
(1) Front part	15.1	15.5
(2) Labeyrie cut	14.9	14.8
(3) Fécamp cut	12.3	16.1
(4) Norwegian Quality Cut	9.0	11.8
(5) Tail part	6.0	9.1

(Table 27.1). The belly tends to accumulate fat due to the high concentration of adipocytes found in this region (Aursand *et al.* 1994; Wathne 1995).

The variation in fat levels found in different parts of the body (over two-fold) clearly demonstrates the importance of standardising sampling routines when describing and communicating fat levels as part of the quality in salmonids.

Proximate composition of salmonids

Salmonids are composed of protein, fat, water and ash, and to a lesser extent, carbohydrate. As an average, fillets of Atlantic salmon muscle contain around 19.1 ± 0.8 % protein and 1.3 ± 0.2 % ash (Nutreco ARC, unpublished data, n = 51), and these levels are relatively constant in fish weighing from 0.5 to 4.0 kg. Fat level increases in general with fish weight (Gjerde & Schaeffer 1989; Shearer *et al.* 1994; Rye & Gjerde 1996). An average fillet fat level of 15% was found in Atlantic salmon with an average weight of 4 kg in Norway in 1995 (Fig. 27.2). However, a high variation was observed between individuals (11% to 19%). As shown in Fig. 27.2, the fat level was negatively correlated with moisture content. Lean fish contained 11% fat and 68.5% water while fat fish had 19% fat and 60.5% water. This means that fat was replaced by water, while protein and ash contents were constant on a wet weight basis.

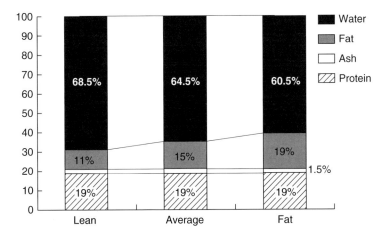

Fig. 27.2   Proximate composition as commonly found in lean, average and fat fillets of Atlantic salmon. (Nutreco ARC results for farmed fish in Norway, 1995).

Analytical methods

The protein level in fish is usually determined by the Kjeldahl method, first described in 1883 (Skoog & West 1976; AOAC 1992). Ash content is measured after burning the sample in a muffle furnace at 540°C to constant weight (AOAC 1923). Water is commonly determined by difference after drying the sample at 103°C to constant weight (AOAC 1990). These methods are widely used and accepted and are therefore recommended as reference methods for the determination of protein, ash and water contents in fish. Thus, no further references will be made to these components in relation to fish quality in this chapter.

Types of fat

Fat, or lipids, are organic substances which are insoluble in water, but soluble in non-polar organic solvents. Lipids are classified in two categories – neutral and polar lipids. The most abundant neutral lipids in salmonids are the tri-acylglycerides (TG) which may constitute more than 90% of the fat in fish (Aursand *et al.* 1994). This group of molecules consists of a glycerol unit where each hydroxyl group is esterified to a fatty acid (Fig. 27.3a). The three fatty acids may differ from each other, with respect to chain length and degree of unsaturation. The TG act as fat depots in the fish. The fatty acid profile in TG in fish is closely correlated to the fatty acid profile of the diet Thomassen & Røsjø 1989; Waagbø et al. 1991; Lie & Huse 1992; Brodtkorb *et al.* 1997), but can be influenced by other factors such as temperature and salinity (Takeuchi *et al.* 1989; Skuladottir *et al.* 1990; Tocher *et al.* 1995).

   Phospholipids (PL) are the dominating polar lipid group in salmonids and constitute about 6 % of the total lipids (Aursand *et al.* 1994). They differ from TG by having one of the primary hydroxyl groups of the glycerol unit esterified to a phosphoric acid (Fig. 27.3b). PL can be sub-divided into several classes, their role depending on the molecule linked to the phosphoryl group. PL therefore vary in size, shape and electric charge. They typically have a

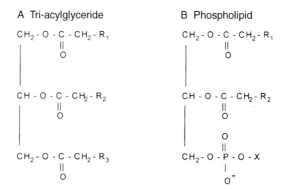

Fig. 27.3 Structural formula of tri-acylglycerides and phospholipids.

polar 'head' and a non-polar hydrocarbon 'tail'. Due to this polarity, they form stronger bonds to the cell matrix than TG, and are thus more difficult to extract. PL generally contain high levels of n-3 fatty acids. Their fatty acid composition is less influenced by the fatty acid profile of the feed (Lie & Huse 1992).

Solvents

Classical fat determination methods use solvents to extract fat from the samples. The fat content is determined as weight after removal of the solvent. Solvent properties play a major role in determining how well fat is extracted from the sample.

   The strength of a solvent is its ability to separate a neutral molecule into a pair of molecules, oppositely charged. The solvent strength is determined by two factors: its separation and its solvation abilities. Separation of ions is enhanced by a high dielectric constant (DEC) because it reduces the electrostatic force of the molecule thereby increasing the ionising power of the solvent. Alcohols, such as methanol and ethanol, are examples of solvents with a high dielectric constant. The solvation ability describes how well the solvent keeps the ions in solution, i.e. how the molecules of the solvent interact or combine with the solute. Solvents such as di-ethyl ether and hexane both have good solvation abilities, particularly for neutral lipids.

Extraction methods

A range of methods exist for extracting fat from tissues. In order to compare variation between methods, Nutreco ARC has used four of the most common methods to determine the fat level in rainbow trout fillet (Table 27.2).

Bligh and Dyer
This method (Bligh & Dyer 1959) uses a mixture of methanol and chloroform as solvents. A higher concentration of methanol at the start of the extraction ensures good separation ability, because of the high dielectric constant of methanol. The presence of chloroform

Table 27.2 Lipid content in rainbow trout fillet determined by four different extraction methods.

Method	Solvent	DEC[1]	Lipid %	Extracted lipids
Blight and Dyer	Methanol and chloroform	32.6 4.8	12.1	Neutral and polar lipids
Ethyl-acetate	Ethyl-acetate	6.0	11.6	Neutral and some polar lipids
Acid hydrolysis	Di-ethylether	4.3	11.2	Neutral and fatty acids of polar lipids
Soxhlet Soxtec	Di-ethylether n-hexane Petroleum ether	4.3 2 —	10.9 10.8 10.6	Neutral lipids Neutral lipids Neutral lipids

[1] DEC: *Dielectric constant*

ensures good solvation ability. As shown in Table 27.2, a high yield (12.1%) was achieved because both neutral and polar lipids were extracted.

Soxhlet

Soxhlet methods apply an extraction technique where the solvent is boiled under reflux, to continuously extract lipids from the sample. Some manufacturers have developed dedicated instruments for this purpose. The analyses reported in Table 27.2 were achieved by using a Soxtec instrument from Foss Tecator AB (1993) according to the method described by Foster and Gonzales (1992). If solvents with low dielectric constants are used, the method will mainly yield neutral lipids. Using petroleum-ether, n-hexane and di-ethyl ether, 10.6%, 10.7% and 10.8% fat, respectively, were extracted from the same sample. These three solvents all have a poor separation ability for polar lipids, but a good solvation ability towards neutral lipids.

Acid hydrolysis

Acid hydrolysis (AOAC 1948) is a method where the sample is boiled in a strong acidic solution in order to hydrolyse, and thereby release, the bound fatty acids. The fatty acids from polar lipids then become free fatty acids which are more easily extracted from the sample. Using di-ethyl ether as the solvent, the fat content in the sample increased to 11.2% because, in addition to the neutral lipids, the fatty acids from the polar lipids were extracted.

Ethyl-acetate

The ethyl-acetate method is the Norwegian reference method for the determination of fat content in fish (Losnegard *et al.* 1979; Norwegian Standard 1994). Ethyl-acetate has a dielectric constant of 6.0. The lipid yield obtained with this method (11.6%) was, as expected, between the Bligh and Dyer and the Soxhlet results (Table 27.2). Due to its dielectric constant, this solvent extracts neutral as well as some of the polar lipids.

Other methods

A wide range of other fat extraction methods are routinely used by many laboratories (e.g. Potter 1955; Folch *et al.* 1957). Many manufacturers of instruments have developed equipment and analytical methods which are safer, use less solvent and are faster than the older classical extraction methods. In addition to the Soxtec instrument from Foss Tecator AB which was mentioned above, another example is the total fat determination with dedicated fat extraction and fat determination units made by Büchi Labortechnik AG (AOAC 1997). This total fat method is of specific interest because it is relatively fast (< 10 minutes analysis time), safe and reliable. The method takes into consideration all fatty acids from C_4 to C_{24} and the fat content is automatically converted to TG content.

The differences between various fat extraction methods, as seen in Table 27.2, show that there is an obvious need to define one method which should be the international reference method for fat determination of fish. Today, there are several methods that act as reference methods in specific geographical areas, e.g. the ethyl-acetate method used in Norway (Norwegian Standard 1994). These local or national reference methods play a very important role in standardising analysis to enable comparison of results. But, in a global market, it would be favourable to agree on one internationally accepted reference method that could be applied in all markets.

The method to be chosen as international reference should extract the lipid components in fish samples that have impact on farmed fish quality – the neutral lipids and the fatty acids from PL. Therefore, acid hydrolysis would be our recommended reference method for determination of fat in fish tissue.

Indirect methods

The industry needs fat results far quicker than any relevant fat extraction method so far can offer. For this reason indirect, instrumental methods are of specific interest. Such methods must be fast, practicable, safe, cheap and sufficiently accurate. In addition, it must be possible to calibrate the method against the reference method.

Torry fat meter

One instrumental non-destructive fat determination method that has been frequently used is the Torry fat meter (Kent 1990). The instrument is portable and analysis time is less than two minutes. It makes use of microwaves to estimate the lipid content of fish flesh. This technique is based on the fact that the dielectric properties of a fish sample depend very much on its water content. Changes in the transmission of microwaves can then be related to the water content of the fish. Since the water content has an inverse linear relationship to lipids in the fish flesh, lipid content can be estimated by calculation (Kent *et al.* 1991). The main disadvantage with the instrument has been reported to be lack of accuracy. For example, the skin and scale thickness, in addition to residual moisture on the fish surface, may affect the results.

Computerised tomography

Computerised tomography (CT) has been used as a non-destructive method for determining fat level and fat distribution in fish (Gjerde 1987; Rye 1991). CT is a technique combining X-ray transmission data with computer calculations to analyse cross-sections of the sample.

CT values represent the density of tissues, where water has a CT-value of zero, while tissues with lower densities (such as fat) have negative CT-values and tissues with higher densities (such as muscle) have positive CT-values. Tomograms can be translated into a picture of the sample where the CT-values are translated into a grey scale, or used in a statistical analysis. The instrument is very expensive and will probably not be used for any routine control purposes. However, it may be promising for selecting fish for breeding purposes (Rye 1991).

Near infrared reflectance

Instruments that meet most of the requirements stated above are those based on near infrared reflectance (NIR). The principle behind NIR is that chemical bonds are able to absorb light of specific energy, i.e. light with specific wavelengths (Mathias et al. 1987; Lee et al. 1992). The sample is illuminated by monochromatic light from a lamp that scans through wavelengths in the range from 1100 nm (for some instrument types from 400 nm) to 2500 nm. The light is partly absorbed by the sample. The amount of absorbed, and thus also the amount of reflected light, depends on which chemical bonds are present in the sample. Using tailored statistical software enables calibration of the reflection of light to the lipid content determined by any fat extraction reference method. However, specific NIR calibrations must be made for each sample type analysed.

An example of a calibration made for NIR analysis of fat in fish is shown in Fig. 27.4. The correlation coefficient and standard error of prediction were $r^2 = 0.97$ and $SEP = 0.7\%$ respectively. The analytical error for determination of fat in fish with the Soxtec method is around 0.5%. As SEP^2 is equal to the sum of variance of the NIR and Soxtec analyses, the calibration in Fig. 27.4 suggests that the analytical error of NIR is comparable to that of the soxtec method.

NIR instruments can be adapted to on-line control systems, and analyses will typically last for less than one minute. During this time the instrument can also analyse a wide range of

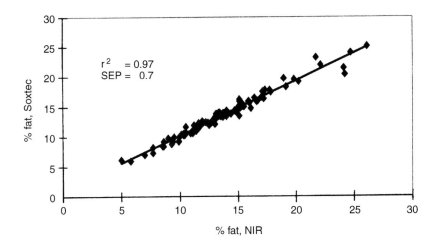

Fig. 27.4 Correlation between fat in Atlantic salmon analysed by Soxtec (Foss Tecator AB 1993) and NIR System 6500 from Foss Tecator AB (400–2500 nm).

other fish quality parameters, such as colour and water. Most NIR instruments are also quite robust and safe, as no chemicals are used. Disadvantages with NIR instruments are the relatively high investment costs, and for some users portable instruments would be preferable.

Determination of fatty acid composition

In addition to fat level, the fat composition also has great impact on fish quality. Fat composition can, to a certain degree, be described by the fatty acid profile. The fatty acid profile is commonly determined by gas chromatography (GC). Because of the complexity and low volatility of the molecules, lipids have to be hydrolysed to fatty acids and further converted into methyl esters of the fatty acids before they can be analysed by GC. From the discussion above, it is obvious that the fatty acid profile will also depend on the type of pre-extraction of fat which has been applied, as analytical methods tend to extract more or less of the different TG, PL and other lipids present.

A time efficient analytical GC method, which is useful when fatty acids from all lipid classes are of interest, has been described by Grahl-Nielsen and Barnung (1985). The method uses a direct esterification of fatty acids from tissue lipids by taking 50 mg of wet fish sample which is hydrolysed in methanolic HCl solution while heated to 90°C for about two hours to form methyl-esters. The methanol is evaporated and water is added to volume. Methyl-esters are extracted by hexane and analysed on GC. The method is relatively quick and does not require any preparation of the fish other than homogenisation. As outlined above, fatty acids from TG as well as PL influence fish quality. This GC method could thus be a good candidate as the reference method for determining the fatty acid profile in fish.

So far, the favourably high *n*-3 fatty acid composition commonly found in salmonids has been highlighted because of the beneficial effects these essential fatty acids have on human health (Carroll 1991; Esterbauer 1992; Grundy 1993; Reddy 1993; Hodgson *et al*. 1996; Pietinen et al. 1997; Rose 1997). However, studies have shown that rancidity of samples expressed as TBARS (Valenzuela 1990; Sørensen & Jørgensen 1995b) increase with higher amounts of polyunsaturated fatty acids (PUFA) present in the sample (Polvi *et al*. 1991; Johansson *et al*. 1995; Olsen & Henderson *et al*. 1997). This was supported by studies made by Nutreco ARC where a sensory panel found significantly less rancid taste and smell for fish which was fed on feed where 50% of the capelin oil was replaced by soya oil. Different quality aspects of fat may thus be conflicting, and in future product development a differentiation of products by varying fish fat quality may be seen.

Conclusions

Harmonised, international standards for sampling and analysis of fish composition are needed in order to define and measure fish quality, because there is significant variation in results between different sampling and analysis techniques.

In relation to fat, the quality of salmonids is determined mainly by the amount of triacylglycerides and the fatty acid composition of the neutral and polar lipids of the fish. A future international reference method for fat measurement should be able to analyse all of these constituents. Acid hydrolysis is thus a preferred candidate for an international reference method for fat extraction.

The fatty acid profile of fish fat also has an impact on fish quality as it is important for human health, fish taste and shelf life of fish products. A reference method is also needed in this area as fat composition will probably become even more important in the future.

Today, fat extraction by acid hydrolysis can be performed efficiently by, for instance, the total fat determination method and instruments from Büchi Labortechnik AG, while NIR instruments, such as, for instance, the NIR System from Foss Tecator AB offer interesting application possibilities for the industry which has to rely on very fast and reliable measurements.

References

AOAC (1923) Association of Official Analytical Chemists Official Method, 923.03 Ash of flour, direct method. *Journal of AOAC International*, 7, 132.

AOAC (1948) Association of Official Analytical Chemists Official Method, 948.16 Fat (crude) in fish meal, acetone extraction method. *Journal of AOAC International*, 31, 98, 606.

AOAC (1990) Official Method 935.29 Moisture in malt, gravimetric method. Official Methods of Analysis, 15th edn, Association of Official Analytical chemists.

AOAC (1992) Association of Official Analytical Chemists Official Method, 981.10 Crude protein in meat block digestion method. *Journal of AOAC International*, 65, 1339.

AOAC (1997) Association of Official Analytical Chemists. Peer-Verified Method No. 4. Total fat determination according to the caviezel method based on a GC technique for food and feed stuff.

Aursand, M., Bleivik, B., Rainuzzo, J.R., Jørgensen, L. & Mohr, V. (1994) Lipid distribution and composition of commercially farmed Atlantic salmon (*Salmo salar*). *Journal of the Science of Food and Agriculture*, 64, 239–48.

Bligh, E.G. & Dyer, W.J. (1959) A rapid method of total lipid extraction and purification. *Canadian Journal of Biochemistry and Physiology*, 37, 911–17.

Brodtkorb, T., Rosenlund, G. & Lie, Ø. (1997) Effects of dietary levels of 20 : 5n-3 and 22 : 6n-3 on tissue lipid composition in juvenile Atlantic salmon (*Salmo salar*) with emphasis on brain and eye. *Aquaculture Nutrition*, 3 (3), 175–87.

Carroll, K.K. (1991) Dietary fats and cancer. *American Journal of Clinical Nutrition*, 53, 1064S–67S.

Einen, O., Waagan, B. & Thomassen, M.S. (1998) Starvation prior to slaughter in Atlantic salmon (*Salmo salar*). I. Effects on weight loss, body shape, slaughter- and fillet-yield, proximate and fatty acid composition. *Aquaculture*, 166, 85–104.

Esterbauer, H. (1992) Lipid peroxidation and its role in atherosclerosis. *Nutrition, Metabolism and Cardiovascular Diseases*, 2, 55–7.

Folch, J., Lees, M. & Sloane Stanley, G.H. (1957) A simple method for the isolation and purification of total lipides from animal tissues. *Journal of Biological Chemistry* 226, 497–509.

Foss Tecator AB. (1993) Solvent extraction using the Soxtec Systems, report AN 301.

Foss Tecator AB, Höganäs, Sweden.

Foster, M.L. & Gonzales S.E. (1992) Soxtec fat analyzer for determination of total fat in meat: collaborative study. *Journal of AOAC International*, **75** (2), 288–92.

Gjerde, B. (1987) Predicting carcass composition of rainbow trout by computerized tomography. *Journal of Animal Breeding and Genetics*, **104**, 121–36.

Gjerde, B. & Schaeffer, L.R. (1989) Body traits in rainbow trout. II. Estimates of heritabilities and of phenotypic and genetic correlations. *Aquaculture*, **80**, 25–44.

Grahl-Nielsen, O. & Barnung, T. (1985) Variations in the fatty acid profile of marine animals caused by environmental and developmental changes. *Marine Environmental Research*, **17**, 218–21.

Grundy, S.M. (1993) Oxidised LDL and atherogenesis: relation to risk factors for coronary heart disease. *Clinical Cardiology*, **3**, 13–15.

Hardy, R.W. & King, I.B. (1989) Variation in *n*-3 fatty acid content of fresh and frozen salmon. *Omega 3 News*, **IV**(4), 1–4.

Hodgson, J.M., Wahlqvist, M.L., Boxall, J.A. & Balazs, N.D. (1996) Platelet trans fatty acids in relation to angiographically assessed coronary artery disease. *Atherosclerosis*, **120**, 147–54.

Johansson, L., Kiessling, A., Åsgård, T. & Berglund, L. (1995) Effects of ration level in rainbow trout, *Oncorhynchus mykiss* (Walbaum), on sensory characteristics, lipid content and fatty acid composition. *Aquaculture Nutrition*, **1**, 59–66.

Kent M. (1990) Hand-held instrument for fat/water determination in whole fish. *Food Control*, January, 47–53.

Kent, M., Lees, A. & Christie, R.H. (1991) In: *Pelagic Fish* (eds H.R. Burt, R. Hardy & K.J. Whittle), pp.157–64. Fishing News Books, Oxford.

Lee, M.H., Cavinato, A.G., Mayes, D.M. & Rasco, B.A. (1992) Noninvasive short-wavelength near-infrared spectroscopic method to estimate the crude lipid content in the muscle of intact rainbow trout. *Journal of Agriculture Food Chemistry*, **40**, 2176–81.

Lie, Ø. & Huse, J. (1992) The effect of starvation on the composition of Atlantic salmon (*Salmo salar*). *Fiskeridirektoratet. Skrifter. Serie Ernæring*, **5**, 11–16.

Losnegard, N., Bøe, B. & Larsen, T. (1979) Undersøkelse av ekstrakjonsmidler for bestemmelse av fett. *Fiskeridirektoratet, rapporter og meldinger*, nr. 1/79.

Mathias, J.A., Williams, P.C. & Sobering, D.C. (1987) The determination of lipid and protein in freshwater fish using near-infrared reflectance spectroscopy. *Aquaculture*, **61**, 303–11.

Norwegian Standard (1994) Norwegian Standard, NS 9402. Atlantic salmon – Measurement of colour and fat.

Olsen, R.E. & Henderson, R.J. (1997) Muscle fatty acid composition and oxidative stress indices of Arctic charr, *Salvelinus alpinus* (L.), in relation to dietary polyunsaturated fatty acid levels and temperature. *Aquaculture Nutrition*, **3**, 227–38.

Pietinen, P., Ascherio, A., Korhonen, P., Hartman, A.M., Willett, W.C., Albanes, D. & Vitamo, J. (1997) Intake of fatty acids and risk of coronary heart disease in a cohort of Finnish men. *American Journal of Epidemiology*, **145**, 876–87.

Polvi, S.M., Ackman, R.G., Lall, S.P. & Saunders, R.L. (1991) Stability of lipids and omega-3 fatty acids during frozen storage and Atlantic salmon. *Journal of Food Processing and Preservation*, **15**, 167–81.

Potter, V.F. (1955) Tissue homogenates. *Methods in Enzymology*, 1, 10–15.

Reddy, B.S. (1993) Dietary fat, calories and fiber in colon cancer. *Preventive Medicine*, 22, 738–49.

Roberts, J.D. & Caserio, M. (1977) *Basic Principles of Organic Chemistry*, 2nd edn.

Rose, D. (1997) Dietary fatty acids and cancer. *American Journal of Clinical Nutrition*, 66 (Suppl.), 998S–1003S.

Rye, M. (1991) Prediction of carcass composition in Atlantic salmon by computerized tomography. *Aquaculture*, 99, 35–48.

Rye, M. & Gjerde, B. (1996) Phenotypic and genetic parameters of body composition traits and flesh color in Atlantic salmon (*Salmo salar* L.). *Aquaculture Research*, 27, 121–33.

Shearer, K.D., Åsgård, T., Andorsdottir, G. & Aas, G.H. (1994) Whole body element and proximate composition of Atlantic salmon (*Salmo salar*) during the life cycle. *Journal of Fish Biology*, 44, 785–97.

Skoog, D.A. & West, D.M. (eds) (1976) In: *Fundamentals of Analytical Chemistry*, 3rd edn, pp.244–5. Whitaker, London.

Skuladottir, G.V., Schioeth, H.B., Gudmundsdottir, E., Richards, B., Gardarssons, F. & Jonsson, L. (1990) Fatty acid composition of muscle, heart and liver lipids in Atlantic salmon (*Salmo salar*) at extremely low environmental temperature. *Aquaculture*, 84, (1), 71–80.

Sørensen, G. & Jørgensen, S. S. (1996) A critical examination of some experimental variables in the 2-thiobarbituric acid (TBA) test for lipid oxidation in meat products. *Zeitschrift für Lebensmittel untersfuschung und Forschung*, 202, 205–10.

Takeuchi, T., Kang, Seok-Joong & Watanabe, T. (1989) Effects of environmental salinity on lipid classes and fatty acid composition in gills of Atlantic salmon. *The Japanese Society Fisheries Science*, 55 (8), 1395–1405.

Thomassen, M.S. & Røsjø, C. (1989) Different fats in feed for salmon: influence on sensory parameters, growth rate and fatty acids in muscle and heart. *Aquaculture*, 79 (1), 129–35.

Tocher, D.R., Castell, J.D., Dick, J.R. & Sargent, J.R. (1995) Effects of salinity on the fatty acid compositions of total lipid and individual glycerophospholipid classes of Atlantic salmon (*Salmo salar*) and turbot (*Scophthalmus maximus*) cells in culture. *Fish Physiology and Biochemistry*, 14 (2), 125–37.

Valenzuela, A. (1990) The biological significance of malondialdehyde determination in the assessment of tissue oxidative stress. *Life Sciences*, 48, 301–309.

Waagbø, R., Sandnes, K., Sandvin, A. & Lie, Ø. (1991) Feeding three levels of *n*-3 poly-unsaturated fatty acids at two levels of vitamin E to Atlantic salmon (*Salmo salar*). Growth and chemical composition. *Fiskeridirektoratet. Skrifter. Serie Ernæring*, 4(1), 51–63.

Wathne, E. (1995) *Strategies for directing slaughter quality of farmed Atlantic salmon* (Salmo salar) *with emphasis on diet composition and fat deposition*. Dr Sc. thesis, Agricultural University of Norway, Ås, Norway.

Chapter 28

Carcass Quality Monitoring at the Farm and Factory

R. Sinnott

Trouw Aquaculture, Wincham, Northwich, Cheshire, CW9 6DF, UK

Introduction

The UK fish farming industry is primarily focused on the production of Atlantic salmon (*Salmo salar*) and rainbow trout (*Oncorhynchus mykiss*). Factors affecting the carcass quality of farmed fish include genetics, dietary composition, food intake, season, health status and life-stage. The farmer is able to influence the quality of his stocks by controlling a number of these factors. Once the fish leave the farm further changes in flesh quality are inevitable as they move through the processing factory and up the value chain. It is therefore important that everybody working in each link of the chain, from those involved in broodstock selection right through to the fish retailer, is fully aware of their potential to influence the quality of the fish they handle.

Many fish farmers are now involved in carcass quality monitoring programmes, often in conjunction with a feed supplier. The monitoring of quality involves time, effort and cost, but the benefits are tangible and most fish farmers now see carcass quality monitoring is an essential part of farming fish. Fish processors are also generally aware of the benefits to be gained from monitoring fish quality at various points in the factory. This chapter will consider the role that carcass quality monitoring can play in ensuring that farmed fish reach the end consumer in prime condition.

Carcass quality monitoring

During their life on the farm and following harvest, various carcass characteristics of farmed fish can be measured at regular intervals. Such an organised schedule of sampling is sometimes referred to as a carcass quality monitoring programme. It is not practical to monitor every carcass quality characteristic but in an ideal situation the following parameters would be included:

- ❏ Taste
- ❏ Texture
- ❏ Succulence

❑ Appearance
❑ Weight
❑ Length
❑ Condition factor
❑ Yield
❑ Sexual status
❑ Carcass fat and oiliness
❑ Flesh colour
❑ Health status.

A typical form for recording carcass quality measurements taken from Atlantic salmon at the farm is shown in Fig. 28.1. There is a section on the form where the history of the fish and their genetic background can be recorded. It is important to know the genetic history of each stock type as most of the important quality parameters are influenced by hereditary factors, for example, flesh colour and carcass fat (Gjerde 1989; Gjerde & Schaeffer 1989). There are further sections on the form for recording information useful to the farmer and fish processor. The increase in average weight of fish from one sample point to another can be used to predict growth rate. The type of feed used and the pigmentation regime employed are other valuable pieces of information worth recording.

In practical terms it is difficult to make an objective assessment of taste, texture and succulence at the farm, although this is possible at the processing factory. Most trout farmers will, however, taste their own fish regularly to check that there are no signs of flesh taint from algal blooms or pollution. Many of the quality parameters listed above will be considered in more detail later in the chapter.

The benefits of monitoring carcass quality

In order to justify the implementation of a carcass quality monitoring programme the benefits must be worthwhile and outweigh the time, effort and cost involved. The next section looks at the benefits for the fish farmer and the fish processor.

Carcass quality monitoring on farm

The benefits of on-farm carcass quality monitoring are comprehensive and include:

❑ Development of a fish quality database
❑ Predicting fish quality
❑ Trend analysis
❑ Feed management
❑ Predictions of stock quality at harvest
❑ Broodstock selection
❑ Modification of carcass quality
❑ Reducing feed costs
❑ Due diligence responsibilities
❑ Third parties

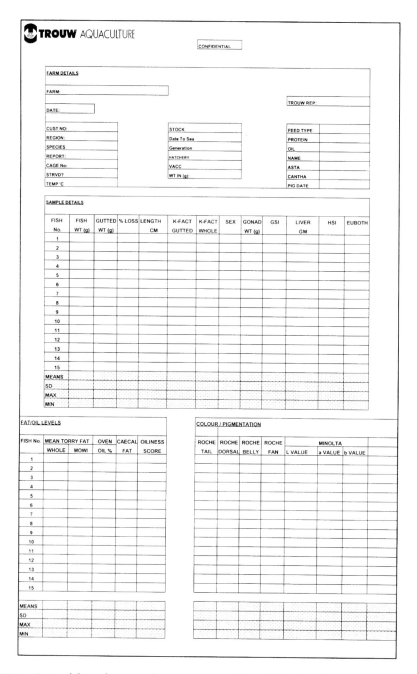

Fig. 28.1   Typical form for recording carcass quality of Atlantic salmon.

Fish quality database

The information collected and recorded from sampling can be used to build up a database valuable for year to year comparisons of stock characteristics and to compare different stocks. Farmers generally accept that different stocks grow and convert feed at different rates but may underestimate the extent of differences in quality characteristics between stocks. The information stored on the database can help the farmer to choose stocks with the best combination of growth, feed conversion efficiency and quality characteristics.

Predicting fish quality

The regular monitoring of important quality parameters in fish stocks will help the farmer to predict the probable carcass quality profile of fish stocks at harvest. The data collected can be compared to the quality specifications of the market. Providing the information is accessible, it is also useful to carefully analyse the data from each sampling point and compare it with past data. Plotting new data alongside data from the previous generation can give a good forecast of the quality of the fish at harvest. If the comparison indicates that target quality parameters might not be reached then the appropriate actions can be taken.

Trend analysis

Seasonal effects influence many carcass quality characteristics. Sampling fish at different times of year can be useful in highlighting critical periods for certain quality parameters and problems. Figure 28.2 highlights the effect of season and weight on fat development in Atlantic salmon. From this it can clearly be seen that regardless of size, flesh fat levels in salmon increase in the autumn of each year and also increase with increasing weight.

Broodstock selection

It is often necessary to kill fish when sampling fish for carcass attributes but it is possible to take a number of useful measurements from anaesthetised fish. Although it is currently not

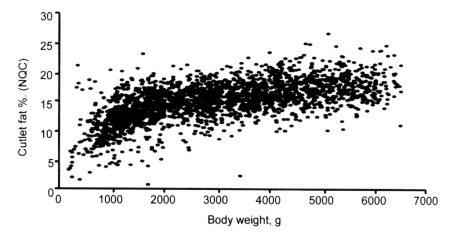

Fig. 28.2 The relationship between weight and fat development in the Norwegian quality cut of farmed Atlantic salmon.

possible to measure flesh colour and yield, most of the other parameters, including carcass fat, can be assessed from live fish. Individual fish showing desirable traits, such as a high growth rate but a lean body mass, can be selected for broodstock. In order to select for flesh colour and yield, a family selection strategy should be used.

Modification of carcass quality

There are focal points for quality assessment in farmed salmon and trout. The key factors are the appearance of the fish, or fish product, and the colour, fat content and texture of the flesh. The farmer can influence these to a degree, by modifying dietary composition, feeding regimes, fish health, environmental conditions, and by selection of stocks. Conversely, monitoring carcass quality will help the farmer assess the impact of changes in diet and feeding practices on fish quality. It should be borne in mind that improving the quality of fish might come at a cost.

Flesh colour

One of the problems facing retailers is that of obtaining supplies of consistently pigmented fish. It could be argued that the consistency of flesh colour is more important than the actual colour of the flesh. A number of factors influence flesh colour in farmed salmonids, in particular the stock type, dietary composition, the amount of pigment in the feed and the time period over which it is fed. Figure 28.3 shows the relationship between visually measured flesh colour and body weight in Norwegian farmed salmon. There is a wide variation in flesh colour for any given size. If the sampled fish have less well developed flesh colour than expected then increasing the level of dietary pigment can generally be used to improve it. To achieve an acceptable flesh colour for table sized trout it is advisable to

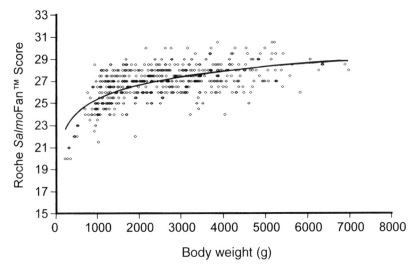

Fig. 28.3 The development of flesh colour in the Norwegian quality cut of farmed Norwegian salmon.

introduce pigmented feed, containing 40–50 mg pigment kg^{-1} of feed when the fish are about 80 g in weight, and continue feeding this until harvest. Salmon are normally fed pigmented diet (70–80 mg pigment kg^{-1} of feed) from seawater transfer until all grilse have been graded out, or until reaching a weight of 2 kg, when dietary pigment levels may be reduced to 30–50 mg pigment kg^{-1} of feed. The savings produced by using this strategy are demonstrated later in the section on reducing feed costs.

When salmonids approach maturity the pigment stored in the white muscle is directed towards the gonads and skin of the fish. Monitoring the gonadal development of sampled fish will provide information that can be used to predict the potential percentage of maturing fish in a season.

Muscle fat

There is a potential for influencing the amount of lipid in the edible flesh of salmonids by modifying dietary composition and feeding regimes. High fat diets grow fish more quickly and can produce a small but significant increase in muscle fat levels compared to fish fed a low fat diet. Feeding fish restricted rations will reduce fat levels on a weight to weight basis, compared to fish fed a high ration, but will also increase the time taken for a fish to reach a given weight. Using normal farming feeding practices, muscle fat will increase with fish size. The development of fat with increasing weight, in Atlantic salmon, is shown in Fig. 28.4.

Table trout are normally harvested before they reach 0.5 kg in weight. Excess muscle fat is rarely a problem at this stage; however, very low muscle fat levels are found in fish that have been held for long periods at constant weight and in fish that have been grown very slowly. Fat can accumulate to high levels in the muscle of market sized Atlantic salmon and may

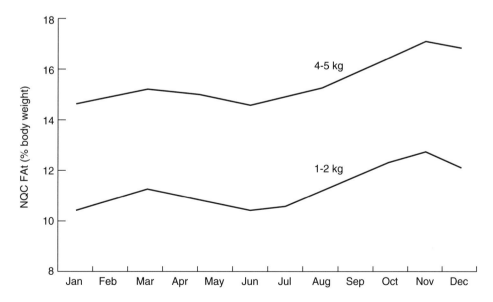

Fig. 28.4 The development of fat in the muscle of farmed Norwegian salmon. The fat is indexed to 1 for a 1.5 kg salmon in January.

exceed 20% of body weight, especially in the autumn and winter of the year before maturity. A carcass monitoring programme can help the farmer to predict the proportion of fish that are likely to mature in a season by plotting the bimodal distribution of growth and by measuring the gonadosomatic index of sampled fish. Salmonids deposit fat in the muscle as a store of energy to be used in times of poor food availability or during maturation. Salmon are slow to release muscle fat and the starvation period required to cause a significant reduction in muscle fat levels can be measured in weeks rather than days. In addition, welfare guidelines in the UK limit the period of starvation, and food deprivation should be viewed as an unacceptable method of reducing flesh fat levels.

Flesh texture

Because of the way connective tissue develops during growth, fast grown farmed fish have fewer cross linkages in the connective tissue, which holds muscle fibres and muscle blocks together, compared to slower grown and therefore older fish of the same size. As well as the energy requirement for growth and metabolism some energy is stored in the muscle as glycogen. Fast growing fish can have significant quantities of glycogen stored in their edible muscle. Poor harvesting techniques, such as the prolonged crowding of fish, will increase stress levels causing this glycogen to break down into lactic acid and also may lead to the fish going into rigor early. Handling fish during rigor can lead to an increase in the incidence and severity of gaping in the flesh. Gaping is an expression used to describe the separation of the myotomes or muscle blocks. This phenomenon can lead to problems in the processing of affected fish. An example of this is when, for example, smoked fillets of salmon are sliced and the sliced sections disintegrate. Rapid chilling of fish to less than 4°C immediately post harvest, and the maintenance of chilling throughout the transport and processing, will help to optimise fish quality and may delay the onset of rigor.

Trout farmers sometimes associate soft texture with 'excess' fat; in fact the opposite is usually true. Muscle fat is inversely proportional to muscle moisture and fish with low fat levels are often 'spongy' for this reason.

Yield

Yield will vary between populations of fish and with the time of year. This is demonstrated for rainbow trout in Fig. 28.5. This contour graph comes from data collected by Nutreco ARC for immature trout. It demonstrates how data collected throughout the year can be used to identify trends in important carcass criteria. This particular data set shows a tendency for the yield to be highest in winter and lowest in the summer. During prolonged starvation salmonids will mobilise visceral energy stores before the bulk of muscle fat is used, resulting in an apparent increase in the yield of starved fish. This may be good for the fish processor but the farmer will lose potential growth and revenue.

Reducing feed costs

By regularly monitoring flesh colour, the farmer of large pigmented trout and salmon can make significant savings on the cost of feed. The pigment used to colour the flesh of farmed salmonids can represent up to 15% of the cost of feed. Once the pigment concentration in the flesh has reached an acceptable level, the amount of pigment added to the diet can be

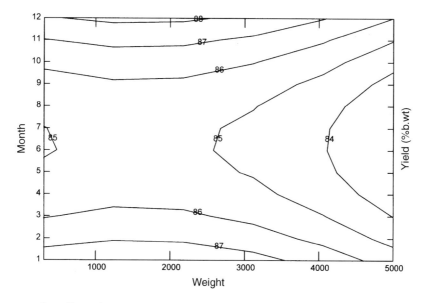

Fig. 28.5   The effect of season and weight on yield in farmed large rainbow trout.

reduced to a 'maintenance level'. This strategy should not be used without a comprehensive carcass monitoring programme to ensure that flesh colour is being maintained at target levels. The flesh of sampled fish should contain at least 7–8 mg of pigment per kg of flesh, before pigment levels in the feed are reduced to a minimum of 30 mg of pigment kg^{-1} of feed. The theoretical savings for a salmon farmer using this strategy for production of 4 kg salmon can be almost one quarter of the cost of pigment.

The dietary pigment level should never be reduced, however, unless a good sample of fish have been examined to check that the visual flesh colour and instrumental colour measured is already above the minimum acceptable level. It is also recommended that a few flesh samples be sent off for high performance liquid chromatography (HPLC) pigment measurement. This is relatively expensive but the cost is a small proportion of the potential savings on pigment costs. Samples of fish on the reduced pigment regime should be checked for flesh colour on a regular basis but particularly a minimum of eight weeks before harvest. If flesh colour is then below that required there may be enough time for an increase in dietary pigment levels to correct the situation. This however is not a realistic option for the producer of trout for the table as the fish are likely to be harvested before optimum pigment levels are reached in the flesh.

Due diligence responsibilities

In the UK it is a fundamental requirement of the major fish retailers that any farmed fish, sampled at random, from one of their shops can be traced back, not only to a particular farm, but also to a specific pen on that farm. It is also becoming customary for the retailer to audit their major fish suppliers and at such times it is useful for the farmer to be able to demonstrate that attention is being directed toward fish quality through the carcass quality monitoring programme.

Third parties

The information on fish quality derived from sampling data can be used to communicate with other interested sectors of the fish industry. Some fish buyers are keen to get regular updates on the development of carcass characteristics from their suppliers. Farms that are able to provide this information may have a distinct advantage over their competitors. The majority of feed suppliers to the fish farming industry offer technical advice in the specialist area of fish quality. If the farmer or fish processor has a good record of carcass quality measurements this can be very useful in helping to find a solution if quality problems develop.

Carcass quality monitoring at the processing factory

Fish are generally purchased according to a specification, which may reflect general quality standards or be specific to the purchaser. The processor will be less interested in the carcass characteristics determined on the farm than the quality of the fish arriving at the factory. The in-factory carcass monitoring programme may include the same items as listed for on-farm monitoring but may be extended to include temperature, processing properties, freshness, odour, sliminess, shelf-life and microbiology.

There are distinct advantages for the fish processor in adopting a carcass quality monitoring programme. The measurement of factors such as flesh fat and colour can be used to compare the purchasing specification with the quality of the material received. The measurement of the physical characteristics of fish, which include their condition factor and textural properties, may be useful in predicting the behaviour of fish during processing. In a similar manner, the fat content and oiliness of fish can influence the quality of cold-smoked salmon and trout. The information collected can be used to compare fish from different sources and to look for seasonal effects on quality. For example, Lavéty *et al.* (1988) found that muscle pH of salmon, one day after death, was at its lowest in mid-summer and that this was associated with an increasing tendency for the muscle to gape.

Figure 28.6 shows the percentage of trout rejected because of poor flesh colour by a fish processor over an eleven-month period. This information can be of great value to the processor and to the supplier in helping to identify the reasons for fluctuations in quality. The processor can also use the information from the monitoring programme to contribute to due diligence requirements.

Sampling systems

It is relatively straightforward to sample fish at the processing factory, where fish are easily accessible and sampling facilities are usually present. The processor may monitor fish quality several times before the fish are dispatched from the premises. The fish may be assessed for quality on arrival, for rigor before processing begins, during processing and finally as an end product. Further testing for the shelf-life of various products may continue.

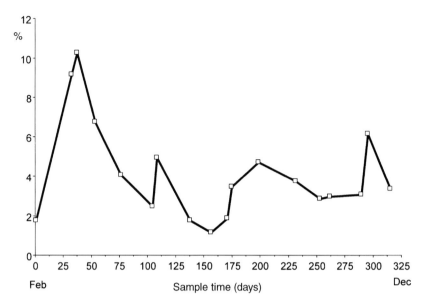

Fig. 28.6   An example of seasonal variation in unacceptably pale-fleshed table sized trout.

On farms, accurate sampling for fish quality can be more difficult, unless a suitable hygienic area with standard light conditions and cleaning facilities is present. The farmer must consider the frequency of sampling, in relation to the cost involved, and balance this against the value of the information collected. The recommended frequency of sampling will be different for each farmed species, mainly because of the differences in the time taken for them to reach market size.

Atlantic salmon
There may be different philosophies regarding the most effective times to sample fish. My own recommendations are given below.

Farmed salmon should be sampled a minimum of three times from transfer to harvest. On a live weight basis, the first sample may be taken when the fish reach 400–600 g, the second when the fish are 1500–2000 g and the third when the fish are 2500–3000 g. For salmon put to sea as smolts in April this equates approximately to the first sample in August/September, the second in December/January and the third in May/June of the following year. Once any maturing fish have been removed, a further sampling point could be considered.

Rainbow trout
Table sized rainbow trout are normally harvested before they reach 0.5 kg in weight, therefore it may be practical to sample them only once, 8–10 weeks prior to harvesting. Sampling harvest-sized fish at different times of the year may emphasise seasonal trends in certain quality parameters.

Sample numbers

To ensure that the fish selected for sampling actually represent the population they originate from, it is important to consider the numbers of fish that should be sampled and the method used to catch them. The 'correct' number of fish is the subject of much discussion. For trout farmers, the unit value of each fish is relatively insignificant but for the salmon farmer each fish may be worth £15 at today's prices. If one considers the numbers of fish on the farm and the natural variation for many carcass characteristics in farmed populations, then the numbers sampled should be high. Against this, one has to consider that fish can be rejected or marked down in value for carcass parameters falling outside purchasing specifications. A population of fish could, for example, have a small but economically significant percentage of fish below the flesh colour specified and the sample population should be large enough to pick this up. If only ten fish are sampled and all of these are of the correct quality does this mean the rest of the population are also of similar quality? Similarly, if one of the fish is poorly pigmented does this mean that 10% of the population will also be pale fleshed?

Statistical tables can be used to find the number of fish that should be sampled. If an incidence of 5% of pale fish in a farm population was economically important, then for a farm with 100 000 fish 58 fish should be sampled. If the same farmer regarded a 2% incidence as important, then the sample number should be increased to 148 fish. From practical experience I would suggest that sampling 30 fish from each population, taken within a narrow size range, would give a reasonable level of security. For routine on-farm monitoring where no problems are anticipated this is often reduced to 10–15 fish from each stock. Whatever the number of fish to be sampled it is always important to try to get a representative sample, i.e. one that represents as closely as possible the population sampled. On a trout farm this may simply involve crowding the fish and scooping out a net full of fish but on salmon farms with large pens this can prove difficult. Salmon farmers will quite often use a few feed pellets to entice fish to the surface, where they can be netted, but these fish are unlikely to represent a cross-section of the entire population. On salmon farms good results have been achieved using a sweep net or a box net for sampling fish.

Practical aspects of carcass quality monitoring

Sample preparation

It is advisable to record all measurements on waterproofed paper or directly on to a computer spreadsheet. Whenever possible the fish to be sampled should have been starved until the gut is completely empty of food and faeces. The gut evacuation rate is temperature dependent and in spring and summer it is generally sufficient to starve salmon and trout overnight. Since the carcass monitoring programme requires the killing of the sampled fish it is recommended that full use be made of them.

Appearance
The appearance of the fish and the presence and location of scale loss should be noted. The opportunity can be taken to carry out a general health audit; fish can be closely examined paying particular attention to the gills, skin and vent region. The presence and location of

skin damage or lesions should be recorded along with a note on the state of the fins. The gill structure should be closely observed and any obvious external parasites removed for more detailed examination. Any bloody discharge from the hindgut may be an indicator of the presence of a bacterial pathogen within the stock and further tests should be undertaken. Trout and salmon are sampled in a similar manner: first the weight is taken and then the length is measured from the tip of the upper jaw to the central edge of the tail fork.

Fat measurement

The fat can be estimated directly at this stage using an instrument, such as the Torry fat meter (see Fig. 28.7), or from a sample taken after the flesh colour has been assessed (see later).

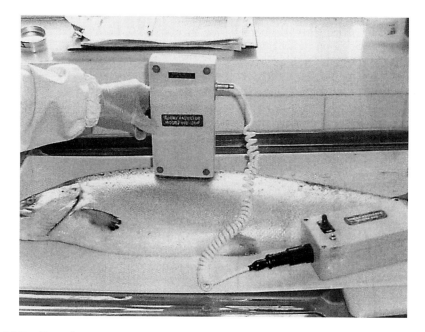

Fig. 28.7   Torry fat meter.

Cleaning the fish

The fish should be gutted and the kidney removed. Farmed trout are often sold gilled and gutted and in order to keep the data collected relevant to real market conditions it is advisable to remove the gills before weighing the fish (gutted weight).

Condition factor (CF)

The condition factor of the fish is calculated from the length and gutted weight of the fish as:

$$CF = \text{gutted weight of fish (g)}^* 100/\{\text{length of the fish (cm)}\}^3$$

Caecal fat index

The amount of fat around the pyloric caecae and between the individual caeca is a useful indicator of condition and can be examined and scored:

No fat = 0
Fat less than the width of a caeca = 1
Fat equal to the width of a caeca = 2
Fat more than the width of a caeca = 3
Fat obscuring the caecae = 4

There are other versions of this index and care should be taken to identify which scoring system has been used when comparing figures from different sources.

Flesh colour measurement

For flesh colour measurement the fish is usually filleted and the pin bones removed. Removal of the pin bones allows a more accurate measurement of the flesh colour of the muscle. In farmed salmon this area around the pin bones can accumulate high levels of lipid, which can be incorporated into distinct striated bands between muscle blocks. Colour measurement is covered in detail in the Chapter 26 by Dave Robb.

The flesh colour can be estimated visually using an industry-accredited method, such as the Roche*Salmo*Fan™ (see Fig. 28.8). Changing light conditions throughout the day will affect the visual assessment of flesh colour. This should therefore take place in standard light

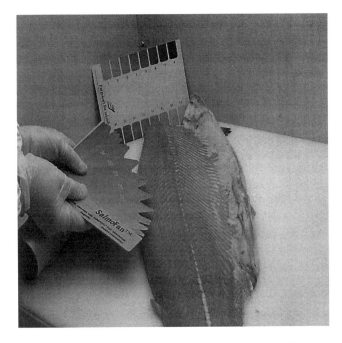

Fig. 28.8   Subjective flesh colour measurement by Roche card and Roche *SalmoFan*™.

conditions, preferably using a light box. Roche recommend that colour assessment be carried out in 'good daylight or under fluorescent lighting with the following characteristics: Ra > 90, colour temperature > 5000 K'. The perception of colour varies between individuals and it is therefore advisable to use trained assessors and whenever possible the same assessors should be used.

Flesh colour can be measured objectively using an instrument such as a light meter. The Minolta chroma meter (see Fig. 28.9) is commonly used for this purpose and this type of instrument can provide valuable information on the colour properties of the flesh. It is important to standardise the position where flesh colour is measured. The three locations shown in Fig. 28.10 are routinely used.

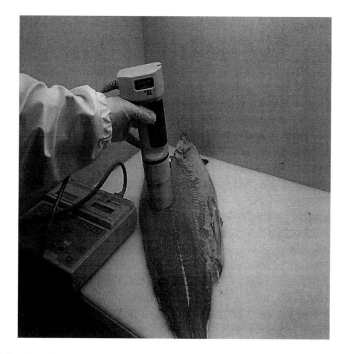

Fig. 28.9   Minolta chroma meter in use.

Fat measurement

The Torry fat meter, mentioned previously, is a non-destructive method for estimating the fat content of the fish muscle. It uses a microwave sensor to determine the moisture content of the muscle, which is inversely proportional to its fat content. The meter is calibrated to estimate the fat content of the muscle from the moisture measurement. The positions used for measuring fat by this method are shown in Fig. 28.9; it is usual practice to take measurements from both sides of the fish.

Other methods of fat measurement require a standard sample of flesh to be removed from the carcass. There are various quality cuts and a number of methods used for analysis of fat. Kurt Fjellanger *et al.* describe these and their respective merits in Chapter 27.

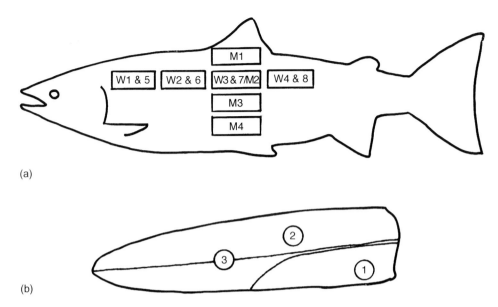

(a)

(b)

Fig. 28.10   Sampling positions for (a) Torry fat meter and (b) Minolta chroma meter colour measurement.

Gonadosomatic index (GSI)

The GSI is a measure of the weight of the gonads, the ovaries or testes, expressed as a percentage of the gutted weight of the fish:

$$\text{GSI (\%)} = \text{weight of both gonads (g)}^* \; 100/\text{gutted weight of fish (g)}$$

It is generally accepted that male salmon can be classified as immature if the GSI is less than 0.10% and to be mature if the GSI is more than 0.14%. Similarly the female salmon is regarded as immature if the GSI is less than 0.26% and mature if it exceeds 0.30%. Fish that fall between these categories cannot, with any confidence, be assigned to either category.

Hepatosomatic index (HSI)

The relationship between the weight of the liver and the weight of the fish can be an indicator of disease or a nutritional imbalance in the fish. The HSI is calculated as:

$$\text{HSI (\%)} = \text{weight of liver (g)}^* \; 100/\text{gutted weight of fish (g)}$$

Yield

The yield of the fish should refer to the proportion of saleable material. For the trout farmer, this could mean the gutted weight of the fish compared to the whole fish. The salmon smoker might regard the yield as the weight of smoked fillet recovered from the whole fish. The yield is normally expressed as:

Yield (%) = weight of saleable material (g)* 100/weight of original raw material (g)

Carcass quality standards

The fish farming industry has few universal objective standards for carcass quality. There are many more detailed quality standards contained in documents drawn up by individual processors and retailers and used by them as purchasing specifications. Trade associations may also supply their membership with a list of fish quality standards.

There are universally accepted 'standards' for carcass characteristics, such as for flesh colour and fat in farmed salmon and trout. These unofficial standards may be too general for the wide range of farmed fish products now sold. A quality standard or specification should help the fish buyer to identify suitable raw material for a particular product. In practical terms a specification must be reasonable and should be agreed by the different links in the value chain. It is of little use, for example, if the fish buyer insists that all 3–4 kg fish have a maximum mid-dorsal fat level of 12% of body weight in October, when Fig. 28.2 confirms that the majority of fish will be out of specification then. If this particular specification is vital to the production of a certain end product then the buyer may have difficulty identifying suitable raw material. If, however, the specification can be relaxed at certain times of year then it will obviously be easier to find supplies of suitable fish. An example of this concerns the difficulty that producers of smoked salmon have with fat fish. The upper level for fat content in salmon destined for smoking is commonly regarded as approximately 14 per cent of body weight. Wild salmon and the earlier generations of farmed salmon generally had fat levels well within this specification but, with improvements in fish health, husbandry and nutrition, farmed salmon now grow more quickly and have higher fat levels on a weight for weight basis. Those salmon smokers who can adapt their process to produce high quality smoked products from these higher-fat fish will have a big advantage over their competitors. It is important that specifications should be flexible and reviewed on a regular basis and be tailored for individual fish products.

It is also important that any proposed quality standard is associated with a strict protocol for its measurement. If, for example, a salmon processor buys fish using a specification which includes a maximum value for muscle fat, then this specification should also include a protocol for measuring the fat level. The muscle fat content of salmon varies down the length of the fish and is much higher at the shoulder than at the tail, therefore it is important to state the exact section of the fish that should be taken for analysis. For example, for the same fish the Norwegian quality cut (NQC) gives a higher measurement for flesh colour and a lower measurement for fat than the Scottish Quality Salmon (SQS) section.

Conclusions

In the future, breeding programmes for farmed fish species will select for desirable characteristics and produce fish of more consistent quality. The price of farmed fish may continue to fall and pressure will then increase on feed suppliers to produce more cost-effective feeds. At the same time further expansion of aquaculture will place heavy demands on the raw materials presently used in farmed fish diets. Further replacement of animal proteins with plant proteins and the introduction of plant oils into salmonid diets will help to ease the pressure on raw materials but could adversely affect carcass quality unless carefully controlled. Against this background of changes the monitoring of fish quality on the farm and in the factory will remain a vital aspect of quality assurance. On a practical level there should be further developments in the on-line measurement of fat and flesh colour at the processing factory and the development of simple methods for the objective measurement of the oiliness, firmness and texture of flesh.

Acknowledgements

Thanks are due to Dr Struksnaes and M. Gisvold of Nutreco ARC for kindly supplying most of the figures used in this chapter and to Edward Branson for his advice and for reading this manuscript.

References

Gjerde, B. (1989) Body traits in rainbow trout. I. Phenotypic means and standard deviations and sex effects. *Aquaculture*, **80**, 7–24.
Gjerde, B. & Schaeffer, L.R. (1989) Body traits in rainbow trout. II. Estimates of heretabilities and of phenotypic and genetic correlations. *Aquaculture*, **80**, 25–44.
Lavéty, J., Afolabi, O.A. & Love, R.M. (1988) The connective tissues of fish. IX. Gaping in farmed species. *International Journal of Food Science and Technology*, **23**, 23–30.

Chapter 29

Sensory Analysis

L. Johansson

Department of Domestic Sciences, Uppsala University, Uppsala, S-75337, Sweden

Introduction

The International Standardisation Organisation (ISO) is a worldwide federation of national standardisation organisations (ISO members) with technical committees established to deal with different subjects. The international standards normally are drafted by these committees, the drafts then being circulated to the ISO members for approval. Publication of the ISO standards is conditional upon their being approved by at least 75% of the ISO members.

ISO standards in the sensory field concern, for example, general methodological guidance of sensory analysis (ISO 1985), methodology of difference tests such as the triangular test (ISO 1983a), the paired comparison test (ISO 1983b), the ranking order test (ISO 1988a) and the 'A' 'not A' test (ISO 1987a), methodology for food evaluation by scales (ISO 1987b), vocabulary of statistics (ISO 1992a), vocabulary of sensory analysis (ISO 1992b), and general guidance for selection, training and monitoring of assessors (ISO 1993). The descriptive test requires a panel specially selected and trained in sensory analysis. The validity and result of profiling depends on the skill of the panel members (assessors). The selection of the assessors therefore needs to be carried out with care, and it should be looked upon as a necessary investment for successful sensory evaluation in product development and in research. In this chapter the recruitment, selection and training of assessors for carrying out a descriptive test (conventional profiling) will be discussed.

Sensory panel – recruitment

Internal recruitment

Candidates for a sensory panel are recruited from amongst the employees of, for example, a plant or a laboratory. The participation in the sensory work is a part of the everyday work. It is therefore crucial to have the support of the management of the establishment. The advantages of an internally recruited panel are that its members are readily available, it has a good stability over time and has a good confidentiality. The disadvantages might be that the product judgements of the candidates could be influenced by their product knowledge, and that selected candidates are difficult to replace when not available.

External recruitment

The most common means of attracting assessors from outside the establishment are advertisements in the local paper or in specialised publications, commercials on local radio and opinion poll institutes and clubs. Interested candidates are then contacted and receive a questionnaire (in some instances coupled with an interview). From the questionnaire and the interview some background information is obtained about the motive for wanting to participate, attitudes to food, health, availability and habits such as smoking. The selected candidates are then invited to participate in various screening tests. The recruitment stage involves the initial selection and the elimination of candidates that are not suitable for undertaking sensory analysis. The final selection is done after the screening tests have been carried out.

Sensory panel – screening

The screening tests provide a basis for final selection of panel candidates with ability to function as proficient assessors in sensory analysis. There are three types of screening tests: tests aiming to determine sensory impairment, tests aiming to determine sensory acuity and those aiming at evaluating the potential for describing and communicating sensory perception.

Sensitivity tests

The candidates are tested for sensitivity to substances which may be present in small concentrations in products, and to detect loss or lack of the sense of smell (anosmia) and taste (aguesia).

Matching tests

A set of samples of taste and smell materials (well above threshold levels) are presented to the candidates. They are given time to become familiar with the samples and are then asked to identify them. Thereafter they are presented with a series of the same materials and asked to match each of them with the original set of samples and describe their sensations. Suggestions of materials and concentrations to be used are given in ISO (1993). Candidates must have more than 80% correct matches to be selected.

Detection tests

To determine the ability to detect differences, a triangular test may be used. Two samples of the test material and one sample of water or other medium are presented. Supra-threshold levels of the test material are used. Suggestions of materials and concentrations to be used are given in ISO (1993). In order to be selected the candidates must have 100% correct responses.

Discrimination tests

To determine the ability to discriminate between levels of stimuli, intensity ranking tests (ISO 1988a) are used. Tests are performed on low concentrations of taste, smell, texture and colour stimuli. Samples of four concentrations of each stimulus are presented to the candidates in random order. The task is to put the samples in order of increasing intensity. Suggestions of materials and concentrations to be used are given in ISO (1993).

Descriptive ability

The aim of the tests is to determine the ability to describe sensory perception. Two tests are recommended: one odour stimuli test and one texture stimuli test.

Odour descriptive test

There are several types of method:

❑ *Direct methods* – the odours are determined by taking a couple of short sniffs from bottles, strips or capsules containing the odour stimulus
❑ *Retronasal methods* (perception of a mouth-originating volatile stimulus) – the odours are determined by ingestion of an aqueous solution, by placing smelling strips in the mouth or from a gaseous medium.

The candidates are presented with a set of olfactory stimuli in the form of five to ten samples. Some of the samples should be taken from common and easily identifiable products, some from products less common and thus difficult to identify. The task is to identify the stimulus correctly or to describe the associated odour. Examples of olfactory materials to be used, associated odour attributes and the grading system are given in ISO (1993). The performance is graded from 0 to 3 points. Candidates must have more than 65% of maximum possible scores to be selected.

Texture descriptive test

The candidates are presented with a set of products with different textural characteristics. Here again some of the samples should be taken from common and easily identifiable products and some from products less common and thus difficult to identify. The task is to identify the stimulus and to describe the characteristics of the texture correctly or associate it with the correct sample. The performance is graded from 0 to 3 points. Examples of products with different texture characteristics to be used, associated odour attributes and the grading system are given in ISO (1993). Candidates must have more than 65% of maximum possible scores to be selected.

Colour vision test

In order to rule out abnormal colour vision of the candidates the effective test of Ishihara (1971) can be used. When matching and judgement of colours are performed normal colour vision of the candidates is necessary.

Sensory panel – training

The training of selected panel members is designed to develop their ability to detect, recognise and describe sensations of odour and taste, and to improve their performance in carrying out sensory assessments. All training must be conducted in test rooms intended for sensory evaluation purposes and equipped according to the recommendations in 'General guidance for the design of test rooms' (ISO 1988b). Before any training takes place it is essential to make it clear to the assessors that the assessments must always be objective and it is never a question of liking or disliking a sample. During training the organiser will discuss the assessments with the assessors. They will also have the opportunity to re-assess samples and to see their scoring of replicates. There are certain instructions that it is necessary to give to the assessors before they arrive at the sensory test laboratory for the first time, such as no using of cosmetics with strong odours, no smoking, and no eating of strongly spiced foods within the 60 min prior to a sensory session.

Assessment procedure

The training must start with an introductory talk to the assessors about assessment procedures. The assessors should during training come to an agreement on how to assess odour, and for mouth assessments should agree on how much of a liquid or solid sample they will put in their mouth, the number of chews to take, when to assess the different attributes etc. They should also agree on minimum time intervals between sample assessments to allow for recovery and on whether samples are to be swallowed or not.

Training in detection and recognition of taste and odours

The aim is to demonstrate tastes and odours in low and high concentrations and to have the assessors recognise and describe the stimuli correctly. The stimuli must be representative of products to be tested and must be served at commonly used temperatures. For guidance about investigating taste sensitivity, about investigating odour sensitivity and about methodology, see ISO 1991, ISO 1992c and ISO 1985 respectively.

Training in the use of scales

The assessors shall be familiarised with the use of rating, classification, interval and ratio scales (see methodology – general guidance, ISO 1985 and evaluation of food products by methods using scales, ISO 1987b).

Descriptive assessments – profiling

The assessors are presented with samples of the actual products to be assessed. Each assessor individually writes down odour, taste and texture attributes that describe the sensory characteristics of the products. The descriptors are listed and similar descriptors are grouped, each group then being treated as one descriptor. These are discussed and the attributes that describe the product the best are agreed on and will constitute the profile to be used for assessing. Thereafter the training takes place in using the intensity line scale on the product profile attributes. Several sessions are used for training, and external standards and samples with particular properties can be used before the actual test sessions are run.

References

Ishihara, S. (1971) *Tests for colour blindness*. Kanahara Shuppan Co. Ltd., Tokyo-Kyoto, Japan.

ISO (1983a) *Sensory analysis – Methodology – Triangular test*. International Organisation for Standardisation, Genève. Ref. No. ISO 4120-1983 (E).

ISO (1983b) *Sensory analysis – Methodology – Paired comparison test*. International Organisation for Standardisation, Genève. Ref. No. ISO 5495–1983 (E).

ISO (1985) *Sensory analysis – Methodology – General guidance*. International Organisation for Standardisation, Genève. Ref. No. ISO 6658–1985(E).

ISO (1987a) *Sensory Analysis – Methodology – 'A' – 'not A' test*. International Organisation for Standardisation, Genève. Ref. No. ISO 8588: 1987.

ISO (1987b) *Sensory analysis – Methodology – Evaluation of food products by using methods using scales*. International Organisation for Standardisation, Genève. Ref.No. ISO 4121–1987 (E).

ISO (1988a) *Sensory analysis – Methodology – Ranking*. International Organisation for Standardisation, Genève. Ref. No. ISO 8587:1988 (E).

ISO (1988b) *Sensory analysis – General guidance for the design of test rooms*. International Organisation for Standardisation, Genève. Ref. No. ISO 8589:1988 (E).

ISO (1991) *Sensory analysis – Methodology – Method of investigating sensitivity of taste*. International Organisation for Standardisation, Genève. Ref. No. ISO 3972:1991 (E).

ISO (1992a) *Statistical analysis – Vocabulary*. International Organisation for Standardisation, Genève. Ref. No. ISO 5492:1992 (E/F).

ISO (1992b) *Sensory analysis – Vocabulary*. International Organisation for Standardisation, Genève. Ref. No. ISO 5492:1992 (E/F).

ISO (1992c) *Sensory analysis – Methodology – Initiation and training of assessors in the detection and recognition of odours*. International Organisation for Standardisation, Genève. Ref. No. ISO 5496:1992 (E).

ISO (1993) *Assessors for sensory analysis Part 1. Guidence to the selection, training and monitoring of selected assessors*. The International Organisation for Standardisation, Genève. Ref. No. 8586-1:1993 (E).

Chapter 30

Preference Testing

G.R. Nute

Department of Clinical Veterinary Science, University of Bristol, Langford, Bristol, BS40 5DU, UK

Introduction

When an individual is introduced to a particular stimulus, a response is generated. When this response is related to preference or acceptance of a product it is known generically as preference testing. There is debate about the effectiveness of consumer testing, be it in-house with a relatively small number of assessors, in the home, or in a hall (hall tests). Hall tests are often known as the 'Mall intercept method'. Most of the argument centres on the number of responses that are required to adequately assess the product in terms of the desired characteristics. Meilgaard *et al.* (1991) give a good overview of sensory evaluation techniques including aspects of preference testing.

Presentation of samples

In relation to the preparation of fish for sensory evaluation, a number of methods have been recommended by various authors. A standardised method was outlined by Kosmark (1986) for the assessment of mid Atlantic fish species, whilst Johansson (1991) published a more detailed methodology for the assessment of rainbow trout. Inevitably there will be differences in culinary procedures between countries. In many circumstances this will affect eating quality. Therefore, it is necessary for the cooking conditions, final internal muscle temperature and general procedures to be described in detail.

Preference tests

Preference tests are often referred to as quantitative affective tests. These are tests whereby decisions are required. There are basically two types of tests. In type 1 tests typical questions are: Which sample do you prefer? Which sample do you like the most? Which sample would you buy? In type 2 tests the consumer is asked to give a rating. Typically, this could be: How much do you like the sample? How acceptable is the sample?

Type 1 tests

Type 1 tests can be classified as: paired preference for two samples (simple choice, A or B), rank preference for three or more samples (order of rank of A, B, C, D), multiple paired preference for three or more pairs of samples (often known as the all pairs test, A *v*. B, A *v*. C, A *v*. D, B *v*. C, B *v*. D, C *v*. D), and multiple selected paired preference (often known as the selected pairs test A *v*. B, B *v*. C, C *v*. D etc.).

The paired comparison test (BSI 1982, ISO 5495), although primarily used to determine differences, can be used to record preferences between two samples. The differences may be directional (e.g. is A tougher than B) or non-directional (e.g. are A and B different), but are generally non-directional in preference tests. Non-directional tests are two-sided (two-tailed tests).

There is debate about the number of assessors required. The British Standards Institution suggests that typically 30 untrained assessors are required, with more than 100 being a more suitable number, whilst other authors state that the numbers required should be between 50 and 400 (Meilgaard *et al.* 1991). It is necessary to balance the order of presentation of the two samples A and B such that successive assessors are presented with each sample first or second. So, assessor 1 receives A then B, assessor 2 B then A, and so on.

A paired preference test was used to compare consumer preference for blue whiting (*Micromesistius poutassou*) and whiting (*Merlangus merlangus*) (Hamilton & Bennett 1981). Twenty-four assessors were presented with samples of plain steamed fish balanced using the scheme described above. The results showed that 15/24 assessors preferred whiting over blue whiting. The critical value for this test using p = 0.5 with *n* repetitions for p < 0.05 is 18/24 for a two-tailed test. The conclusion was that there was no significant preference between the two samples of fish.

Further work on consumer preferences for fish species (Hamilton & Bennett 1983) included a selected pairs test as a way of assessing pairs of fish species. Assessors were allowed a 'no preference' response and those that indicated no preference were not counted in the total number of responses in the p = 0.5 tables. Nine species of north Atlantic fish were compared in pairs, haddock versus whiting, cod versus haddock, cod versus whiting, cod versus ling, cod versus saithe, whiting versus blue whiting, lemon sole versus plaice, lemon sole versus dab and plaice versus dab. The pairings used were based on a matching test, where three sets of fish were used based on similarities in size and type of fillet and in flake size. The sets were: (1) cod, ling and saithe, (2) haddock, whiting and blue whiting and (3) lemon sole, plaice and dab. All nine permutations of order within each set were presented an equal number of times to the panel. The results from these tests showed that assessors were able to significantly differentiate the samples within each set. However, the results of the preference tests showed that of the nine pairs of samples presented, only in lemon sole versus plaice and plaice versus dab were there any significant preference effects; in the former, lemon sole was significantly preferred over plaice (p < 0.01) and in the latter, plaice was significantly preferred over dab (p < 0.05).

In a study of consumer preference of smoked salmon colour (Gormley 1992), eight samples, which in reality were four pairs, were presented to consumers, who were asked to select the sample they would buy. Samples were placed in one of four categories based on descriptors identified by a trained panel. The categories were 1 – deep orange, 2 – orange pink, 3 – lightish pink, not orange and 4 – dark, kipper-like. Of 386 respondents, 43% selected a sample that was pink. This sample had a Hunter L value of 1.90, a Hue angle of 35 and a Roche card value of 14.5. The author concluded that consumers preferred smoked salmon which had a light pink colour. However, frequent purchasers (38% of the total) of smoked salmon preferred their salmon to have a deep orange colour.

Preference when about to purchase a product can be influenced by many factors, not necessarily based on the quality of the product but more on the perceived potential quality. Consumers can be influenced by information appearing on the package containing the product. Work by Wessells *et al.* (1994) investigated this effect as part of a large study investigating consumer preferences and concerns for North-Eastern Aquaculture Products and included work on salmon, trout, hybrid striped bass and tilapia as well as shellfish , mussels, clams and oysters.

Consumers were given nine examples of labels that describe salmon and were asked to rank the labels in order of preference. The examples given were:

(1)   Farm-raised fresh salmon $5.49/lb
(2)   Wild caught fresh salmon $4.99/lb
(3)   Fresh salmon $4.99/lb Inspected by the US Food and Drug Administration
(4)   Wild caught salmon $4.49/lb Inspected by the US Department of Agriculture
(5)   Fresh salmon $5.49/lb Inspected by the US Department of Agriculture
(6)   Wild caught fresh salmon $5.49/lb Inspected by the US Food and Drug Administration
(7)   Fresh salmon $4.49/lb
(8)   Farm-raised fresh salmon $4.49/lb Inspected by the US Food and Drug Administration
(9)   Farm-raised fresh salmon $4.99/lb Inspected by the US Department of Agriculture

One thousand, one hundred and fifteen consumers took part in the test. The greatest number of replies (402) ranked sample 8 their most preferred. The second most preferred was sample 9 with 219 responses. Consumers' rating for their second most preferred sample was sample 8 with 256 responses. The least preferred was sample 7 with 461 replies. The authors concluded that the label most frequently ranked first was one that contained information stating the product was farm raised, had the lowest price and had been inspected by the US Food and Drug Administration.

The link between preference and acceptance

Acceptance tests are effectively the 'affective status' of the product and using relative acceptance ratings, preference is often inferred. Therefore, the assumption is that higher ratings mean higher preference. This assumption occurs primarily through the use of hedonic scales. The most common was developed by Peryam and Girardot (1952) who devised a nine-point scale which defines the psychological states of 'like' as shown in Table 30.1.

Other scales, such as the 'smiley, facial expression scales' show the degree of liking by representations of facial expressions. The smile, if turning down at the corner of the mouth, is equivalent to dislike, whereas if the smile is upturned it is equivalent to liking. Various degrees of smile are shown and the consumer selects the face that most represents their feeling about the sample.

Whitefish is a freshwater species of economic importance in the Great Lakes area of North

Table 30.1   Hedonic scale devised by Peryam and Giardot (1947).

Scale labels	Values assigned after assessment
Like extremely	9
Like very much	8
Like moderately	7
Like slightly	6
Neither like or dislike[*]	5
Dislike slightly	4
Dislike moderately	3
Dislike very much	2
Dislike extremely	1

[*] neutral point that can be omitted.

America. In a study by Wesson *et al.* (1979) on the discrimination of whitefish quality by laboratory consumers (effectively a hall test) seven point hedonic scales were used to reveal differences in consumer preferences for fresh of frozen whitefish. A trained sensory panel comprising 21 assessors was asked to rate texture and oxidised flavour and also preference.

The panel showed that fresh fish had less oxidised flavour and was more tender than fish that had been previously frozen. The ratings for preference were 4.8 and 2.5 for fresh versus frozen fish respectively. The laboratory consumer test comprising 208 consumers also revealed a significant preference for fresh fish with means of 5.3 and 4.3 for fresh and frozen respectively.

These results contrast somewhat with those found by Bennett and Hamilton (1986) using different species. These authors studied consumer acceptance of cod and whiting after chilled storage, freezing and thawing. Using a seven point hedonic scale, where 1 = dislike very much and 7 = like very much, 66 consumers were given samples of cod and whiting that was fresh, quick thawed and slow thawed. Results showed that there were no significant differences between the three treatments when analysed using the Wilcoxon matched pairs signed rank test.

Shorter hedonic scales are often used in the belief that consumers may be confused if given a large number of categories from which they must select just one. This debate is outside of the scope of this current chapter but it is useful to see what has been achieved in studies with five point scales. A study by Ostrander *et al.* (1976) compared consumer preference for salmon and trout. Samples of trout, sockeye salmon and chinook salmon were baked and served to 137 consumers in a canteen. Consumers were asked to rank the samples on a five point 'smiley' scale on the basis of preference. Using analysis of variance it was concluded that there was no difference between the samples with means of 4.1, 4.0 and 3.7 for sockeye salmon, trout and chinook salmon respectively.

Overall liking, which is effectively a summation of all the characteristics of the fish that give rise to liking, does not always provide sufficient information to understand which of the individual characteristics of the fish are most preferred by consumers. In a paired test with 104 families, Rounds *et al.* (1992) compared the eating quality of farm-reared rainbow trout versus brown trout. Consumers were asked to cook each pair of fish and using a nine point hedonic scale, rate their liking for odour, colour, flavour, texture and overall liking. The results showed that there was no significant difference in liking for odour, colour or texture between the fish, but there were significant differences for flavour and overall liking where brown trout were preferred over rainbow trout.

Preference testing can be extended to identify those characteristics of fish which lead to the selection by consumers of one fish species over another. These will have important implications for consumption patterns and hence production.

In an extensive study of factors which affected consumer choice, Sawyer *et al.* (1988) conducted a survey that was concerned with identifying the consumer evaluations of the sensory properties of fish. Two hundred and ninety consumers were asked how frequently they ate fish, which fish from a list of 32 species they had eaten at least once, and the degree of liking for that species. Consumers were also asked their major reason for their hedonic rating. Results showed that those consumers who rated fish greater than five (increased liking) on a nine point hedonic scale were mainly concerned with flavour. This was also true of those who rated fish less than five (increasing disliking) although these consumers also showed a tendency to be affected by texture in their judgements. More consumers were familiar with haddock, flounder and swordfish than with pollock, catfish, whitefish, whiting, carp and hake.

When consumers are presented with a number of different types or species of fish all of good quality, then simple comparison of means for each fish may not reveal the true preferences of individual consumers. Interest in the way individual consumers behave, and whether or not they form part of a sub-group of consumers, has led to the development of more advanced preference testing techniques.

One such technique is 'preference mapping' which is a form of multidimensional scaling (MDS). There are basically two types. PREFMAP uses an external analysis (i.e. a sensory space from a trained panel) into which are mapped consumer data. MDPREF uses an internal analysis performed on data sets that consist exclusively of preference data which can be derived from paired comparison, rank or rating data. MDPREF configures subjects (consumers) and objects (products) solely on preference data.

In a study by MacFie and Thomson (1984) six samples were used comprising blue whiting, scad (good quality), scad (poor quality), white fish in parsley sauce, white fish fingers and cod mince. The MDPREF solution explained 85% of the variance and consumers were shown to be segmented, such that no single product satisfied all consumers. There were consumers whose region of highest preference was related to whole fish and consumers whose region of preference was centred on fish products. This approach is becoming more widespread as consumers are presented with more choice and, although they might find all products acceptable, they will still show a preference for some products over others. Trying to understand the underlying reasons for consumer preference has led to the development of the concept of 'appropriateness'.

Appropriateness

The concept of appropriateness as described by Schutz (1988) is based on several assumptions:

(1) Simple affect or preference is not the same as appropriateness for use
(2) What is appropriate to the consumer varies widely and is affected by:
   ❑ Who is consuming the item
   ❑ The situation in which the item is to be consumed and the other items to be consumed at the same time

For example, if a consumer is presented with a choice of fresh salmon, fish fingers, fish cakes or fish in a sauce, it is likely that one of the fish products would be selected over the fresh salmon if the consumer was in a rush to prepare a meal. However, if the intention was to produce a meal in which time was not important, and maybe there was something special about the meal, then the fresh salmon might be selected over the products.

Marshall (1988) showed that there are both positive and negative features of fish that may affect preference and the link with appropriateness. Positive features were: healthy, light, quick to prepare, nutritious, non-fattening, no waste and a variety of species. Negative features of fish were associated with: display of the whole fish, smell, dislike of handling raw fish, fiddly to eat, bones, few cooking methods, unusable leftovers and confined to one meal.

Conclusions

Preference testing covers a very wide range of comparisons, from simple choice of one item over another to complex reasons for preference selection. It is important to understand that there are subtle differences between the terms liking, preference and acceptability. The type of test selected should be based on an understanding of these terms and how they are related to consumer choice.

References

Bennett, R. & Hamilton, M. (1986) Consumer acceptability of cod and whiting after chilled storage and freezing and thawing. *Journal of Food Technology*, **21**, 311–17.

BSI (1982) *Sensory analysis of food. Part 2. Paired comparison test*. BS 5929,1982. British Standards Institution, Milton Keynes, UK.

Gormley, T.R. (1992) A note on the consumer preference of smoked salmon colour. *Irish Journal of Agricultural and Food Research*, **31**, 199–202.

Hamilton, M & Bennett, R. (1981) A comparative study of consumer preference for blue whiting (*Micromesistius poutassou*) and whiting (*Merlangius marlangus*) before and after incorporation into products. *Journal of Food Technology*, **16**, 655–9.

Hamilton, M & Bennett, R. (1983) An investigation into consumer preferences for nine fresh white fish species and sensory attributes which determine acceptability. *Journal of Food Technology*, **18**, 75–84.

Johansson, L (1991) Effects of starvation on rainbow trout. *Acta Agriculture Scandinavia*, **41**, 207–16.

Kosmark, J.J. (1986) Standardizing sensory evaluation methods for marketing fish products. In: *Seafood Quality Determination* (ed. D.E. Kramer & J.Liston) pp.99–107. Elsevier Science Publishers BV, Amsterdam, Netherlands.

MacFie, H.J.H & Thomson, D.M.H. (1984) Multidimensional scaling methods. In: *Sensory Analysis of Food* 2nd edn (ed. J.R. Piggot), pp.351–75. Elsevier, London.

Marshall, D.W. (1988) Behavioural variables influencing the consumption of fish and fish products. In: *Food Acceptability*, (ed. D.M.H. Thomson) pp. 219–31. Elsevier, London.

Meilgaard, M., Civille, G. & Carr, B.T. (1991) *Sensory Evaluation Techniques*, pp.210–18. CRC Press, Inc. London.

Ostrander, J., Martinsen, C., Liston, J. & McCullough, J. (1976) Sensory testing of pen-reared salmon and trout. *Journal of Food Science*, **41**, 386–90.

Peryam, D.R. & Girardot, H.F. (1952) Advanced taste-test. *Food Engineering*, **24**, 58–61, 194.

Rounds, R.C., Glenn, C.L & Bush, A.O. (1992) Consumer acceptance of brown trout (*salmo trutta*) as an alternative species to rainbow trout (*Salmo gairdneri*). *Journal of Food Science*, **57**, 572–4.

Sawyer, F.M., Cardello, A.V. & Prell, P.A. (1988) Consumer evaluation of the sensory properties of fish. *Journal of Food Science*, **53**, 12–18, 24.

Schutz, H.G. (1988) Beyond preference: appropriateness as a measure of contextual acceptance of food. In: *Food Acceptability*. (ed. D.M.H. Thomson pp.115–34. Elsevier, London.

Wessells, C.R., Morse, S.F., Manalo, A & Gempesaw II, C.M. (1994) *Consumer preferences for north-eastern aquaculture products*. Report on the results from a survey of north-eastern and mid-Atlantic consumers, Publication 40, 36–42. Department of Resource Economics, University of Rhode Island.

Wesson, J.B., Lindsay, R.C & Stuiber, D.A. (1979) Discrimination of fish and seafood quality by consumer populations. *Journal of Food Science*, **44**, 878–82.

Chapter 31
Value-added Products: Surimi

M. Tejada

Instituto del Frío (CSIC). C/Ramiro de Maeztu, s/n. Ciudad Universitaria, 28040 Madrid, Spain

Introduction

Fish farming is one of the food producing sectors with the greatest growth potential and is considered the best option for maintaining present levels of fish consumption at a time when wild captures are stagnant. In order to get the highest profit from fish farming, the aquaculture industry cultivates species that are commercialised whole or as processed products with a high market value. When the fish is filleted, a large percentage of the muscle adheres to the fish frames after filleting and can be extracted as minced muscle. For some fish, it is possible to recover up to 60% of minced meat from the filleted frames. Although the nutritional and functional value of this muscle is high, it is frequently used to make products not for direct human consumption. It is therefore important for the fish farm industry to become aware of the possibility of manufacturing high value products for direct human consumption. One such possibility that the fish farming industry has so far failed to consider is the use of fish which are not readily marketed whole, or frames and trimmings left from filleting, as raw material for 'surimi' production.

Surimi is a Japanese term defining washed and refined ground fish meat, which is used mainly for its functionality, and especially its capacity to form elastic fish gels. Surimi is essentially a concentrate of myofibrillar proteins, mainly natural actomyosin, since the manufacturing process eliminates most of the fat, stromal proteins and water-soluble muscle components. Surimi gels, known generically as 'kamaboko' gels, are thermally irreversible, and when cooked, retain their original shape with practically no change in their sensory characteristics. Surimi has been used for centuries in Japan to make products with a range of specific flavours and textures. Traditionally, surimi production did not exist as a separate process; it was a previous step to producing fish gels. Mass production of frozen surimi began in Japan in the 1960s as a means of marketing the muscle of Alaska pollock (*Theragra chalcogramma*), which was treated as a by-product of roe (Suzuki 1981; Toyoda *et al.* 1992). Since then the use of surimi as a raw material has spread to western countries, chiefly for the manufacture of surimi-based shellfish analogues such as crab legs, scallops, shrimps, lobster tails, etc. These products are better suited to western tastes and have been in increasing demand in the last two decades. Most surimi is now marketed in frozen form. In 1996 the total world production of kamaboko and frozen surimi was 1,378,609 metric tonnes, more than half of which was produced in Japan (FAO 1998).

Manufacture of surimi

The most important requirement of fish for processing into surimi is freshness. The best quality surimi generally comes from factory vessels, as fish can be processed sooner after catching than at land-based plants. The process is basically the same with minor modifications (Fig. 31.1). To obtain surimi, the fish is washed and scaled. Large fish are filleted and small fish are headed and gutted so that the gills and viscera do not pass into the surimi as the high proteolytic activity of digestive enzymes can hydrolyse the myofibrillar proteins and thus reduce gel forming ability. In gadoid fish (cod, hake, saithe, hoki, etc.) kidneys have to be removed as kidney tissue has a very high trimethylmine N-oxide demethylase (TMAO-ase) activity, and, if mixed with the muscle, high levels of formaldehyde (FA) from TMAO are formed in these species during frozen storage. The increase of FA accelerates denaturation and aggregation of myofibrillar proteins (Del Mazo *et al.* 1994; Tejada *et al.* 1997) reducing gel forming ability during frozen storage. All peritoneum and scales residues must be removed as these are considered impurities for the purposes of commercial surimi quality. The flesh is forced through a meat separator (deboner machine), where a belt presses the meat against a perforated drum (holes 3–5 mm diameter) to produce mince. Where skin-on fillets are used, the skin side should face the belt.

The minced fish is put through various cycles of washing in water–dewatering–sieving to remove fat and water-soluble substances (sarcoplasmic proteins, digestive enzymes, inorganic salts, TMAO etc). Washing removes compounds that shorten the frozen storage life of surimi, improves the colour, attenuates the fishy flavour and enhances gel forming ability. The water:mince ratio and the number and duration of washings depends on the freshness, composition of the fish and on the agitation time. As a rule, the fresher the fish the fewer washings and less soaking time and water required. The hardness of the wash water affects the swelling tendency of the meat during the washing process; the greater the ionic strength of the water, the easier it is to remove the water from the meat. When water with low ionic strength is used it is advisable to add sodium chloride at 0.15–0.25% (w/v) to the last washing solution to increase ionic strength and make dewatering easier (Suzuki 1981).

The use of calcium or magnesium chloride in the washing water is controversial. In Alaska pollock the Ca^{2+} ions are thought to increase firmness, reduce gel cohesiveness and cause denaturation of actomyosin during frozen storage (Tamoto 1971). However, in ultra-fresh cod the use of calcium or magnesium chloride improves the firmness and elasticity of gels but can also activate proteolysis resulting in loss of surimi quality when the fish is in post-rigor mortis condition (Martinez *et al.* 1991).

The pH of the washing water is generally not a problem in white fish. For red meat fish attaining a low final pH (5.7–6.0), alkaline agents are used to adjust the pH to 6.5–7. The water temperature must be less than $10°C$ to prevent denaturation of the actomyosin during processing and the decrease of gel forming ability. On the other hand, higher temperature favours water removal and separation of sarcoplasmic proteins. It is therefore important to ascertain the heat stability of the functional proteins in different species in order to favour the removal of water and water soluble compounds without this affecting the gel forming ability of the myofibrillar proteins. Any impurities that may have remained through the washing of the mince are removed by sieving. Industrially the process is done continuously with appropriate equipment (Toyoda *et al.* 1992). Surimi manufacture requires a large amount of

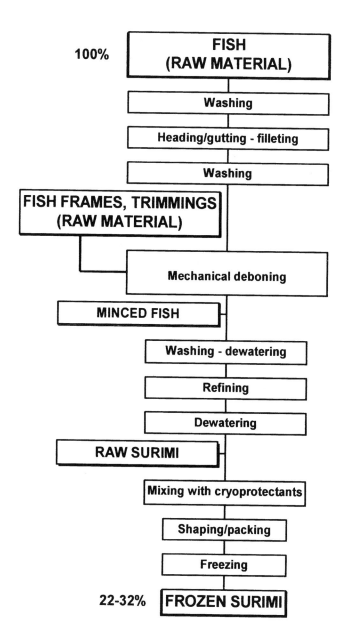

Fig. 31.1   Surimi production.

fresh water that creates a disposal problem, as the wastewater is rich in soluble and suspended organic matter. A number of effluent treatments have been designed to separate the suspended particles, the water-soluble proteins and the lipids. The most widely used methods are air flotation, electric or chemical coagulation, settling, etc. followed by microbiological treatment to comply with effluent standards (Okada 1992).

Refining consists of passing the washed mince through a perforated cylindrical screen (holes 1–3 mm diameter) to remove most of the bones, scales or connective tissue that have passed through the meat separator and have not been removed in the washing cycles. It is generally carried out before the last dewatering to prevent temperature increase during refining. In the final dewatering, a screw press is generally used to adjust the final moisture of the surimi.

Washed minced fish meat is easily denatured during frozen storage so decreasing or destroying its gel forming ability. In order to stabilise proteins in surimi during frozen storage, protective substances are added. From the 1960s, Japanese researches studied which proteins were mainly denatured during frozen storage, how their functional or enzymatic properties were altered, what substances would act as cryoprotectants and what was the actual protective mechanism. From these studies it was concluded that cryoprotectants worked by binding to the protein molecules resulting in an increased hydration of the protein molecules, reducing ice formation during freezing and hence reducing the degree of protein unfolding or denaturation during freezing and frozen storage (Matsumoto & Noguchi 1992). Other mechanisms have been proposed in which cryoprotectants are considered to have a stabilising effect by favouring preferential hydration of the proteins (Arakawa *et al.* 1990). Nevertheless the cryoprotectants initially proposed by the Japanese researchers, a mixture of saccharose-sorbitol (4 + 4% w/w) and polyphosphate (0.2–0.3%), are still the ones most widely used in the surimi industry. Cryoprotectants are blended with the washed minced muscle using various kinds of equipment. Blending time should be long enough for even distribution and short enough not to raise the temperature.

Surimi yield varies with the species and the technology employed, ranging between 20 and 33% depending on the starting material (species, size, season, freshness, etc.), the final water content and the technological parameters applied during the process.

Most surimi is marketed frozen ($\leq -25°C$) in 10 kg blocks, packed in corrugated cardboard cartons, two blocks per carton. Frozen surimi blocks are easily stored, transported and handled, and as surimi is graded according to its quality, frozen surimi can be used as a standard raw material available all the year around, is easy to store and manage and does not cause problems of waste disposal in the surimi-based product plants.

There are differences between surimi and minced fish that make surimi a versatile raw material with high functionality. Concerning composition, the percentage composition of the minces is quite similar to that of the original muscle. However, surimi is a concentrate of myofibrillar proteins; lipid content ranges between 1 and 3% depending on the fish species, and the final water content depends on the process. The percentage of carbohydrates in surimi is much higher than in minces (around 8%) as these are added to the surimi as cryoprotectants. There are also sensory differences in colour, odour and flavour. In minces this depends on the species, but the compounds producing colour, odour and flavour are partially eliminated by surimi processing. Minces deteriorate faster than whole muscle in frozen storage, as mincing accelerates denaturation and aggregation of the myofibrillar proteins (Tejada *et al.* 1998), whereas surimi keeps well due to the addition of cryo-protectants.

Frozen surimi is generally made from under-utilised wild white-flesh saltwater species (mainly Alaska pollock) or unused parts of commercially valuable species (FAO 1998) as whiteness is one of the quality parameters required by the industry for surimi-based products. However, surimi is also made from small pelagic species with a high proportion of red muscle (Davis 1988). Fish with high proportion of red muscle have higher fat content, a stronger fishy odour and flavour, higher concentrations of heme pigments even in the white muscle, a higher concentration of sarcoplasmic proteins, less stable proteins, higher proteolytic activities, more TMAO and lower final pH. Sardine (*Sardina* spp.) menhaden (*Brevoortia* spp) and horse mackerel (*Trachurus murphyi*) have been studied in depth for industrial surimi manufacture (Ishikawa *et al.* 1979; Suzuki & Watabe 1986; Dondero *et al.* 1991, 1994; Shimizu *et al.* 1992; Alvarez *et al.* 1995; Alvarez & Tejada 1997).

Although gel forming ability can be comparable to that of white flesh species, surimi from dark fleshed species is considered inferior because of darker colour. Its use is restricted to surimi products for which final whiteness is not a requirement. The objectives pursued in industrial preparation of surimi from dark fleshed species are to eliminate the red muscle, raise the pH, remove fat and prevent the muscle minces from being contaminated with skin. In addition, the fish must be fresh and must have been rapidly chilled when caught, as proteolytic activity is higher and pH lower than in white fish species. For storage purposes, as well as the cryoprotectants used for surimi from white flesh species, an antioxidant must be used as these species tend to become rancid in frozen storage. There are various procedures used by the industry to make surimi from these species (Nonaka *et al.* 1989; Shimizu *et al.* 1992). In all species prolonged chilled or frozen storage of the fish before the surimi is made generally decrease the gel forming ability of the surimi.

Farmed species

Seawater and fresh water farmed species have been studied to assess the possibility of using them to make surimi. These species include tilapia (red hybrid) (Tejada *et al.* 1995), channel catfish (*Ictalurus punctatus*) (Lim & Sessa 1995), African sharptooth catfish (*Clarias gariepinus*) (Hoffman & Prinsloo 1996) and hybrid Clarias catfish (*C. macrocephalus* and *C. gariepinus*) (Nongnuch 1996). Although there are numerous studies on wild salmon surimi, there are not specific studies on farmed salmon (Yasunaga *et al.* 1998a,b). In Japan, 1% of the salmon catch is used to make surimi (approximately 3000 MT) (Arai *et al.* 1998). There is no evidence to show that surimi is currently being made industrially from farmed species, although industrial manufacture of surimi from fresh water fish has been attempted in China.

Surimi gels

Gelling is the major functional property of surimi and the reason for the spectacular growth in this industry over the last few years. Gels consist of strands or chains cross-linked to create a continuous network immersed in a liquid medium (Tanaka 1981). Gelation is the formation of a three dimensional continuous network which exhibits a certain degree of order. In contrast, coagulation is a random aggregation of proteins. Fish protein gels are hydrogels because they are formed when proteins dispersed in water cross-link to form a continuous matrix which entraps water within.

Solubilisation

Myofibrillar proteins are responsible for the formation of the homogeneous, thermally-irreversible kamaboko type gels made from surimi. Myofibrillar proteins constitute 70–79% of muscle proteins, and of these, actomyosin is the one responsible for gel forming ability. The gelling characteristics of actomyocin are associated with the myosin portion and the species specificity to the heavy-chain portion of the myosin molecule. Elimination of sarcoplasmic proteins during surimi manufacture enhances gel forming ability due to the interference of the sarcoplasmic proteins in the network formed by the myofibrillar proteins (Okada 1980; Shimizu *et al.* 1992), or to an increase in the concentration of myofibrillar proteins (Nishioka *et al.* 1990; Toyoda *et al.* 1992). Other myofibrilar proteins seem to have no effect on gelation (Careche *et al.* 1995).

Myofibrillar proteins form myofibrils and myofilaments, which are the contractile structures in muscle and give the skeletal muscle the striated appearance when observed by microscope. However, for gel making the structure has to be destroyed so that the proteins can rearrange and cross-link to form the matrix of the gel. Although the microstructure of intact muscle is partially degraded during the mincing and comminution steps in surimi manufacture, the surimi is ground with sodium chloride (2–3% w/w) in order to destroy the structure and make soluble the actomyosin and the myosin filaments. The myofibrillar proteins are partially solubilised because when the salt ions are bound to acidic and basic amino acid residues, intermolecular ionic bonds are ruptured and the proteins are dispersed in water forming a viscous paste (sol) (Niwa 1985). In the sol the proteins are partially denatured and cross-linked so that they can form the continuous matrix of a gel. It is considered that if salt is not present the sol will not be obtained even if grinding is carried out for a long time (Niwa 1992). Although gel formation has also been described in conditions where there was hardly any solubilisation of myofibrillar proteins (Chang 1997), grinding with salt is the procedure normally used in the industry. As fish proteins are very unstable, especially when solubilised with salt, grinding with salt must be performed at low temperature to prevent excessive denaturation of the actomyosin, since gel forming ability declines when conditions for gel formation are not present. When proteins have become previously aggregated by heating or during frozen storage, it is not possible to form high quality gels.

Gelation

Gelation involves a number of changes in the myofibrillar proteins, resulting in the aggregation of these proteins as a firm, elastic network. Kamaboko gels may be obtained by heating the sol directly at final temperature ($\geq 80°C$) (direct cooking) or in two steps after setting at temperature below $50°C$. The final texture of gels varies according to a number of factors including species, grinding conditions, and setting and cooking time and temperature. Gel characteristics are also influenced by habitat temperature, season of capture and sexual cycle of the fish, and variations in moisture, lipids, protein, nitrogen or pH in the muscle (Suzuki 1981; Shimizu 1985; Nishimoto *et al.* 1987, 1988; Kato *et al.* 1989; Akahane & Shimizu 1990; Yamazawa 1990; Couso *et al.* 1992; Alvarez *et al.* 1993, 1995). Changes of texture reflect changes in the gel microstructure. In sardine surimi a predominant fibrous image is associated with good textural properties, while a globular image is associated with lower gel strength (Couso *et al.* 1992; Alvarez *et al.* 1999a).

Setting

When solubilised proteins are heated, at the setting temperature they lose their natural conformation and begin to unfold. The hydrophobic amino acid residues emerging at the surface of the molecule as a result of unfolding increase protein–protein hydrophobic interactions. In the transition from sol to gel induced by heating at setting temperatures, ionic bonds, hydrogen bonds, hydrophobic interactions, disulphide bonds between adjacent cystein residues, and covalent bonds other than disulphide formed through the action of transglutaminases, are formed among the polymer segments (Suzuki 1981; Sano 1988; Niwa 1992; Careche *et al.* 1995; Arai *et al.* 1998). When setting is correctly applied it ensures a gradual sol–gel transition to form an initial protein mesh. The fish species, the quality of the surimi and the heating conditions all influence the type of bonds formed by the protein molecules, resulting in gels with different texture characteristics. Setting determines the texture and structure of the final kamaboko gels (Alvarez & Tejada 1997; Alvarez *et al.* 1999a). The set gels (suwari gels) formed at relatively low temperature are highly translucent, quite elastic but very soft (Niwa 1992; Alvarez & Tejada 1997).

Heating of the set gel

When set gels are heated at higher temperatures (cooked, > 80°C) gel strength increases through the enhancement of gel elasticity and firmness. This is due to the fact that in set gels conditions are adequate for orientation of the proteins to form the required bonds to produce elastic and firm gels (Suzuki 1981; Sano 1988; Alvarez *et al.* 1999a). Upon heating the protein, α-helixes unfold due to the heat instability of hydrogen bonds, exposing greater numbers of hydrophobic amino acids, and more extensive hydrophobic bonding occurs between molecules. There is also further oxidation of sulphydryl groups and disulphide bonds are formed. The rigidity of the gel increases dramatically and the increased light-scattering caused by the coagulation of proteins results in a whiter, more opaque gel (Niwa 1992). In some species prolonged setting time or higher temperatures have been found to cause a change in the ultrastructure and a decrease in gel strength in set (suwari) and cooked (kamaboko) gels (Alvarez & Tejada 1997; Alvarez *et al.* 1999b).

Direct heating

Heating of the sol without previous setting (direct cooking) results in a different gel structure due to the rapid unfolding of proteins, so that massive hydrophobic interactions and disulphide bonds are formed without prior formation of a mesh. As a result, coagulation is more intense, more water is released from the gel and the protein dispersion becomes less uniform.

The resulting gel is quite white and opaque with lower elasticity and water holding ability than pre-set gels, producing more brittle kamaboko gels (Niwa 1985; Alvarez *et al.* 1999b). In some cases there is rapid coagulation of actomyosin and no gel is formed.

Modori

Exposure of a sol or a set gel to temperatures near 60°C for any significant period of time causes irreversible degradation of the gel mesh. This thermal degradation is known as 'modori'. Some species are highly susceptible to modori (Tsukamasa & Shimizu 1989). In sardine surimi modori has been observed when setting was performed at 50°C (Alvarez *et al.* 1999b).

Three main mechanisms responsible for modori have been described: proteolytic degradation due to enzymes, thermal coagulation of myofibrillar proteins during heating, and involvement of specific non-enzymatic modori inducing proteins (Tsukamasa & Shimizu 1991; Niwa 1992). Enzymatic degradation of the gel with breakdown of myosin heavy chains has been shown to depend on species and season. For the same species there are different types of heat activated proteolytic enzymes of myofibrillar and/or sarcoplasmic origin that produce modori at different temperatures (Itoh *et al.* 1995). Another type of modori-inducing enzyme is involved in the degradation of fish species infested with parasites of *Myxosporidia* spp. Hydrophobic interactions and disulphide bridges have been found to increase among myofibrillar proteins and surimi sols at modori temperatures (Sano 1988; Careche *et al.* 1995). Although the mechanism whereby modori-inducing non-enzymatic proteins act on the myofibrillar protein network is not completely understood, it has been suggested that this could be due to a non-enzymatic interaction between such proteins and myofibrillar proteins (Iwata *et al.* 1977).

Ingredients

It is common practice to add other ingredients to surimi-based fish gels in order to achieve different textures, reduce costs or improve shape. The ingredients most widely used in surimi gel manufacture include starches (native or modified), proteins (egg white, soy protein isolate, whey protein concentrate, gluten, gelatine, etc.) and hydrocolloids (carrageenan, alginate, methyl cellulose, etc). In most cases the ingredients are entrapped within the protein matrix 'filling' the gel. The ingredients exert their effect by modifying the formation of the continuous surimi gel matrix during heat induced gelation or due to changes on pH and water (Alvarez *et al.* 1992, 1997; Lee *et al.* 1992; Couso *et al.* 1998). Among the factors that influence the type of structure formed in gels with added ingredients are the thermodynamic compatibility of the components, the mutual reactivity and the potential for interaction (Ziegler & Foegeding, 1990).

Surimi quality: standards

Surimi is graded according to its characteristics as a raw material and the quality of gels made in specified conditions from SA (super class) to C (off graded) (Suzuki 1981).

Standardised methods for measuring surimi quality vary according to the country. The required tests on raw material in the Japanese Standards for frozen surimi include moisture, pH and detection of impurities such as small bones, black membranes, etc. which are evaluated according to size (≥ 2 mm in diameter counts as one; < 2 mm counts as one half) on a scale from 10 (0 impurities) to 1 (≥ 31) (Lanier 1992). Optional tests on raw material include whiteness, brightness, drip under pressure and viscosity. Functional properties are measured by making a gel in standardised conditions in which starch may be added or not. This kamaboko test measures the texture characteristics (gel or jelly strength) by a penetration test, whiteness, brightness, folding test and sensory assessment. The sensory tests measure mainly the strength and elasticity of the gel when chewed, on a scale from 10 (extremely strong) to 1 (fragile). The folding test involves folding a round-shaped slice, 30 mm in diameter and 3 mm thick, and evaluating the way it breaks when folded. The maximum score, 'AA' or 5 points, is awarded when no cracks are visible after folding twice, and the minimum score, 'D' or 1 point, when the slice breaks under finger pressure. In this case the surimi is judged to have no gel forming ability.

In the USA the test has been related to compositional and functional properties (Lanier 1992). The compositional properties considered in raw material are protein, moisture and fat content, pH and visual contaminants such as particles of skin, belly lining etc. Functional properties are measured in gels made under standardised procedures in which a range of setting temperature/time conditions are established for low, median or high temperature setting ability; cooking conditions are fixed at 90°C for 15 min. Colour, flavour, odour and gel-forming ability (torsion test) are measured. The optional requirements for frozen surimi in the draft prepared by the Codex Committee on Fish and Fishery Products (Codex Alimentarius Commission 1999) evaluate moisture, pH, 'objectionable matter' such as particles of skin, bones and other than fish meat in raw material. The standardised methods for gels measure whiteness, expressible moisture, folding test on a scale from 5 (no crack when folded in four) to 1 (splits into two when folded in two), and sensory assessment (biting test) on a scale from 10 (extremely strong) to 1 (incapable of forming a gel). Gel strength and deformability, measured in standardised conditions by puncture or torsion test, have to be decided between the buyer and the seller.

Conclusions

Until now, the manufacture of surimi from farmed fish species was not considered an industrial option in western countries. However, as the ratio of farmed to wild catches increases, the fish farm industry will have more top-grade muscle available for processing in high-priced products for which there is a growing consumer demand. The producers and the processing industry should therefore be prepared to grasp every opportunity to utilise individual fish of inadequate size or quality for commercialisation, or to recover muscle from fish frames and trimmings generated by filleting in products that are sold ready-processed. The characteristics of surimi products depend on the quality of the surimi and the gelation conditions. As they vary according to intrinsic differences between species and differences in protein stability, it is essential to establish optimum parameters for surimi manufacture and gelation for farmed species in different habitats and production systems.

References

Akahane, Y. & Shimizu, Y. (1990) Effects of setting incubation on the water-holding capacity of salt-ground fish meat and its heated gel. *Nippon Suisan Gakkaishi*, 56, 139–46.

Alvarez, C. & Tejada, M. (1997) Influence of texture of *suwari* gels on *kamaboko* gels made from sardine (*Sardina pilchardus*) surimi. *Journal of the Science of Food and Agriculture*, 75, 472–80.

Alvarez, C., Couso, I., Tejada, M., Solas, M.T. & Fernández, B. (1992) Action of starch and egg white on the texture, water holding capacity and microstructure in surimi gels. In: *Quality Assurance in the Fish Industry* (eds H.H. Huss, M. Jakobsen, J. Liston), Developments in Food Science, 30. Elsevier Science Publishers, Amsterdam.

Alvarez, C., Couso, I., M., Solas, M.T. & Tejada, M. (1993) Influence of manufacturing process conditions on gels made from sardine surimi. In: *Food Proteins. Structure and Functionality* (eds K.D. Schwenke & R. Mothes), pp.347–53. VCH Weinheim, Germany.

Alvarez, C., Couso, I. & Tejada, M. (1995) Sardine surimi gels as affected by salt concentration, blending, heat treatment and moisture. *Journal of Food Science*, 60, 622–6.

Alvarez, C., Couso, I., Solas, M. & Tejada, M. (1997) Waxy corn starch affecting texture and ultrastructure of sardine surimi gels. *Zeitschrift für Lebensmittel Untersuchung und Forschung*, 204, 121–8.

Alvarez, C., Couso, I. & Tejada, M. (1999a) Microstructure of suwari and kamaboko sardine surimi gels. *Journal of the Science of Food and Agriculture*, 79, 839–44.

Alvarez, C., Couso, I. & Tejada, M. (1999b) Thermal gel degradation (*modori*) in sardine surimi gels. *Journal of Food Science*, 69, 633–7.

Arai, K.I., Yasunaga, K. & Abe, Y. (1998) Characteristic quality of surimi-based products from chum salmon with use of food additive containing microbial transglutaminase or bovine plasma. *Fourth International Symposium on the Efficient Application and Preservation of Marine Biological Resources*, Kangnung, Korea, pp.283–303.

Arakawa, T., Bhat, T. & Timasheff, S.N. (1990) Why preferential hydration does not always stabilize the native structure of globular proteins. *Biochemistry*, 29, 1024–1931.

Careche, M., Alvarez, C. & Tejada, M. (1995) Suwari and kamaboko sardine gels: effect of heat treatment on solubility of networks. *Journal of Agricultural and Food Chemistry*, 43, 1002–1010.

Chang, H.S. (1997) *Solubility and gelation of chicken breast muscle proteins as affected by salts.* PhD dissertation, University of Massachusetts, Amherst, MA.

Codex Alimentarius Commission (1999) *Draft of code of practice for fish and fishery products.* Codex Committee on fish and fishery products. FAO/WHO-Food-Standards-Programme 1999.

Couso, I., Alvarez, C., Solas, M.T., Tejada, M., Fernández, B. & Barba, C. (1992) Ultrastructure of sardine surimi gels influence of final moisture and setting conditions. In: *EUREM 92. Electron Microscopy 92.* (eds L. Megías-Megías, M.I. Rodriguez García, A. Ríos & J.M. Arias). Vol. III, pp.913–14. Secretariado de Publicaciones, Universidad de Granada.

Couso, I., Alvarez, C., Solas, M.T. & Tejada, M. (1998) Morphology of starch in surimi gels. *Zeitschrift für Lebensmittel Untersuchung und Forschung*, 206, 38–43.

Davis, N. (ed.) (1988) Fatty fish utilization: upgrading from feed to food. *Proceedings of a national technical conference in Raleigh, N.C.* UNC Sea Grant Publication 88-04, 405pp.

Del Mazo, M.L., Huidobro, A., Torrejón, P., Tejada, M. & Careche, M. (1994) Role of formaldehyde in formation of natural actomyosin aggregates in hake during frozen storage. *Zeitschrift für Lebensmittel Untersuchung und Forschung*, **198**, 459–64.

Dondero, M., Yañez, G., Curotto, E. & Canto, M. (1991) Efecto de crioprotectores en surimi de jurel (*Trachurus murphyi*). *Revista Española de Ciencia y Tecnología de Alimentos*, **31**, 539–49.

Dondero, M., Gandolfo, M. & Cifuentes, A (1994) Efecto crioprotector de Maltodextrina 25DE, suero y mezcla de aditivos en surimi de jurel (*Trachurus murphyi*). *Revista Española de Ciencia y Tecnología de Alimentos*, **34**, 389–408.

FAO (1998) *Fishstat plus*. An electronic database of fisheries statistics. Food and Agriculture Organization, Rome, Italy.

Hoffman, L.C & Prinsloo, J.F. (1996) The potential of freshwater fish in South Africa. *Food Industries*, **49**, (7) 30.

Ishikawa S., Nakamura, K., Fujii, Y., Yamano, G., Sugiyama, T., Shinozaki, K., Tobita, K. & Yamaguchi, Y. (1979) Fish jelly product (Kamaboko) and frozen mince meat (frozen surimi) made of sardine. III. Influence of the treatment methods for material just after catch on the kamaboko forming ability of sardine meat. *Bulletin of the Tokai Regional Fisheries Research Laboratory*, **99**, 31–42.

Itoh, Y., Maekawa, T., Pantip Suwansakornkul, P. & Obatake, A. (1995) Seasonal variation of gel-forming characteristics of three lizardfish species. *Fisheries Science*, **61**, 942–7.

Iwata, K., Kobashi, K. & Hase, J. (1977) Studies on muscle alkaline protease. VI. Purification of proteins which induce the 'modori' phenomenon during kamaboko production and of cathepsin A from carp muscle. *Nippon Suisan Gakkaishi*, **43**, 181–93.

Kato, N., Nakagawa, N. & Terui, S. (1989) Changes in myofibrillar protein in surimi during grounding with NaCl in relation to operating condition of a continous mixer. *Nippon Suisan Gakkaishi*, **55**, 1243–51.

Lanier, T. (1992) Measurement of surimi composition and functional properties. In: *Surimi Technology* (eds T.C. Lanier & C.M. Lee), pp.123–63. Marcel Dekker, Inc., New York, USA.

Lee, C.M., Wu M.C. & Okada, M. (1992) Ingredient and formulation technology for surimi-based products. In: *Surimi Technology* (eds T.C. Lanier & C.M. Lee), pp.272–302. Marcel Dekker, Inc., New York, USA.

Lim, C.E. & Sessa, D.J. (eds) (1995) *Nutrition and Utilization Technology in Aquaculture*. AOCS Press, Champaign, IL, USA.

Martinez, I., Solberg, C. Lauritzen, K. & Ofstad, R. (1991) Effect of the addition of $CaCl_2$ and $MgCl_2$ during the washing procedure on the protein content of cod surimi and water soluble fraction. 6pp. *21st annual WEFTA meeting*. Copenhagen.

Matsumoto, J.J. & Noguchi, S.F. (1992) Cryostabilization of protein surimi. In: *Surimi Technology* (eds T.C. Lanier and C.M. Lee), pp.357–88. Marcel Dekker, Inc., New York, USA.

Nishimoto, S., Hashimoto, A., Seki, N., Kimura, I., Toyoda, K, Fujita, T. & Arai, K.I. (1987) Influencing factors on changes in myosin heavy chain and jelly strength of salted meat paste from Alaska pollack during setting. *Nippon Suisan Gakkaishi*, **53**, 2011–20.

Nishimoto, S.I., Hashimoto, A., Seki, N. & Arai, K.I. (1988) Setting of mixed meat paste of two fish species in relation to cross-linking reaction of myosin heavy chain. *Nippon Suisan Gakkaishi*, 54, 1227–35.

Nishioka, F., Tokunaga, T., Fujiwara, S. & Yoshioka, S. (1990) Development of new leaching technology and a system to manufacture high quality frozen surimi. In: *Chilling and Freezing of New Fish Products*, pp.73–80. International Institute of Refrigeration (ed.), Paris, France.

Niwa, E. (1985) Functional aspect of surimi. In: *Proceedings of the International Symposium on Engineered Seafood including Surimi* (eds R.E. Martin & R.L. Collete), 19–21 November, Seattle WA, USA, pp.141–7. National Fisheries Institute, Washington DC, USA.

Niwa, E. (1992) Chemistry of surimi gelation. In: *Surimi Technology* (eds T.C. Lanier & C.M. Lee), pp.389–427. Marcel Dekker, Inc., New York, USA.

Nonaka, M., Hirata F. Saeki, H. Nakamura, M. & Sasamoto, Y. (1989) Gel forming ability of highly nutritional fish meat for food stuff from sardine. *Bulletin of the Japanese Society of Scientific Fisheries*, 55, 2157–62.

Nongnuch, R. (1996) Processing of hybrid Clarias catfish. *INFOFISH International*, 3, 33–6.

Okada, M. (1980) Utilization of small pelagic species for food. In: *Proceedings of the Third National Technical Seminar on Mechanical Recovery and Utilization of Fish Flesh* (ed. R.E. Martin), pp.265–82. National Fisheries Institute, Washington, DC.

Okada, M. (1992) History of surimi technology in Japan. In: *Surimi Technology* (eds T.C. Lanier & C.M. Lee), pp.3–21. Marcel Dekker, Inc., New York, USA.

Sano, T. (1988) *Thermal gelation of fish muscle proteins*. Doctoral thesis, Laboratory of Biochemistry, Department of Chemistry, Faculty of Science and Technology, Sophia University, Tokyo, Japan, 344pp.

Shimizu, Y. (1985) Biochemical and functional properties of material fish. In: *Proceedings of the International Symposium on Engineered Seafood Including Surimi* (eds R.E. Martin & R.L. Collete), pp.148–67. 19–21 November, Seattle, WA.

Shimizu, Y., Toyohara, H & Lanier, T.C. (1992) Surimi production from fatty and dark-fleshed fish species. In: *Surimi Technology* (eds T.C. Lanier & C.M. Lee), pp.181–207. Marcel Dekker, Inc., New York, USA.

Suzuki, T. (1981) *Fish and Krill Protein: Processing Technology*, 260pp. Applied Science Publishers Ltd, London.

Suzuki, T. & Watabe, S. (1986) New processing technology of small pelagic fish protein. *Food Reviews International*, 2, 271–307.

Tamoto, T. (1971) Effect of leaching on the freezing denaturation of surimi proteins. *New Food Industry*, 13, 61–9.

Tanaka, T. (1981) Gels. *Scientific American*, 244, 124–38.

Tejada, M., Alvarez, C., Martín, O. & Barbosa Cánovas, G.V. (1995) Infuencia del tratamiento térmico y la humedad en la calidad de los geles de surimi de tilapia. *Revista Española de Ciencia y Tecnología de Alimentos*, 35, 297–314.

Tejada, M., Torrejón, P., Del Mazo M.L. & Careche, M. (1997) Effect of freezing and formaldehyde on solubility of natural actomyosin isolated from cod (*Gadus morhua*), hake (*Merluccius merluccius*) and blue whiting (*Micromesistius poutassou*). In: *Seafood from Producer to Consumer. Integrated Approach to Quality* (eds J.B. Luten, T. Børresen & J. Oehlenschläger), pp.265–80. Elsevier Ed., Amsterdam.

Tejada, M., Torrejón, P., Del Mazo, M. & Careche, M. (1998) Extractability of myofibrillar protein from cod (*Gadus Morhua*) frozen stored at different temperatures and muscle integrity. *Twenty-first annual conference, Tropical and Subtropical Seafood Science and Technology Society of the Americas*. Florida Sea Grant Program, University of Florida, Gainesville, FL, Sea Grant Report 118, pp.36–45.

Toyoda, K. Kimura, I., Fujita T., Noguchi, S.F. & Lee, C.L. (1992) Surimi manufacturing from whitefish. In: *Surimi Technology* (eds T.C. Lanier & C.M. Lee), pp.79–112. Marcel Dekker, Inc., New York, USA.

Tsukamasa Y. & Shimizu, Y. (1989) The gel-forming properties of the dorsal muscle from Clupeiformes and Salmonoidei. *Bulletin of the Japanese Society of Scientific Fisheries*, **55**, 529–34.

Tsukamasa, Y. & Shimizu, Y. (1991) Another type of proteinase-independent *modori* (thermal gel degradation) phenomenon found in sardine meat. *Bulletin of the Japanese Society of Scientific Fisheries*, **57**, 1767–71.

Yamazawa, M. (1990) Effect of high temperature heating on physical properties of kamaboko-gel. *Nippon Suisan Gakkaishi*, **56**, 497–503.

Yasunaga, K., Abe, Y., Nishioka, F. & Arai, K.I. (1998a) Effect of bovine plasma on heat-induced gelation of salt ground meat from walleye pollack and chum salmon. *Nippon Suisan Gakkaishi*, **64**, 685–96.

Yasunaga, K., Abe, Y., Nishioka, F. & Arai, K.I. (1998b) Change in quality of preheated gel and two step heated gel from walleye pollack and chum salmon on addition of microbial transglutaminase. *Nippon Suisan Gakkaishi*, **64**, 702–709.

Ziegler, G.R & Foegeding, E.A. (1990) The gelation of proteins. In: *Advances in Food and Nutrition Research* (ed. J.E. Kinsella), 34, pp.203–98. Academic Press, New York.

Chapter 32
Safety of Aquaculture Products

P. Howgate

26 Lavender Row, Stedham, Midhurst, West Sussex, GU20 0NS, UK

Introduction

The world production of animal products from aquaculture in 1996 was 23.3 million tonnes (FAO 1998). In 1984, when FAO starting recording aquaculture production, the amount was 6.9 million tonnes and the increase since then has been approximately 10% compound a year, perhaps making aquaculture the fastest growing food production system in the world. Since the mid-1980s, catches of fish for human consumption from wild stocks have remained approximately the same at around 57 million tonnes and the fact that total supplies of fish for an increasing world population have increased since then is due to the increase in aquaculture production. On a world-wide basis, aquaculture provided 29% of total fish and shellfish supplies for human consumption in 1996.

Aquaculture products are food products and an important aspect of quality of foods is that they must be safe to eat. There is no suggestion from epidemiological evidence that products of aquaculture are more hazardous than their counterparts caught from the wild, but aquaculturists must ensure that their practices do not introduce hazards into their products or increase the risk of food poisoning. Indeed, given the controlled conditions under which aquaculture products are produced, there should be opportunities to make cultivated products even safer than counterparts caught in the wild.

Aspects of the public health safety of aquaculture products have been reviewed (Ward 1989; Reilly and Käferstein 1997; Alderman & Hastings 1998; Dalsgaard 1998; Howgate 1998; Rim, 1998) and this paper will highlight particular features of aquaculture production that have a bearing on the hazards and risks of the products. A hazard is 'a biological, chemical or physical agent in, or condition of, food that may cause a food to be unsafe for consumption'; a risk is 'the likelihood that an adverse health effect will occur within a population as a result of a hazard in a food' (National Advisory Committee on Microbiological Criteria for Foods, 1998a,b). A more detailed overview of microbiological problems in the salmon processing industry is given in Chapter 8 by Tony Laidler.

General features of food poisoning from fishery products

Though there is widespread and increasing concern about food poisoning, the data on sources and causes of food poisoning are not as comprehensive and as detailed as one would wish (Guzewich *et al.* 1997). The quality of surveillance of food poisoning incidents varies considerably among countries, and is often poor. Summaries are seldom published, and when they are they often lack important information. Even in countries where surveillance and reporting are good, experts consider that perhaps only about 10% of cases of food poisoning are reported.

The most comprehensive summaries of food poisoning relate to the USA (Bean & Griffin 1990; Ahmed 1991). In that country, in the 15 years 1973–1987, the food commodity responsible for outbreaks was identified in only about half of the incidents, and of these, 15% involved vertebrate fish, and 6% shellfish, predominately bivalve shellfish. In the two other countries for which summaries are available, Canada (Todd 1992), and the Netherlands (Beckers 1986; Simone *et al.* 1997), the incidence of food poisoning from fish and shellfish combined was around 7%. There do not appear to be any data that distinguish aquacultured products and products caught from the wild as vehicles of food poisoning.

Aquaculture production

Hazards in aquaculture products are associated primarily with the environment in which they are cultured, and with farming practices. To put these factors in context of aquaculture production, in 1996, 57% of production was cultivated in freshwater, 37% in the marine environment, and 6% in brackish waters. Of the geographical regions listed in the statistics, Asia accounted for 89% of production. The breakdown by species groups was 55% freshwater species, 32% bivalve shellfish, 6% diadromous species, 4% crustacean shellfish and 2% miscellaneous marine species.

Hazards and risks of some aquaculture products systems

Bivalve molluscs

In the few published country reports of food poisoning, and from hearsay, it appears that bivalve shellfish are often involved in food poisoning incidents from consumption of aquatic foods. Bivalves concentrate bacteria and virus particles from their surroundings and the presence of pathogenic bacteria and viruses is a hazard in these organisms (Guzewich & Morse 1986). The shellfish are safe for consumption when they are cooked or heat processed, but the risk of causing illness is considerably increased by the practice of eating them raw, or after only light cooking. Viruses in bivalve shellfish can withstand more heat processing than can the pathogenic bacteria that might be present (Abad *et al.* 1997), and thereby pose a greater risk of causing illness than do bacteria. Rose and Sobsey (1993) have calculated a risk of 1 in 100 of illness from virus infection from consumption of 60 g of raw bivalve shellfish based on epidemiological and contamination data for the USA. Official control of the safety of bivalves is exercised by a combination of regulation of the quality of the water from which they are harvested and by a requirement for depuration of harvested shellfish if the waters do not meet the required standard. These controls are not adequate to ensure safety because in many incidences of outbreaks of food poisoning by shellfish the product has come from approved waters or has been depurated (Gerba 1988). Counts of enteric bacteria, the criterion for approval of waters, are not reliable indicators of presence of pathogenic viruses, and viruses depurate slowly from shellfish so that depuration times that are adequate to clear bacteria do not clear viruses (Vaughn & Landry 1984; Beril *et al.* 1996; Abad *et al.* 1997).

Of the total world production of bivalve molluscs of 9.05 million tonnes in 1996, 7.3 million tonnes, 81%, came from aquaculture. Summary epidemiological statistics do not differentiate between incidents involving aquacultured products and wild products but it is inevitable that aquacultured shellfish must be responsible for incidents of food poisoning. However, it cannot be concluded that aquacultural practices influence the presence of hazards in the product; illness from consumption of raw bivalves has been recognised well before aquaculture made an important contribution to supplies. In the USA, where food poisoning from consumption of bivalve shellfish is higher than in other countries for which data are available, only about 15% of supplies comes from aquaculture. Whether or not cultivation practices increase or decrease the hazard in the shellfish does not seem to be known. Bacteria and viruses accumulate on debris, and viruses attached to sediment have a slower rate of inactivation than the same virus in the water column. It might be concluded from these observations that shellfish harvested from the wild, that is from contact with the sea bottom, could be more contaminated than shellfish grown on racks or on ropes, but shellfish can accumulate viruses and bacteria efficiently from the water column, and there might be no significant effect of cultural and harvesting practices.

Mariculture

Several species of bacteria pathogenic to man are indigenous to the marine environment, and can be expected to be found on fish harvested from that environment (Ahmed 1991; Ward & Hackney 1991; Gibson 1992; Huss *et al.* 1997; Dalsgaard 1998). Epidemiological data show that vertebrate fish are rarely the vehicle for bacterial food poisoning. One reason for this is that the fish are commonly cooked before consumption and the main risk comes from consumption of raw fish products. For example, there is a high incidence of poisoning from *Vibrio parahaemolyticus* in Japan because of the custom of eating raw fish. Marine fish from inshore or estuarine waters can be contaminated by human enteric organisms from sewage, but again the incidence will be low, and such organisms are readily killed by cooking.

There is no reason to expect the bacterial hazards of aquacultured fish from the marine environment to be any different from those of wild populations of fish caught in similar environments. Aquacultured fish are less likely to be contaminated by enteric organisms because it is the usual practice to site fish farms in unpolluted waters.

Marine fish, including those species that are farmed, can harbour nematode parasites that are pathogenic to man, the main species of concern being *Anisakis simplex* and *Pseudoterranova decipiens* (Deardorff & Overstreet 1991). The definitive hosts of the parasites are piscivorous marine mammals such as seals, and marine invertebrates and fish are the two intermediate hosts. Humans are infected when they consume raw or minimally processed fish, or more rarely, the invertebrate host. Anisakiasis, the disease in humans caused by the parasite, is an uncommon disease because the parasite is killed by normal cooking, and by freezing. There is a risk from fishery products consumed raw – the incidence of anisakiasis is high in Japan – or after only mild processing like salting at low concentrations or cold smoking. Many countries now require that fish used for these mildly processed products be frozen before processing.

The parasite infects the fish through its feed, and farmed fish which are fed processed, pelleted, feeds – the common practice for farmed marine fish – should not become infected. It has been reported following surveys that salmon farmed in the USA, Norway and Scotland at least do not harbour nematodes (Deardorff & Kent 1989; Bristow & Berland 1991; Angot & Brasseur 1993), and this freedom from nematodes is recognised by the European Commission in that farmed salmon are excused from the provisions of the hygiene regulations that require minimally processed fish intended for consumption without cooking to be frozen before sale. However, it must not be assumed farmed marine fish are necessarily free of nematodes as in some systems the cultured fish are fed with raw fish and the farmed product can thereby become infected. This has been reported for farmed yellowtail in Japan, and also trout in freshwater (Wootton & Smith 1975). Trash fish and offal for fish feed should be heated or frozen before being used.

Bacterial hazards in products from fresh and brackish waters

Species of pathogenic bacteria indigenous to the marine environment can also be found in fresh and brackish waters, though the composition of the flora is not the same. For example, vibrios are salt-tolerant organisms and are associated with the marine and brackish water environments, but some species can be found in fresh water. The human pathogens *Edwardsiella tarda*, *Aeromonas hydrophila* and *Pseudomonas shigelloides* are more common in brackish and freshwater environments than in marine environments, and have been associated with farmed fish. For example, all have been isolated from channel catfish ponds in the USA (Wyatt *et al.* 1979; Leung *et al.* 1992).

Clostridium botulinum has been found in fish and in sediments in freshwater trout farms in Britain and Denmark (Huss *et al.* 1974; Cann *et al.* 1975), though given the widespread occurrence of the organism in both the aquatic and terrestrial environments, it is undoubtedly much more widely spread in fish farms than in just those countries. The organism is not infective and illness arises from the toxin formed when the organism grows in the fishery product. The risk factors are more associated with aspects of the processing, packaging and storage of the product than with the presence of the organism in the fish, but in any analysis of the public health hazards of farmed fish it must be assumed that the fish will carry *C. botulinum* spores.

The aquatic environment of onshore aquaculture establishments has a greater risk of being contaminated by enteric organisms than do waters of offshore installations because faecal material from animals, birds and humans can enter water bodies directly or from run-off from the land. There are only a few reports on the presence of enteric pathogens in farmed fish cultivated in unfertilised systems (Wyatt *et al.* 1979; Saheki *et al.* 1989; Leung *et al.* 1992; Nedoluha & Westoff 1993), and they show that the numbers of enteric organisms are low and comparable to the incidence in fish caught from the wild. Nematode parasites are not present in the freshwater environment, and the trematode parasites to be described later are unlikely to be present in unfertilised systems. The risks to human health from fish cultivated in unfertilised freshwater systems will be very low.

Of much more concern are the widespread practices of using human and animal excreta as fertilisers in pond aquaculture and of raising fish in wastewaters. A number of reviews and reports have described the use of excreta in aquaculture and the raising of fish in sewage wastewaters, and most have drawn attention to the health hazards of these practices. They are reviewed in Howgate (1998). Though the amount of data is small, the evidence is that fish can be cultured in wastewater treatment ponds without their posing a significant risk to public health as long as some safeguards are met. For example, it would not be advisable to raise fish for human consumption in ponds early in the sequence of treatment. The available data on the bacterial flora of fish raised in sewage treatment systems shows that the fish can be carriers of enteric pathogens on the skin and in the gut, but a comparison with data on the microbial flora of fish from natural waters suggests the risk to public health from septic-raised fish might not be greater than that posed by fish harvested from the wild.

Cultivation of fish in ponds directly fertilised by animal and human excreta is widespread in Asia and South East Asia, and is practised elsewhere. Unfortunately, there are very little published data on the bacteriology of these systems (Howgate 1998). Reilly & Twiddy (1992) and Dalsgaard (1998) have reviewed the occurrence of *Salmonella* in cultured tropical shrimp. The organism is often recovered from the pond system or from the shrimp, but it cannot be assumed that *Salmonella* is a normal constituent of the aquatic environment of tropical shrimp ponds.

It is disappointing that there is so little information on hazards of products from aquaculture systems using excreta as a fertiliser bearing in mind the amount of fish globally that must be raised in such systems and the obvious risks to health. There is certainly a need to investigate the bacteriological hazards of fish and crustacean shellfish cultured in contact with sewage and manures in more detail, particularly in systems in tropical climates. The available evidence is that products from such systems will be contaminated by enteric pathogens on the skin and in the guts, and in some circumstances the organisms can be present in the flesh as well. There is some, but rather inconclusive, evidence from epidemiological studies (Blum & Feachem 1985) of an increased incidence of diarrhoeal illnesses in communities consuming fish from excreta-manured ponds. The pathogenic bacteria are killed in normal cooking, and the health risks are associated with the consumption of raw or lightly preserved products, and with cross contamination of other foods, hands, and food preparation surfaces.

Parasites in fish cultivated in freshwater

Fish, including shellfish, can be infected by a variety of parasites, though only a few species pose a risk to humans. One group that is of serious concern is the trematodes. A recent WHO report (WHO 1995) has reviewed the public health significance of foodborne trematode infections and estimated that somewhere in the order of 40 million people worldwide are affected by fishborne trematode infections. The species responsible for illnesses are prevalent in freshwater fish in warm climates, and all have similar life cycles involving a definitive host and two intermediate hosts (Malek 1980). The definitive host is man and other mammals. The first intermediate host is a snail, and the second, depending on genus, a fish or a crustacean shellfish. Man and animals are infected by eating raw, or minimally processed, infected fish or shellfish. The most important parasites, so far as numbers of people affected is concerned, are species of the genera *Clonorchis*, *Opisthorchis*, and *Paragonimus*.

Clonorchiasis, the disease caused by *Clonorchis,* is endemic in some countries in East Asia – China, Hong Kong, Macao, Republic of Korea, Laos and Vietnam – though not confined to those countries. Only one species, *C. sinensis,* is reported to be involved. The geographical distribution of clonorchiasis within countries is associated with the distribution of its main, but not sole, snail host, *Parafossaulus manchouricus.* A wide range of freshwater fish can act as the second intermediate host, including the cyprinid species that are so commonly used in aquaculture in the affected countries.

Opisthorchiasis is endemic in Thailand, Laos, the Russian Federation, Ukraine and Kazakhstan, though again not confined to those countries. Two species of parasite are involved, *Opisthorchis viverrini* and *O. felineus.* The former is found mainly in Laos and Thailand, the latter mainly in Russia, Ukraine, Kazakhstan, and Northern Korea. A range of species of snails can act as first intermediate hosts for *Opisthorchis* species, and a range of species of fish as second intermediate hosts, again including species that are farmed in the affected areas. *Clonorchis* and *Opisthorchis* parasites affect the liver and can cause illness ranging from debility to death.

Paragonimiasis is endemic in some countries in east Asia – China and Korea – and is present in Japan, South America and West Africa, but again, not confined to those countries or regions. Some 40 species of *Paragonimus* have been reported to cause disease, but by far the most common agent is *P. westermani.* The first intermediate host is a snail, the species depending on locality and species of parasite, and the second a crustacean shellfish from a range of species. The parasite causes disease in the lung which is often confused with tuberculosis. The WHO (1995) report shows more people being affected by paragonimiasis than by clonorchiasis and opisthorchiasis combined. However, there is only a very small production of aquacultured freshwater crustacea in the affected areas, and, by default, aquacultured products will hardly contribute to infection.

Heterophydiasis is an enteric disease caused by intestinal trematode parasites of the family *Heterophydiae.* Several species have been reported to cause disease, but the two most important are *Heterophyes heterophyes* and *Metagonimus yokogawi.* The former is more prevalent in China and Egypt, the latter in Japan, Korea and the Russian Federation, though the disease is known in other countries. The intermediate hosts are freshwater and brackish water snails and fish, including species like mullet, that are farmed.

The literature on trematode infections in humans, reviewed by Howgate (1998), describes in some detail the epidemiology of fishborne trematode diseases, and the incidence of infection of various species of fish in affected areas. However, none of the reports explicitly state that the parasites have been found in farmed fish or shellfish, though an association between disease and aquacultural practices is sometimes made. The WHO report on foodborne trematode infections (WHO 1995) makes the link on more than one occasion in the text without providing any examples or data. Naegel (1990), in a review of health problems relating to the use of animal excreta in aquaculture, writes that there is potential for spread of trematodes by such practices, but does not give any examples. Cross (1991) attributed an increase in the incidence of clonorchiasis in Hong Kong to the importation of pond-reared fish from China. Chen (1991) reported that clonorchiasis was common in farmers in parts of Taiwan, and associated the high incidence of infection with the use of pig manures in fish farming in pond systems.

It is only to be expected that farmed freshwater fish will be infected with trematode parasites where fish-borne trematode diseases are endemic. The diseases are maintained by transmission through fish and the fact that a disease is endemic must mean that some fish at least in the region, wild, cultured or both, are infected. However, quantitative, or even qualitative, data on the relative importances of aquaculture products and of products from the wild in maintaining the transmission are lacking. Until there is evidence to the contrary, it must be assumed that trematodes are present in farmed freshwater fish where trematode diseases are endemic. Judging from the epidemiology of trematode diseases, the risks of illness from consuming infected fish are high. Normal cooking will kill the parasites, but raw and lightly preserved fish products are common items in the diet in countries where these parasites are endemic.

The control over environmental conditions possible with aquaculture could provide means to reduce, if not eliminate, the hazard of trematode parasites in freshwater fish. Treatment of excreta, for example by composting (Strauss 1985), could prevent primary infection of the water, and elimination of snails from the pond will break the transmission cycle. Treatment of the fish by antihelminthics like praziquantel just prior to harvesting for sale could eliminate the parasite from the flesh (Mitchell 1995). These actions are only possibilities and will be difficult to apply in the context of existing systems of cultivation of freshwater fish in the regions where trematode diseases are endemic. There is an obvious need for further research into the control of trematode hazards in farmed fish.

Conclusions

All foods have a risk of causing food poisoning, but there is no evidence to suggest that, overall, aquacultured products pose more risk than similar products caught from the wild. In some aquaculture systems, the products might be, or have the potential to be, safer than the wild counterparts. Bivalve shellfish, cultured or harvested from the wild, are associated with a significant risk of causing food poisoning, particularly when consumed raw, but being cultured does not seem to be a factor in increasing or reducing the risk. Fish products from mariculture are safer than similar species from the wild because of the freedom from nematode parasites. The hazard from enteric-derived pathogens might be increased in farmed crustacean shellfish in tropical environments compared with shellfish caught in the wild, but the evidence for this is equivocal, and there might not be an increased risk of causing food poisoning. Farmed freshwater fish in temperate climates pose a very low risk of causing illness when cultivated in unfertilised systems. Fish cultivated in systems fertilised by human and animal excreta have the potential for an increased risk of causing illness of bacterial origin, but there are few data on which to base an estimate of the risk. Freshwater fish in tropical countries are associated with a high risk of causing illnesses from trematode parasites, and farmed fish must share this risk.

Aquaculture products are making an important, and increasing, contribution to fish supplies for human consumption, and there is a need for more studies into the safety of aquaculture products, and the effects of cultural practices on them.

References

Abad, F.X., Pintó, R.M., Gajardo, R. & Bosch, A. (1997) Viruses in mussels: public health implications and depuration. *Journal of Food Protection*, **60**, 677–81.

Ahmed, F.E. (ed) (1991) *Seafood Safety*. Committee on Evaluation of the Safety of Fishery Products, Food and Nutrition Board, Institute of Medicine, National Academy Press, Washington, DC.

Alderman, D.J. & Hastings, T.S. (1998) Antibiotic use in aquaculture: development of antibiotic resistance – potential for consumer health risks. *International Journal of Food Science and Technology*, **33**, 139–55.

Angot, V. & Brasseur, P. (1993) European farmed Atlantic salmon (*Salmo salar* L.) are safe from anisakid larvae. *Aquaculture*, **118**, 339–44.

Bean, N.H. & Griffin, M. (1990) Foodborne disease outbreaks in the United States, 1973–1987: pathogens, vehicles, and trends. *Journal of Food Protection*, **53**, 804–17

Beckers, H.J. (1986) Incidence of foodborne disease outbreaks in The Netherlands: annual summary, 1981. *Journal of Food Protection*, **53**, 924–31.

Beril, C., Crance, J.M., Leguyader, F., Apaire-Marchais, V., Leveque, F., Albert, M., Goraguer, M.A., Schwartbrod, L. & Billadeul, S. (1996) Study of viral and bacterial indicators in cockles and mussels. *Marine Pollution Bulletin*, **32**, 404–409.

Blum, D. & Feachem, R.G. (1985) *Health Aspects of Nightsoil and Sludge Use in Agriculture and Aquaculture. Part III. An Epidemiological Perspective*. International Reference Centre for Waste Disposal, Dübendorf, Switzerland.

Bristow, G.A. & Berland, B. (1991) A report on some metazoan parasites of wild marine salmon (*Salmo salar* L.) from the west coast of Norway with comments on their interactions with farmed salmon. *Aquaculture*, **98**, 311–18.

Cann, D.C., Taylor, L.Y. & Hobbs, G. (1975) The incidence of *Clostridium botulinum* in farmed trout raised in Great Britain. *Journal of Applied Bacteriology*, **39**, 331–6.

Chen, E.R. (1991) Current status of food-borne parasitic zoonoses in Taiwan. In: *Emerging problems in food-borne parasitic zoonosis: impact on agriculture and public health*. Proceedings of the 33rd SEAMEO-TROPMED Regional Seminar, Chiang Mai, Thailand, 14–17 November 1990 (ed. J.H. Cross), pp.62–4. SEAMEO Regional Tropical Medicine and Public Health Project, Bangkok, Thailand.

Cross, J.H. (ed.) (1991) *Emerging problems in food-borne parasitic zoonosis: impact on agriculture and public health*. Proceedings of the 33rd SEAMEO-TROPMED Regional Seminar, Chiang Mai, Thailand, 14–17 November 1990. SEAMEO Regional Tropical Medicine and Public Health Project, Bangkok, Thailand.

Dalsgaard, A. (1998) The occurrence of human pathogenic *Vibrio* spp. and *Salmonella* in aquaculture. *International Journal of Food Science and Technology*, **33**, 127–38.

Deardorff, T.L. & Kent, M.L. (1989) Prevalence of larval *Anisakis simplex* in pen-reared and wild-caught salmon (*Salmonidae*) from Puget sound. *Journal of Wildlife Diseases*, **25**, 416–19.

Deardorff, T.L. & Overstreet, R.M. (1991) Seafood-transmitted zoonoses in the United States: the fishes, the dishes, and the worms. In: *Microbiology of Marine Food Products* (eds D.R. Ward & C. Hackney), pp.211–65. Van Nostrand Reinhold, New York.

FAO (1998) *Fishstat plus*. An electronic database of fisheries statistics. Food and Agriculture Organization, Rome, Italy.

Gerba, C.P. (1988) Viral disease transmission by seafoods. *Food Technology*, **42**, 99–103.

Gibson, D.M. (1992) Pathogenic microorganisms of importance in seafoods. In: *Quality Assurance in the Fish Industry* (eds H.H. Huss *et al.*), pp.197–209. Elsevier Science Publishers, Amsterdam.

Guzewich, J.J. & Morse, D.L. (1986) Sources of shellfish in outbreaks of probable viral gastroenteritis: implications for control. *Journal of Food Protection*, **49**, 389–94.

Guzewich, J.J., Bryan, F.L. & Todd, E.C.D. (1997) Surveillance of foodborne disease. I. Purposes and types of surveillance systems and networks. *Journal of Food Protection*, **60**, 555–66.

Howgate, P. (1998) Review of public health safety of products from aquaculture. *International Journal of Food Science and Technology*, **33**, 99–125.

Huss, H.H., Pedersen, A. & Cann, D.C. (1974) The incidence of *Clostridium botulinum* in Danish trout farms. I. Distribution in fish and their environment. II. Measures to reduce contamination of the fish. *Journal of Food Technology*, **9**, 445–58.

Huss, H.H., Dalgaard, P. & Gram, L. (1997) Microbiology of fish and fish products. In: *Seafood from Producer to Consumer: Integrated Approach to Quality* (eds J.B. Luten, T. Børrensen & J. Oehlenschläger), pp.413–30. Elsevier, Amsterdam.

Leung, C-K., Huang, Y-W. & Pansorbo, O.C. (1992) Bacterial pathogens and indicators in catfish and pond environments. *Journal of Food Protection*, **55**, 424–7.

Malek, E.A. (1980) *Snail-transmitted Parasitic Diseases*, Vol II. CRC Press, Inc., Boca Raton, USA.

Mitchell, A.J (1995) Importance of treatment duration for praziquantel used against larval digenetic trematodes in sunshine bass. *Journal of Aquatic Animal Health*, **7**, 327–30.

Naegel, L.C.A. (1990) A review of public health problems associated with the integration of animal husbandry and aquaculture, with emphasis on southeast Asia. *Biological Wastes*, **31**, 69–83.

National Advisory Committee on Microbiological Criteria for Foods (1998a) Hazard Analysis and Critical Control Point Principles and Application Guidelines. *Journal of Food Protection*, **61**, 1246–59.

National Advisory Committee on Microbiological Criteria for Foods (1998b) Principles of risk assessment for illness caused by foodborne biological agents. *Journal of Food Protection*, **61**, 1071–4.

Nedoluha, P.C. & Westhoff, D. (1993) Microbiological flora of aquacultured hybrid striped bass. *Journal of Food Protection*, **56**, 1054–60.

Reilly, A. & Käferstein, F. (1997) Food safety hazards and the application of the principles of the hazard analysis and critical control point (HACCP) system for their control in aquaculture production. *Aquaculture Research*, **28**, 735–52.

Reilly, P.T.A. & Twiddy, D.R. (1992) *Salmonella* and *Vibrio cholerae* in brackish water cultured tropical prawns. *International Journal of Food Microbiology*, **16**, 293–301.

Rim, H-J. (1998) Field investigations on epidemiology and control of fish-borne parasites in Korea. *International Journal of Food Science and Technology*, **33**, 157–68.

Rose, J.B. & Sobsey, M.D. (1993) Quantitative risk assessment for viral contamination of shellfish and coastal waters. *Journal of Food Protection*, **56**, 1043–50.

Saheki, K., Kobayashi, S. & Kawanishi, T. (1989) Salmonella contamination of eel culture ponds. *Nippon Suisan Gakkaishi*, **44**, 675–9 (in Japanese).

Simone, E., Goosen, M., Notermans, S.H.W. & Borgdorff, W. (1997) Investigations of foodborne diseases by food inspection service in the Netherlands, 1991 to 1994. *Journal of Food Protection*, **60**, 442–6.

Strauss, M. (1985) *Health aspects of nightsoil and sludge use in agriculture and aquaculture. Part II. Pathogen survival*. Report No. 04/85, International Reference Centre for Waste Disposal, Dübendorf, Switzerland.

Todd, E.C.D. (1992) Foodborne disease in Canada – a 10 year summary from 1975–1984. *Journal of Food Protection*, **55**, 123–32.

Vaughn, J.M. & Landry, E.F. (1984) Public health considerations associated with molluscan aquaculture systems: human viruses. *Aquaculture*, **39**, 299–315.

Ward, D.R. (1989) Microbiology of aquaculture products. *Food Technology*, **43**(11), 82–6.

Ward, D.R. & Hackney, C. (1991) *Microbiology of Marine Food Products*. Van Nostrand Reinhold, New York.

WHO (1995) *Control of Foodborne Trematode Infections*. WHO Technical Report Series 849. World Health Organization, Geneva.

Wootton, R. & Smith, J.W. (1975) Observational and experimental studies on the acquisition of *Anisakis* sp. larvae (Nematoda: Ascaridida) by trout in fresh water. *International Journal for Parasitology*, **5**, 373–8.

Wyatt, L.E., Nickelson, R. & Vanderzant, C. (1979) *Edwardsiella tarda* in freshwater catfish and their environment. *Applied and Environmental Microbiology*, **38**, 710–14.

Chapter 33

A Practical Approach to Controlling and Improving Product Quality for the Farmed Fish Industries at the Point of Retail Sale

A.R. Greenhalgh

'Woodlea', Hill Drive, Failand, Bristol, BS8 3UX, UK

Introduction

This is a non-academic chapter, but one which it is hoped will be interesting and thought provoking. It will provide some insight into the commercial world by bringing into focus details of the steps that need to be taken to put into practice in the marketplace ways to improve farmed fish quality. This chapter will focus on one of the best independent quality assurance schemes in the UK food industry. This is the Product Certification Scheme for Scottish Quality Farmed Salmon, more commonly known as the SQS (Scottish Quality Salmon) Scheme, or the TQM (Tartan Quality Mark) Scheme, because products meeting the standards of the scheme may carry a specific quality symbol.

The objective of the scheme is to provide a means by which a range of fresh farmed salmon which has been produced in Scotland, including the islands off the West Coast, Orkney and the Shetland Isles, can be supplied to the market to a standard that is demonstrated by a certificate of approval and where appropriate a mark of conformity. This shows that they have been produced and prepared for marketing by approved producers complying with standards, operational procedures and practices approved by the governing board of the certifying body, as specified in the quality manual for the scheme. The quality scheme currently covers 25 salmon farming companies growing salmon at 133 farm sites, 14 packing stations and 12 further processing plants and retail packers (excluding smokers). A number of producers in Shetland have joined the scheme, even though they have their own quality scheme on Shetland run by Shetland Seafood Quality Control (SSQC).

History of the scheme

The scheme was initiated in the early 1980s by members of the Scottish Salmon Growers Association (SSGA), a trade association whose main role is to represent the interests of Scottish salmon farmers on technical and political matters and to promote Scottish salmon in the UK and overseas. Membership of the Association is on a purely voluntary basis and the association has no statutory powers. Income is generated by subscription and a levy based on annual salmon production. The SSGA, who were at that time producing some 4000 to 5000 tonnes of farmed salmon per year, began to realise that as volumes of farmed salmon increased, their product, which in its wild state had been traditionally recognised as a low volume, high quality, gourmet food which sold at a high price, was in some danger of degenerating into a mediocre, low priced, commodity item. The precedent for this had been set some years before by the uncontrolled expansion of the broiler chicken industry. Not only were volumes of farmed salmon growing in Scotland, but also in a number of other countries such as Norway, the Faroe Islands and Ireland, much of which was deemed to be inferior in taste and quality to the best Scottish product.

In 1983 the SSGA decided to set up a quality scheme to develop and promote the Scottish product (Table 33.1). At about the same time the government had set up Food From Britain (FFB), one of its functions being to improve the quality standard of British food, particularly that which was unprocessed or only lightly so. The two groups, one experienced in salmon production and the other having the skills to produce workable quality systems, came together and produced the initial Product Certification Scheme for Scottish Quality Farmed Salmon. For the next eight years the scheme was administered jointly by the two organisations, but all the inspection and auditing of the scheme was done by FFB using its own full time inspectors, thus providing independence. During this period Scottish Quality Salmon (SQS) was set up as the marketing arm of SSGA and in 1992 the quality scheme was given a further boost when it was accepted by Qualité France. In 1994 when the government closed the technical side of FFB, the scheme reverted initially to the SSGA/SQS, but it was in danger of being perceived to have lost its independence. In 1996/7 the whole structure of the Salmon Group in Scotland was revised.

Figure 33.1 shows the current structure and the interaction of the various committees and groups involved. The quality scheme is now run by Food Certification (Scotland) Ltd, an independent certification body accredited in 1997 by UKAS (United Kingdom Accreditation Service), one of the first product certification companies to achieve this

Table 33.1   Significant dates in the development of the product certification scheme for Scottish quality farmed salmon.

Date	Achievement
1983	Salmon scheme first established
1986	Incorporated into Food From Britain's (FFB) quality certification scheme
1986–94	Scheme jointly operated through SSGA and FFB's quality council
1992	Label Rouge certification acceptance
1994	FFB quality schemes transferred to MAFF. Most schemes allowed to close but salmon scheme continued by SSGA/SQS
1997 Jan	Food Certification (Scotland) Ltd established
1997 Nov	UKAS accreditation granted to Food Certification (Scotland) Ltd

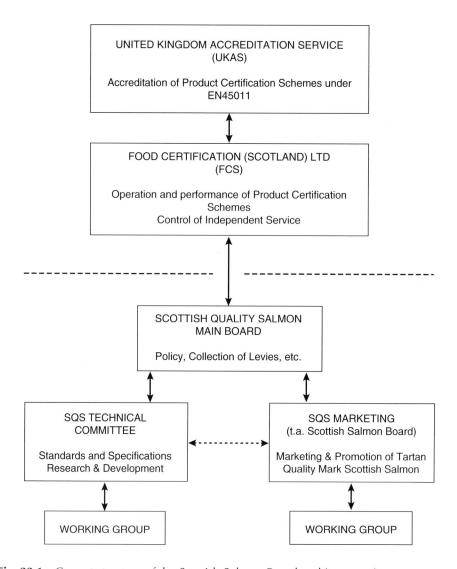

Fig. 33.1   Current structure of the Scottish Salmon Board and its committees.

accreditation. Throughout the 16 year history of the scheme many detailed changes have been made. Some have resulted from scientific advances either in the academia or as a result of research funded by the members of the scheme. Others have resulted from changes in legislation, particularly in the areas of medicines, veterinary treatments, food hygiene and labelling. These factors and the consolidation and overhaul of the management structure ensure that there is an ongoing dynamism which not only supports the quality scheme, but which invests in research and development requested by members. It also injects funds into marketing and advertising the scheme to keep its quality mark at the forefront of the retail market, so that it is recognised as a reliable indicator of a high level of quality.

The requirements of a quality scheme

For any product quality certification scheme to gain wide customer acceptance, five things are essential:

(1) Standards set must be acceptable, consistent and regularly achievable, and must be at a perceived level higher than the rest of the market
(2) Products meeting the standards must be clearly identifiable by customers
(3) Products failing to meet the standard in any way, must not be marketed as a 'quality scheme' product
(4) Time must be allowed for trade and customer recognition and acceptance that the 'quality scheme' identified product will be of a high and consistent standard compared to non-marked product
(5) Financial support must be sufficient to bring the quality standard and mark to the attention of purchasers.

These five rules are by no means easy to enforce even with legal backing. It is particularly difficult to do in a large diverse group of independent producers ranging from small single farms, largely one man operated, to large multisite national companies with a full range of specialist staff.

The written documentation of the scheme must be detailed and comprehensive. Even if this is achieved, the goodwill and honest involvement of the whole industry is essential to ensure that the scheme works. In any situation such as this there will always be non-compliers who will try to work the system in their favour. They must be identified and either reformed or removed, whether they be individual employees or companies. There must be no exceptions. One producer, or even one salmon grader, who flouts the rules will ruin the system for all the other members because customer confidence will ultimately be affected.

Details of the Product Certification Scheme for Scottish Quality Farmed Salmon

The full details of the scheme are published in a detailed manual consisting of an operating manual and a quality manual. Both are subject to full amendment control standards.

The operating manual contains the following information:

(1) Introduction and definitions: describes the reasoning behind the scheme and defines the terminology used in the manual.
(2) Administrative structure: details the responsibilities of the governing board, executive director and staff of Food Certification (Scotland) Ltd.
(3) Details of the quality manuals and the inspection service: briefly describes the content of the quality manual and establishes the rights of access.

(4)   The operating procedure: describes the method of application, the assessment inspection, method of reporting, certificate of approval, appeal procedure and sur- veillance inspections. The latter is one of the major strengths of the scheme. The inspections are:

❑   Two per year, for farms and on-farm packing stations packing less than 500 tonnes per annum, when harvesting operations are being carried out
❑   Four per year, two of which shall be unannounced, for on-farm packing stations packing 500–1000 tonnes per annum
❑   Six per year, for preparation and processing establishments
❑   Eight per year, for large packing stations packing salmon sourced from a number of farms.

The programme for inspections provides for:

❑   A preliminary meeting to discuss the arrangements for the inspection
❑   For farms, an inspection of the husbandry practices and harvesting procedures
❑   For packers and processors, inspection of a sample of products selected at random
❑   Detailed examination of the production environment, operational procedures and practices, product handling, storage, transport, hygiene and quality control
❑   Examination of relevant records and how they are maintained
❑   Taking of samples for submission to independent laboratories for analysis covering veterinary residues, chemical composition, microbiological quality or organoleptic characteristics
❑   From farms a sample of the feeding stuff may also be taken for analysis
❑   Progress on previous non-compliances will be checked and any new non-compliances detailed
❑   During an inspection, observance that the product does not conform to the relevant product specification or that there are major non-compliances with other criteria defined in the manual which may affect the quality of the product. The inspector has the authority to require the batch or production run to be segregated as non-conforming product which may not carry the quality mark
❑   On completion of the inspection the findings will be discussed with the manage-ment and completion dates will be fixed for rectification of non-compliances
❑   A written report will be forwarded to the company and to the executive director of Food Certification (Scotland) Ltd, who will review it and notify the outcome, in writing, to the company.

(5)   The rules governing certification: specify the rules of the scheme, which cover items such as validity of certificates, compliance, maintenance of a quality management system, access, authorised management representatives, correction of any deficiencies, use of quality marks and payments for the service. Details of the standards to be expected from staff of Food Certification (Scotland) Ltd and their obligations are also detailed.

(6)   The complaints procedure: specifies action on complaints both for product and the scheme.

(7)   Appendices containing:

❏ Application forms for farms, preparation and processing establishments
❏ Non-compliance forms; certificate of approval
❏ Monitoring programmes for detection of residues and determination of oil content
❏ Composition of feeding stuffs.

The quality manual contains the following information:

(1) Introduction and definitions: describes the manual and defines terms used.
(2) Product specifications: provides a set of full specifications covering such areas as har-vesting, transport to packing station, preparation, selection, organoleptic character-istics, packing, labelling, and storage and distribution of whole and gutted salmon and basic specifications for portions and further processed products such as 'ready meals'.
(3) Standards for salmon production on farm: covers all farming operations, welfare, stocking, stocking densities, feeding and feed formulation, animal welfare and the use of veterinary medicines, harvesting and transportation.
(4) Standards for preparing, processing and packing salmon: details minimum standards for facilities and equipment; raw materials; preparing, processing, packing, storage and transport operations; high risk areas; general hygiene requirements; prevention of cross contamination; hygiene requirements applicable to staff; monitoring of hygiene pro-cedures (microbiological testing); shelf-life monitoring and traceability systems.
(5) Quality management system: details a basic structure for a suitable quality manage-ment system following the outline of the requirements of the European system EN 9002.
(6) Appendices containing:

❏ SSGA guidelines for the humane slaughter of farmed Atlantic salmon
❏ Salmon harvesting form
❏ Protocol for determining the fat content of salmon
❏ EU freshness categories and rating scale
❏ Colour matching procedure – Roche colour card
❏ Protocol for organoleptic testing of salmon
❏ Procedure for monitoring salmon for residues of veterinary medicines
❏ List of records required to demonstrate achievement of criteria defined in the manual.

Membership of the scheme necessitates applicants passing through a number of stages which are summarised in Fig. 33.2. The rules of the scheme require initial examination of the product, the production process, the production environment and the distribution facilities, assessment of the quality management system and acceptance by the certifying authority. Acceptance is followed by regular surveillance, which takes into account the quality man-agement system and the examination of samples at the point of production and from the open market. In practice, after participants have been initially inspected and approved they are expected to comply with the standards which are set out in the quality manual.

The standards are policed by three inspectors, two full time and one part time, who visit production sites, packing stations, processing plants and the wholesale markets to check compliance. The same inspectors also cover the parallel Product Certification Scheme for Smoked Scottish Quality Salmon.

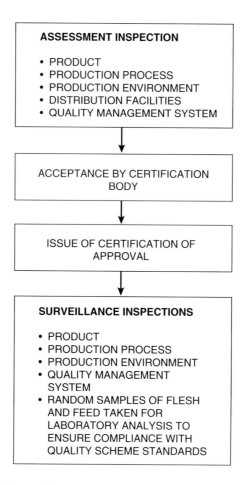

Fig. 33.2   Stages in the certification procedure.

Members who are registered with the scheme and who continue to comply with the standards are encouraged to identify their conforming salmon and salmon products with the Tartan Quality Mark which is incorporated onto gill tags, box labels and pre-pack labels, each of which bears a unique number to allow traceability of product back to the packer and from their records back to production farm and ultimately the cage.

The traceability which the labelling system allows, coupled with the enforcement of extremely high standards of production and handling, places the members of the scheme in a unique position in the UK in what has now become a very competitive marketplace. A major indication of the international recognition which the scheme has received was its acceptance by 'Label Rouge' in 1992. Since this acceptance, salmon produced under the scheme sold in France has been eligible to carry the 'Label Rouge' quality mark. It is a unique honour because this is the first occasion on which a foreign product has been recognised in this way by the French authorities.

Acknowledgement

The author would like to acknowledge help and advice from Scottish Quality Salmon, particularly Dr J.L. Webster; also Mr C.G. Absolon, Executive Director of Food Certification (Scotland) Ltd, for permission to extract information from their relevant manuals.

Poster Abstracts

Slaughtering of Atlantic Halibut (*Hippoglossus hippoglossus*): Effect on Quality and Storing Capacity

L. Akse and K. Midling

Norwegian Institute of Fisheries and Aquaculture Ltd, N-9291 Tromsø, Norway

Through 15 years of intensive biological research, Atlantic halibut is now established as the most promising species in Norwegian marine aquaculture. Annual production is still small, but is expected to grow rapidly. However, knowledge on muscle quality and what factors affect the product in this species are scarce. In this project a number of experiments were conducted on farmed Atlantic halibut, including:

❑ *Slaughtering stress and storing capacity* – 32 of 64 Atlantic halibut were stressed prior to slaughter. The two groups were identically slaughtered and thereafter compared during 28 days of storage on ice regarding pH, total volatile nitrogen (TVBN), microbiology and sensory attributes.

❑ *Anaesthetic and killing methods* – Anaesthesia with CO_2 or clove oil-derived Eugenol were compared to killing the halibut by a blow to the head. The groups were compared regarding initial pH, onset and development of *rigor mortis* and residues of haem iron in the muscle after bleeding.

❑ *Bleeding and gutting* – One group was gutted directly and then bled for 20 minutes in water, the other group was bled for 20 minutes and then gutted.

Except for the initial pH, we were not able to detect major differences in post-mortem quality between the stressed and not stressed halibut. Compared to several other species, Atlantic halibut has exceptional storing capacity both in sensory, chemical and microbiological terms. In this experiment, this is demonstrated by almost no development in pH, TVBN and viable bacteria values until after day 21. Farmed Atlantic halibut is likely to be accepted in the market as fresh fish at least one week longer than other white fish species (e.g. cod).

CO_2 performed poorly as an anaesthetic and resulted in rapid onset of *rigor mortis* and high hardness values. Eugenol postponed onset of *rigor mortis* by 14 to 16 hours compared to CO_2 and gave a significantly lower measured hardness. Killing by a blow to the head resulted in increased variation in onset time and development of *rigor mortis*. Small differences were found in haem iron muscle residues among the groups, but halibut killed by a blow to the head were significantly better bled than fish anaesthetised with CO_2.

Nucleotide Variation in Cod (*Gadus morhua*)

G. Cappeln and F. Jessen

Ministry of Agriculture and Fisheries Research, Department of Seafood Research, Technical University of Denmark, Building 221, DK-2800 Lyngby, Denmark

An investigation of the variation in nucleotide composition throughout the fillet of cod (*Gadus morhua*) was carried out. Furthermore it was investigated whether the sampling place affects the rate of decomposition and synthesis of nucleotides. Knowledge of this is important when sampling for K-value analysis or rigor examination.

Samples were taken from 16 different positions on the fillet (Fig. 1). These positions were grouped into four different areas: front piece, loin, belly and tail. All samples were analysed for ATP, ADP, AMP, IMP, Ino and Hx. The results are expressed as a mean of the samples within each area.

The ATP content in newly killed cod seems to be lowest in the front piece and highest in the loins. After 20 h at 0°C the degradation of ATP had been slowest in the belly area. ATP degradation occurs fastest in the front piece but the level in loin was almost the same. A similar picture was found for the IMP synthesis rate. No significant variation in ADP, AMP, Ino and Hx content was found.

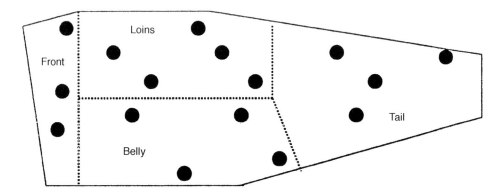

Fig. 1 Sampling positions. The fillet was divided into four areas: Front piece, loin, belly and tail. All analyses were performed on two fish just after killing and on two fish after storage at 0°C for 20h.

Effects of Dietary Fat on Growth and Body Composition of Atlantic Salmon (*Salmo salar*)

M. Ekli, S.J.S. Johansen and M. Jobling

Norwegian College of Fishery Science, University of Tromsø N-9037 Tromsø, Norway

Fat content in the muscle is a quality criterion of farmed Atlantic salmon, so farmers should aim for a combination of low muscle fat content, good food conversion and fast growth. In the present experiment, growth and body composition of Atlantic salmon were examined to provide indications of the ways in which dietary fat levels influence patterns of nutrient partitioning.

Post-smolt Atlantic salmon (initial weight 50 g) were fed to excess on either a commercial dry pellet (high-fat), or this feed without the fat top dressing (low-fat). The fish were held at 10°C under constant light, and were fed for 4 h each day. The experiment was run for 13 weeks and food consumption was monitored every fourth week using the X-radiographic method. Samples of fish were taken for body composition analysis at the start and end of the experiment.

At the beginning of the experiment food consumption was similar for both dietary groups (0.83% body weight). At the end of the experiment food consumption was significantly higher in the low-fat group than in the high-fat group (1.02 versus 0.73% bwt, respectively). This shows that the fish were able to compensate for the lower energy content in the low-fat diet by consuming more food.

There were no significant differences between the two dietary groups in final body weights. Body composition, however, was significantly influenced by the fat content of the diet. When fish were sampled at the end of the experiment, fish in the high-fat group had a higher dressing out loss than those fed the low-fat diet (7.6 versus 6.2% respectively). This was a reflection of the higher proportion of visceral fat amongst the salmon fed the high-fat diet (23.4 versus 12.8%). The relative fat content of the carcass was significantly higher in the fish fed the high-fat diet (7.5 versus 4.9%), whereas the proportion of protein was lower (18.0 versus 18.5% in the fish fed the low-fat diet).

Combining the data for carcass and viscera resulted in a significant difference in relative fat content in whole fish between the two dietary groups, with a higher proportion in the high-fat group (9.3 versus 5.6%). The low-fat group, however, contained a higher proportion of protein (18.3 versus 17.5%).

The results of this experiment show that feeding a dry pellet with a high fat content may not result in faster growing fish. A low fat diet may result in fish growing just as fast as those fed with a high-fat diet, with the advantage of producing a fish with a low fat content in the muscle.

The Method Chosen for Salting and Smoking Atlantic Salmon (*Salmo salar*, L.) affects the quality of the fillet

M. Espe[1], R. Nortvedt[1], O.J. Torrissen[2] and Ø. Lie[1]

[1] *Institute of Nutrition, Directorate of Fisheries, P.O.B. 185 Sentrum, N-5804 Bergen, Norway;* [2] *Institute of Marine Research, N-5392 Storebø, Austevoll, Norway*

The effects of different salting and smoking methods on the quality of Atlantic salmon (*Salmo salar*, L.) have been studied in a project financed through the FAIR-program. The general objective is to improve the quality of Atlantic salmon by characterising interactions between raw material characteristics and different parameters in the traditional smoking process.

Fish of $4\,kg \pm 100\,g$ were filleted and processed by one of seven methods. Fish were individually tagged, filleted and then the left fillet was analysed when fresh, while the right fillet was processed and analysed after smoking. The processing methods used were: dry salting followed by smoking at either $20°$ or $30°C$, brine salting and smoking at either $20°$ or $30°C$, salt injection followed by smoking at either $20°$ or $30°C$ and finally dry salting followed by an electrostatic smoking procedure. Following processing, 15 fresh and smoked samples from each group were analysed for fatty acid composition, vitamin E and vitamin C content. Oxidation status was measured as TBARS to evaluate the effect of the different processing methods on the quality changes through the processing step. The HunterLab tristimulus colour instrument measured the fillet redness and yellowness.

Following rejection of individual outlier values, the group mean values of the percentual differences between fresh and smoked fillets were pretreated by autoscaling before a principal component analysis was carried out. Three significant principal components explained 92.2% of the total variance. The biplot from the principal component analysis showed that the dry salted ($30°C$), the salt injected ($30°C$) and the dry salted and electrostatic smoked groups clearly separated from the other groups. Atlantic salmon from the salt injected ($30°C$) group showed a close correlation to high loss (8.4%) of polyenic fatty acids, which completely overlapped the sum of n-3 and n-6 fatty acids and their ratio in the plot. Atlantic salmon from the electrostatic smoked group were negatively correlated to the change in vitamin C from fresh to smoked fillet, meaning that low changes (8.1%) appeared in this group, compared to the other groups (80–88%). The electrostatic smoked group showed a positive correlation to the change in TBARS, meaning that the oxidation status in the fillet decreased after smoking in this group, but increased in the other groups. The electrostatic smoked group, however, lost more of the colour during smoking, indicated by a higher correlation to the change in the Hunter (a) (redness) and (b) (yellowness). This group also showed the highest loss in vitamin E (33.5%) during processing (overlapping with Hunter (a) in the plot), meaning that vitamin E probably had an antioxidative function in the electrostatic smoked fillets. Both the dry salted (30°C) and the electrostatic smoked groups were characterised by loss (10–11%) of saturated fatty acids.

Compensatory Growth: A Tool for Product Tailoring?

S.J.S. Johansen, M. Ekli and M. Jobling

Norwegian College of Fishery Science, University of Tromsø, N-9037 Tromsø, Norway

A prerequisite for successful product tailoring is a means of manipulating quality parameters towards clearly stated goals. Among quality parameters of farmed Atlantic salmon, fillet fat content has received much attention. The problem of reducing fillet fat deposition in salmon is complicated by the use of lipid-rich feed formulations, combined with high production rates aimed at low feed:gain ratios. A cyclic feeding regime that alternates periods of restricted and satiation feeding may be used to combine the high overall growth rates achieved on high energy diets with a low rate of fat deposition.

When a period of restricted feeding is followed by unlimited feed availability, fish may show a dramatic increase in feed intake (hyperphagia). This leads to recovery growth, so that body weight may catch up with, and sometimes exceed, that of fish fed continuously to satiation. During the recovery phase there seems to be a shift towards deposition of lean body mass in preference to fat, so a cyclic feeding regime is predicted to produce a leaner fish than satiation-feeding without any marked reduction in growth. To fully exploit the potential of cyclic feeding it is crucial to obtain an even feed distribution amongst fish during periods of restricted feeding. Unequal feed acquisition during the restriction phase results in heterogeneous growth, and will also influence the ability of individuals to display the recovery growth response. The challenge is therefore to impose a restricted feeding regime that leads to reduced rates of growth without increasing heterogeneity, thereby allowing the perceived benefits of the recovery growth response to be fully exploited.

Post-smolt Atlantic salmon were exposed to restricted feeding for eight weeks to reduce growth rates to about 60% of those of satiation-fed controls. An attempt was made to ensure that all individuals were given the opportunity to feed, the success of the feed distribution regime being confirmed by a lack of increase in heterogeneity amongst the restricted fish compared to the controls. After eight weeks the fish were then switched to satiation-feeding, and growth and body composition changes studied.

The restricted-fed fish displayed a compensatory response, and achieved the same weight as the controls within eight weeks of the switch to satiation-feeding. Thus, the salmon post-smolt displayed a complete compensatory response, and there were indications of an over-compensation as time progressed. At equal body weight the feed restricted group had a lower body lipid content (by 9%) than the controls. Thus, it seems that a cyclic feeding regime has potential as a production tool, as long as feed can be distributed equally amongst the fish. A follow-up is planned on a commercial scale, where satiation-feeding will be compared with continuous feed-cycling, starting with post-smolt salmon and terminating with harvesting of market size fish.

Seasonal Variations in Quality Characteristics Within and Between Age Classes of Farmed Atlantic Salmon (*Salmo salar*)

T. Mørkøre and K.-A. Rørvik

AKVAFORSK, Institute of Aquaculture Research,1432 Ås-NLH, Norway

The experiment was conducted in sea cages (125 m^3) from July 1995 to July 1996 at the Norwegian west coast (AKVAFORSK, Ekkilsøy). LT-fishmeal based feeds (25.1 MJ/kg) were given to triplicate groups of 1+ salmon (IBW 0.2 kg) and 0+ salmon (IBW 0.4 kg). All fish were weighed in bulk every second month, and at each sampling, ten salmon per net-pen were used for analyses of fat content, colour and texture in raw fillets. The body weight at slaughter was 5.1 kg for 0+ salmon and 3.3 kg for 1+ salmon.

The fillet fat content increased substantially in both fish groups during autumn (from 3–4% in July to 14–17% in November). During winter the fat content decreased slightly, thereafter there was a further increase during spring. Colour analyses (Roche colour card, RCC) were performed from November and onwards. RCC-scores for the 1+ salmon were stable throughout the year, with a significant increase from May to July. RCC-scores for the 0+ salmon increased during the period from November to January, thereafter the RCC-readings were relatively stable. Within both fish groups, the fillet colour tone became more reddish and less yellowish during the experiment (Minolta hue-value decreased). No significant difference was observed between 1+ and 0+ salmon in fillet fat content or RCC-score, when the results were corrected for differences in fish size. The colour tone of the 1+ salmon fillets, however, seemed more yellowish compared with the 0+ salmon. The breaking strength, measured instrumentally, showed significant variations within both fish groups, but no significant difference was found between the fish groups. Variations in breaking strength were more closely related to water temperature and growth rate, than to fish size.

Field Observations of Intraperitoneal Vaccination of Atlantic Salmon (*Salmo salar* L.)

H. Moynihan and D.B. Garforth

Aquaculture Consultants, Muighinis, Carna, Co. Galway, Ireland

Intraperitoneal vaccination of salmon smolts is an effective method of disease immunisation. Protection can be enhanced with the use of adjuvants which help the fish respond better to the vaccine. Prolonged protection can be secured, particularly with the use of mineral oil adjuvanted vaccines. Possible disadvantages of i.p. vaccination can be stress and mortality caused by anaesthetising and injecting fish, the formation of intra abdominal adhesion and granulomas within the body cavity and arguably, a consequential reduction in growth.

Severe lesions of the cavity wall can also disrupt flesh quality with the result of poor market acceptability and downgrading of the entire carcass. In this field trial on the west coast of Ireland, cage groups of Atlantic salmon post smolts were assessed for (1) growth performance and (2) level of adhesion/granulomas after i.p. injection with three commercially available vaccines all containing mineral oil adjuvants (labelled A–C). Vaccination was by hand and undertaken by the same vaccine contractor. Adhesions were assessed on 50 fish using a 0–6 score of severity (0 = absent; 6 = severe). As there was no control group (unvaccinated fish) the authors make observations rather than draw conclusions. All other conditions were site related and similar for all groups. This abstract presents interim results of (1) growth performance (mean weight and specific growth rate (SGR)) and (2) level of adhesion 34 weeks post vaccination.

Table 1 Cage data for each vaccine group and mean adhesion score.

Vaccine	Mean wgt at vaccination (g)	Mean wgt at sea transfer (g)	No. of fish per cage	Mean wgt at sample date (16.9.98)	Mean adhesion score ± SD at sample date	Percentage adhesion occurrence (%)
A	66	66+	29731	1105.80	0.00±0	0
B	48	90	37035	745.08	1.59±1.13	78.4
C	40	87	36500	903.27	1.04±1.08	55.8

+ last batch sample pre-transfer

The severity of adhesions in any group 34 weeks post vaccination was less than two. Differences in mean adhesion score were observed between all groups with a complete absence of adhesions from the group A sample. The occurrence of any adhesion ranged from zero (A) to 78% (B) (Table 1). The highest mean monthly weight and SGR was observed in Group A throughout the trial period with a maximum of 1.82% bw/day in May. SGRs for Group B and C were similar for each month.

Impact of Dietary Lipid Source on Muscle Fatty Acid Composition and Sensory Evaluation of Atlantic Salmon (*Salmo salar* L.)

A. Obach, E. Åsgard Bendiksen, G. Rosenlund and M. Gisvold

Nutreco Aquaculture Research Centre, P.O. BOX 48, 4001 Stavanger, Norway

Salmon farming employs diets with different gross composition and raw materials. The source of dietary fat will determine the fatty acid profile of the fillet and thereby affect the quality of the end product. The objective of this trial was to study the impact of three different dietary lipid sources on the fatty acid composition and the sensory characteristics of salmon fillets. The three diets had the same basal composition but were coated with either pure capelin oil (CO diet), pure Peruvian anchovy oil (PO diet) or a mixture of 50% capelin oil and 50% soybean oil (SO diet).

Atlantic salmon with an initial weight of 873 g were fed the SO diet, for 107 days. Fish were sampled every third week to determine the fatty acid composition of fillets and viscera. At the end of this period, the fish (average weight of 1666 g) were divided in three groups and either fed the SO, the PO or the CO diets for 93 days. During this second period, fish were also sampled at regular intervals to determine fatty acid composition of fillet and viscera. At the end of the trial (average weight of 2837 g), a sensory evaluation of cooked and smoked fish was performed at the Norwegian Food Research Institute (Matforsk, Ås, Norway).

During the first period, the main changes in the muscle lipid composition affected the *n*-6 family, with an increase in linoleic acid (18 : 2 *n*-6, the main fatty acid in soybean oil) from 3.8% to 14.5%, whilst the level of saturates, monoenes and *n*-3 fatty acids decreased slightly. In the second period, a major decrease in the level of *n*-6 fatty acids was observed in the PO and CO groups, which was compensated by increased levels of monoenes in the CO group and of *n*-3 and saturated fatty acids in the PO group. The fatty acid profile in the SO group remained almost constant during this period. The changes in viscera fatty acid composition followed those in the fillet.

The sensory evaluation was performed according to standard procedures by a panel of 12 trained judges. The intensity of each sensory attribute was judged on a scale from 1.0 (= no intensity) to 9.0 (= high intensity). The CO group was used as a reference. In smoked salmon, significant differences were found in 2 out of the 26 sensory attributes that were tested: colour tone and firmness. Although the SO group scored lowest and the PO group highest for both attributes, the differences in average scores were only 0.40 and 0.33. In cooked salmon, significant differences were found in 4 out of the 19 attributes tested. Salmon taste, acidic taste and juiciness were significantly higher in the CO group compared to both the PO and the SO group, while the PO group had significantly higher intensity of bitter taste. Again the differences in average scores were small, varying between 0.34 and 0.44.

Although dietary fatty acid composition was clearly reflected in the fillet, this resulted in only minor differences in the sensory characteristics of both smoked and cooked salmon. In general, the differences between the two groups fed fish oil diets were of the same magnitude as the differences between these groups and the group fed the diet containing 50% soybean oil.

Effect of Starving Carp (*Cyprinus carpio* L.) before Slaughter on Parameters of Meat Quality

M. Oberle, F.J. Schwarz

Bavarian State Institute of Fishery, Substation for Carp Pond Culture, D-91315 Höchstadt; Working-group Aquaculture of the Technical University of Munich, D-85350 Weihenstephan, Germany

Carp commonly spend many weeks without food in commercial suppliers' premises while waiting to be sold on. In the cold winter months in Central Europe they also consume little food if kept in ponds. There are very few investigations on the effect of starving carp before slaughter on the chemical, physical and sensory parameters of their meat. This experiment investigated these parameters in carp kept in identical conditions but having different fat content and fatty acid patterns.

In a three-factorial experiment 144 carp (initial live weight 0.54 kg) were kept in 18 aquaria (water temperature 23°C) and were given diets with 44% (I), 33% (II) or 22% (III) crude protein, in which a protein pre-mix (soya protein, fish meal and casein) was replaced by increasing amounts of a wheat-maize mixture. Diets of each protein level were given with or without added fish oil (8.1%). The carp were fed to an average end weight of 1.06 kg. At this point 12 fish on each diet were slaughtered after keeping them without food for 24 h or starving them for 42 days (kept at 12°C). The slaughtered fish were dissected into fillet (skinless), viscera and remainder.

Starving the fish for 42 days resulted in a small weight loss of 0.028 kg and also had little effect on the relative weight of the fillet or its fat content or the pattern of fatty acids, whilst the viscera percentage (13% versus 12%) and fat content of the viscera decreased significantly (17.5% versus 14.1%). Some physical parameters of meat quality were significantly affected by starving the fish. The pH values (measured at 0, 24 and 48h after slaughter) of the meat of the starved fish were significantly higher (pH_0 6.98, pH_{24} 6.50 and pH_{48} 6.39) than those not starved (pH_0 6.80, pH_{24} 6.25 and pH_{48} 6.24). The shear strength (N) of both the raw and the cooked meat was significantly increased by starving the fish (raw; 257 versus 178 N, cooked 149 N versus 116 N). A significant correlation between the pH value post-mortem and shear strength of raw meat was found. Starvation led to a significant decrease of the b value (meat colour, Hunter L, a, b) of the raw muscle and of the L value of the cooked meat. Water binding capacity was not affected. The meat of the starved fish scored slightly higher in smell, taste, juiciness and overall impression in taste panel tests.

The Influence of Deep Frozen Storage on Fat Oxidation and Sensory Properties of Carp Meat

M. Oberle

Bavarian State Institute of Fishery, Substation for Carp Pond Culture, D-91315 Höchstadt, Germany

Carp is marketed almost exclusively fresh, practically never as a frozen product. The following experiment attempts to establish how far deep frozen storage affects the quality of the carp meat.

The investigation was carried out on carp with a live weight of about 1.2 kg, which had been raised without supplementary feeding. Three groups were compared: deep frozen for four months at $-18°C$ and vacuum-packed (Group II) or deep frozen but not vacuum-packed (Group III) and freshly slaughtered carp of the same origin, which remained alive during the winter (Group I) in which Groups II and III were in deep frozen storage. The nutrient composition of the skinned fillets was established. A sensory test was carried out by a taste panel of six by means of an evaluation scale. In addition the fat oxidation (TBARS) of the carp meat was measured by means of the thiobarbituric acid method.

The nutrient composition of the fillets from the three groups did not vary. The average fat content was 1.9%, protein content was 18.1% and ash content 1.2%. The fat oxidation (TBARS) of the vacuum-packed frozen carp fillets (0.98 mg/kg) was nearly identical to that of the fresh carp fillets (1.10 mg/kg). The fat oxidation of the carp fillets which had been in deep frozen storage without vacuum packing (Group III) (3.28 mg/kg) was about three times higher than that of the fresh or the vacuum-packed frozen fillets. The higher fat oxidation had a definite effect on the sensory quality. The smell, taste and overall impression of the fish from Group III was judged significantly inferior to that of the fish from Groups I and II. The taste panel frequently considered the fish from Group III to be 'rancid' and 'nauseating'. The fish which had been vacuum-packed in deep frozen storage (Group II) were noted to be of considerably lower sensory quality in comparison with the fresh carp (Group I), but the deterioration in quality was by no means so noticeable as in the fish which had not been vacuum-packed (Group III). The vacuum-packed carp were also judged significantly inferior in taste to the fresh fish. They tasted less aromatic than the fresh carp, but not bad. Only slight differences in smell and overall impression were noted between Groups I and II. The meat of the frozen fish (Groups II and III) was judged to be appreciably dryer and tougher. The deep freezing had no significant influence on the colour of the cooked fish.

Effect of Packing, Glazing and Salting on the Oxidative Stability of Frozen Half-shelled Pacific Oysters

N.M. O'Gorman, J.P. Kerry, D. Gilroy, D.J. Buckley, T.P. O'Connor and R.G. Fitzgerald

Department of Food Science and Technology and Aquaculture Development Centre University College Cork, National University of Ireland, Cork, Iceland

Due to consumer demand, increasing quantities of shellfish and fish are being distributed to markets in their freshest forms. Therefore, more attention is being focused on novel methodologies and technologies which might improve the quality of fish products and by so doing, extend their shelf-lives. Cold storage has been a common approach taken to prevent shellfish and fish deterioration; however, new approaches such as alternative packaging and the use of synthetic and natural preservatives have been applied to extend the shelf-life of fish products. The objective of this study was to investigate the effect of packaging, glazing and salting on the shelf-life of frozen Pacific oysters as determined by oxidative stability.

Fresh farmed Pacific oysters were depurated in UV sterilised water tanks for 24 hours. Oysters were removed and (a) frozen to $-18°C$ and top shell removed, (b) chilled at $4°C$ for 7 h, frozen to $-18°C$ and top shell removed, (c) chilled to $4°C$ for 7 h, frozen to $-18°C$ and top shell removed and glazed with a solution containing 1% sodium ascorbate and 3% salt. Oysters were held in heat-sealed packs, stored at $-18°C$ and sampled over a period of 42 days. The trial was repeated to observe the effect of the individual components of the glaze under two packaging systems. Fresh depurated oysters were (a) untreated (control), (b) glazed with a solution of 1% sodium ascorbate, (c) glazed with a solution of 3% salt, (d) glazed with a solution containing 1% sodium ascorbate and 3% salt. Oysters from each treatment group were subsequently held under aerobic conditions in heat-sealed plastic packs or vacuum packed in similar plastic pouches. All samples were stored at $-18°C$ and sampled over a period of 120 days. Lipid oxidation was determined by measuring thiobarbituric acid reacting substances (TBARS).

The use of sodium ascorbate significantly ($p < 0.05$) reduced lipid oxidation in glazed oysters compared to the unglazed treatment groups. TBARS numbers were higher in oysters that were allowed to rest prior to freezing than in fresh-frozen oysters; however, these trends were not significantly different. The use of 1% sodium ascorbate as a glazing agent in combination with vacuum packaging significantly ($p < 0.05$) reduced lipid oxidation in oysters compared to all other treatments. In general, samples held in heat-sealed packs had higher oxidation levels compared to similar treatments held under vacuum packaging conditions. The use of a salt glaze in the absence of 1% sodium ascorbate had pro-oxidative effect on oysters stored under aerobic conditions and were similar to control oysters held under similar packaging conditions. All other treatments had lower TBARS values compared to control oysters held in aerobic packaging conditions. Similar results were detected in oysters which had been blanched prior to glazing or packaging. These results show that the use of sodium ascorbate glaze in combination with vacuum packaging provides an effective method of extending the shelf-life of oysters.

Use of High Pressure Steam and Addition of Herbs to Cooking Water and their Effects on Mussel Shelf-life Quality

N.M. O'Gorman, J.P. Kerry, D.J. Buckley, T.P. O'Connor and R.G. Fitzgerald

Department of Food Science and Technology and Aquaculture Development Centre
University College Cork, National University of Ireland, Cork, Ireland

Shellfish have a limited shelf-life due to oxidative rancidity of highly unsaturated *n*-3 fatty acids. Natural antioxidants are capable of protecting lipids from oxidation in foods. The objective of this study was to investigate the effect of a number of fried herbs on the shelf-life of mussels as determined by oxidative stability.

Fresh farmed mussels were depurated in tanks with UV sterilised water for 24 h. Mussels were cooked in a pressure cooker at 15 psi for 120 s with 0 (control), 2, 4, 6, 8 and 10% dried sage (% w/v of cooking water). On determining optima for form of use and working concentration from the initial sage screening experiment, mussels were pressure cooked with herbs: basil, oregano, marjoram, rosemary, sage and thyme using determined experimental parameters. All samples were held in heat sealed plastic packs and chilled (4°C × 8 days) and/or frozen (-20°C × 40 days). Lipid oxidation was determined by measuring thiobarbituric acid reactive substances (TBARS).

A concentration of 8% whole sage was most effective at reducing TBARS values in fresh mussels compared to all other treatments ($p < 0.05$). Mussels were pressure cooked with 8% whole basil, oregano, marjoram, rosemary, sage and thyme and held under chilled and frozen conditions in heat-sealed packs. Basil, oregano, marjoram, rosemary, sage and thyme reduce ($p < 0.05$) TBARS in both chilled and frozen storage samples compared to controls. Thyme and oregano were the most effective herbs in reducing TBARS. The addition of herbs to cooking water prior to pressure cooking appeared to be effective in inhibiting lipid oxidation in mussel samples, thereby extending shelf-life.

Quality Differences in Sea Bass (*Dicentrarchus labrax*) from Different Rearing Systems

E. Orban, A. Ricelli, G. Di Lena, I. Casini, R. Caproni

National Institute of Nutrition, Unit of Study on the Quality of Fish as Food, via Ardeatina 546, 00178 Roma, Italy

The development of different rearing techniques, intensive, extensive and semiextensive, has given a great contribution to the increased demand for fish products. The objective of this research was to compare the chemical composition, fatty acid profiles and content of some unsaponifiable components, particularly α-tocopherol, important for its antioxidant and vitamin action, in fillets from intensively and extensively reared sea bass (*Dicentrarchus labrax*) of commercial size (350–500 g), caught in the same month. The intensively reared sea bass were raised on artificial diets in tanks of salted water collected from the lagoon, while extensively reared fish were bred and raised without an artificial diet in the lagoon. Results on the proximate composition of fish fillets evidence no significant difference, probably due to the high variability of lipid content in fish from the extensive plant (Table 1). The unsaponifiable lipid fraction composition of fillets from differently reared fish showed no significant difference (Table 2). On the contrary by comparing the fatty acids profiles significant differences emerged in some fatty acids (Table 3).

Table 1 Nutrient content (g/100 g) of *D. labrax* from different rearing systems.

	Extensive	Intensive		Extensive	Intensive
pH	6.60 ± 0.11	6.37 ± 0.04	Non protein N	0.32 ± 0.01	0.32 ± 0.01
Moisture	76.48 ± 2.11	70.63 ± 0.34	Lipid	2.75 ± 1.78	7.88 ± 1.19
Protein	18.71 ± 0.22	19.09 ± 0.00	Ash	1.22 ± 0.17	1.16 ± 0.01

Table 2 Unsaponifiable lipid fraction composition of *D. labrax* fillets from different rearing systems (mg/g lipid).

	Extensive	Intensive		Extensive	Intensive
Cholesterol	29.68 ± 12.18	8.25 ± 0.25	All-trans retinol (µg)	3.74 ± 1.44	6.67 ± 0.11
Squalen	0.18 ± 0.03	0.13 ± 0.01	α-tocopherol	0.25 ± 0.07	0.10 ± 0.02

Table 3 Fatty acid profiles of *D. labrax* fillets from different rearing systems (% of total fatty acids).

Fatty acids	Extensive	Intensive	Fatty acids	Extensive	Intensive
14:0	2.31 ± 0.31	3.90 ± 0.18*	22:1	0.60 ± 0.1	2.25 ± 0.34*
15:0	0.90 ± 0.12	0.36 ± 0.00*	18:2	3.04 ± 0.8	8.84 ± 0.19**
16:0	23.84 ± 0.57	20.32 ± 0.10*	18:3	0.50 ± 0.2	1.55 ± 0.08*
17:0	0.74 ± 0.56	0.26 ± 0.07	18:4	0.46 ± 0.1	1.25 ± 0.11*
18:0	5.81 ± 0.79	3.99 ± 0.16	20:2	0.58 ± 0.0	0.39 ± 0.05*
20:0	0.51 ± 0.19	0.52 ± 0.11	20:4	3.60 ± 0.3	0.61 ± 0.09*
16:1	6.72 ± 1.25	5.40 ± 0.18	20:5	5.47 ± 0.7	6.77 ± 0.39
18:1 ω9	20.62 ± 1.23	25.15 ± 0.50	22:5	2.85 ± 0.4	0.99 ± 0.08*
18:1 ω7	4.42 ± 0.69	2.73 ± 0.09	22:6	11.61 ± 3.6	7.99 ± 0.09
20:1	0.87 ± 0.10	2.66 ± 0.26*			

$n = 2$ $*p \leq 0.05, **p \leq 0.01$

Do Killing Methods Affect the Quality of Atlantic Salmon?

H. Ottera, B. Roth and O.J. Torrissen

Institute of Marine Research, Department of Aquaculture, P.O. Box 1870 Nordnes, N-5817 Bergen, Norway

The methods used for stunning and killing fish species in aquaculture have recently received a lot of attention, from an ethical point of view – does the fish suffer unnecessary pain during the process – and also from a product quality point of view. These two aspects were the rationale for the EU-project 'Optimisation of harvest procedures of farmed fish with respect to quality and welfare – FAIR CT97-3127 FAQUWEL'. Here we present some of the preliminary results on product quality of Atlantic salmon (*Salmo salar*) as affected by killing method.

We are evaluating four methods for killing the salmon:

- ❑ Sedation by CO_2, followed by gill-cutting
- ❑ Electro-stunning, followed by gill-cutting
- ❑ Brain destruction by a pin-bolt machine, followed by gill-cutting
- ❑ Direct gill-cutting

Sedation by CO_2 or direct gill-cutting are the most commonly used methods in the aquaculture industry, but recently the interests in alternative methods have evolved, and commercial use of electro-stunning and various types of brain destruction techniques are in development. We also did a simple trial on using laughing gas N_2O as a sedative, but that apparently had no effect on the salmon.

Major evaluation criteria included development of *rigor mortis* and pH during storage on ice, cortisol measurements as an indicator of stress during slaughter, and various product quality measurements taken on raw fish stored four days on ice.

As expected, fish killed by the methods supposed to be most 'brutal', use of CO_2 or direct gill-cutting, also went into *rigor mortis* first, and had the highest rigor index. Similarly, they seemed to have the most rapid initial drop in pH. Both these factors indicate that the use of these traditional killing methods for salmon may be inferior to new methods like electro-stunning and pin-bolting. Differences in ultimate flesh quality, measured on raw fish stored four days on ice are, however, more difficult to find. Preliminary data analysis does not indicate differences between killing methods on fillet colour; on the other hand there are indications that fish bled to death after gill-cutting had softer fillets (measured as Warner-Bratzler shear force). Further analysis and experiments will go into more detail in evaluating salmon quality as a function of killing method.

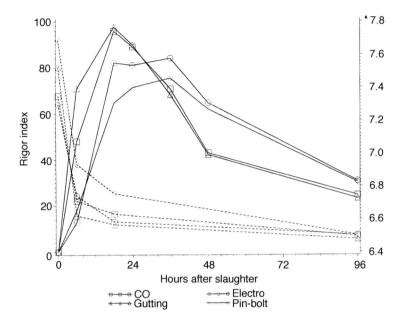

Fig. 1   Rigor index (solid lines) and pH (broken lines) of salmon stored on ice as a function of killing method. On rigor index 100 represents full rigor. pH was measured by spear electrode in the musculature. Each point represents the average of 20 individuals.

Flesh Quality in Seawater Reared Brown Trout (*Salmo trutta*) as Affected by Dietary Fat Levels and Pre-harvest Starvation

C. Regost , J. Arzel, M. Cardinal, M. Laroche and S.J. Kaushik

Fish Nutrition Laboratory, Unité mixte Inra-Ifremer, Centre de Brest IFREMER, 29280 Plouzané, France

Three isoproteic (crude protein content 56%) diets with different fat levels (11, 20 and 26%), designated respectively as LF, MF and HF were fed to triplicate groups of brown trout, *Salmo trutta* (initial average body weight of 1.5 kg), reared in seawater. At the end of an initial period of three months of feeding, fish fed diet HF were split into two groups : a triplicate group of fish received diet LF and the other triplicate group was kept unfed for a further two month period. Fish initially fed diet LF during the first phase continued to be fed the same diet. At the end of each phase, besides comparative carcass analysis, sensory and instrumental analyses were made on fresh as well as smoked fillets.

After the first three months, while the dietary fat levels had no significant effect on growth rate nor feed utilisation, an increase in dietary fat induced high lipid levels in the whole body (14.6 to 17.9%, on wet weight basis) and in the muscle (8.3 to 11.0%, on wet weight basis). During the second phase, while fish fed diet LF had similar growth performance irrespective of previous nutritional history, a significant effect of starvation was observed. Whole body fat composition was similar in all groups (around 15%) at the end of phase 2. Both starvation and feeding a low fat diet (LF) led to muscle lipid mobilisation in fish initially fed the high fat diet (HF).

Fillet yield was found to decrease significantly ($p < 0.05$) in unfed fish. Sensory analyses revealed few differences between treatments, in terms of visual aspects, for both cooked and smoked fillets at the end of phase 1. A positive relationship between instrumental colour analyses and dietary fat levels was observed, but no differences were observed between groups in terms of instrumental texture measurements. At the end of phase 2, a significant effect of starvation was observed on colour as well as on texture.

Both treatments applied two months before slaughtering, starvation or feeding a low fat diet led to modifications on the flesh quality criteria. However, fish fed during this last period presented the advantage of conserving growth performance.

Effect of Dietary Fatty Acid Profile on Fatty Acid Composition in Salmon (*Salmo salar*) when Replacing Fish Oils with Vegetable Oils

G. Rosenlund, A. Obach, E.Å. Bendiksen, M. Gisvold and B. Ruyter

NUTRECO Aquaculture Research Centre, P.O.Box 48, N-4001 Stavanger, Norway

The nutritional value of the fat in salmon is related both to the amount of fat and to its fatty acid composition. Both parameters are affected by feeding. Fish lipids typically contain long chain fatty acids ($\geq$ 20 carbons) of which the polyunsaturated fatty acids of *n*-3 family are of special interest. In comparison, vegetable oils principally contain shorter fatty acids typically with 18 carbons. Consumption of eicosapentaenoic acid (EPA, 20 : 5 *n*-3) and docosahexaenoic (DHA, 22 : 6 *n*-3) is known to have a positive effect in relation to arteriosclerosis and cardiovascular diseases. No agreement has been reached regarding recommended daily intake in this respect, but the British Nutrition Foundation goes as far as recommending a daily intake of 1.25 g of *n*-3 fatty acids as EPA and DHA. Surveys in England have shown that the actual daily intake of EPA and DHA is 0.25 g.

The objective of the present study was to investigate how different combinations of fish oils and vegetable oils will affect the lipid composition of farmed salmon. In a multivariate designed trial, fish with an initial weight of 275 g were fed 30 different diets for 90 days. The trial was carried out in sea water (7°C) in indoor tanks (1 × 1 m) at Nutreco ARC's research station (Stavanger, Norway). The feeds were based on a common base pellet coated with 30 different oils which were either pure fish oils produced from capelin, anchovy or jack mackerel, or these fish oils replaced by up to 75% soybean, rapeseed or linseed oil. This resulted in a wide variation in dietary fatty acid profiles, i.e. the ratio of *n*-6 and *n*-3 fatty acids was between 0.2 and 2.9 and the level of EPA and DHA varied from 7.5 to 27% of the diet fat. The diets contained 45.8 ± 0.1% crude protein and 32.1 ± 0.1% crude fat.

The fish grew to an average final weight of around 600 g. No major differences were found in gross body composition of the fish analysed at the end of the trial. However, a very close correlation was found between the ratio of *n*-6 and *n*-3 fatty acids in feed and fish muscle ($y = 0.8758x + 0.073$, $R^2 = 0.9671$). Analyses of individual fatty acids showed that the level of C18-fatty acids in the fish could be described by linear equations whereas the drop in the level of EPA and DHA was better described by a polynomial equation ($y = 0.0367x^2 - 0.3973x + 9.192$, $R^2 = 0.9208$). The results may indicate that salmon at low dietary levels of EPA and DHA will conserve or form these fatty acids from 18 : 3 *n*-3. The latter is in agreement with other reports.

The level of EPA and DHA varied between 11.6 and 22.1% of total fatty acids in the muscle fat of fish fed different fish oils, whereas the corresponding range in fish fed the different oil combinations was 7–20%. Assuming a fat level of 15% in harvest size salmon, these results suggest that a portion of 100 g of salmon is enough to provide the recommended daily intake of EPA and DHA even at high levels of vegetable oils in the diet.

The Effect of Slaughtering Procedures on Bleeding in Atlantic Salmon (*Salmo salar*) and Rainbow Trout (*Oncorhynchus mykiss*)

B. Roth, O.J. Torrissen and E. Slinde

Institute of Marine Research, Bergen, P.O. Box 1870-Nordnes, 5817 Bergen, Norway

After being sedated with CO_2, Atlantic salmon (*Salmo salar*) and rainbow trout (*Oncorhynchus mykiss*) are slaughtered by bleeding followed by gutting. Bleeding is often practised with a gill or throat cut combined with a 30 to 60 min bleeding period ensuring that the fish are bled and are dead before gutting. Accordingly this is done under the circumstances that the heart plays an active role in forcing the blood out of the muscle tissue. From a theoretical point of view this practice is difficult to explain, since the circulation in fish is maintained by the heart generating a flow through the fish. A gill or throat cut would cause drop of pressure eliminating the circulatory flow and the bleeding would become a passive action. This would not empty the blood vessel in the fillets. The aim of this study was therefore to investigate different bleeding and gutting procedures and their effect on bleeding by analysing the amount of blood clots in smoked fillets.

During our experiments we used a total of 80 rainbow trout that were sedated with CO_2 followed by a blow to the head, and slaughtered in four different ways:

(1) Gill cut following 30 minutes bleeding period before gutting
(2) Gill cut only, no gutting
(3) No gill cut, but directly gutted
(4) Neither gill cut nor gutted.

Within 45 minutes after slaughter all fish were placed on ice and stored for three days before filleting and transport to a smoking facility. After smoking, the remaining blood in the fillet became visual and the size and abundance of the blood clots were measured. In addition the fillets were given a visual bleeding grade ranging from 1 to 5. The same experiments were also performed on salmon, but in this case the fish was only slaughtered as described in (1) and (3). In addition 10 rainbow trout from groups (1) and (4) received rough treatment to determine if treatment during *rigor mortis* would affect the bleeding. Rough treatment consisted of a 1 m drop at three hours post-mortem, and one bend on each side four times within the next 24 hours.

The results show that the heart does not contribute to the bleeding of fish fillets. Sufficient bleeding was obtained by cutting one or several of the blood vessels at an early stage, allowing a passive drainage of blood from the muscle tissue. Poor bleeding and major blood clots were clearly observed in fish that were neither bled nor gutted. Rough handling did not influence the amount of blood clots in fish that were bled and gutted after stunning, but the handling did influence fish that was neither bled nor gutted. This shows that bleeding had already occurred before the fish entered *rigor mortis*.

If one can assume that future stunning methods meet the ethical demands required for direct gutting, we conclude that gill cutting is not necessary.

Cold Shortening and Drip Loss in Atlantic Salmon (*Salmo salar*), Rainbow Trout (*Oncorhynchus mykiss*), Halibut (*Hippoglossus hippoglossus*), Cod (*Gadus morhua*), Haddock (*Melanogrammus aeglefinus*) and Saithe (*Pollachius virens*)

E. Slinde, B. Roth and O.J. Torrissen

Institute of Marine Research, Bergen, P.O. Box 1870-Nordnes, 5817 Bergen, Norway

The aim of this study was to investigate the degree of muscle contraction and drip loss due to *rigor mortis* and cold shortening in Atlantic salmon (*Salmo salar*), rainbow trout (*Oncorhynchus mykiss*), halibut (*Hippoglossus hippoglossus*), cod (*Gadus morhua*), haddock (*Melanogrammus aeglefinus*) and saithe (*Pollachius virens*). Pieces of fillet approximately $10 \times 2 \times 1$ cm were cut from each side of the fish, and the lengths were measured and the pieces were weighed. The pieces were either stored on ice (control group) or frozen overnight before thawing. To estimate the degree of muscle contraction and drip loss, length and weight were measured at approximately 30 hours post-mortem on the control group, and 15 hours after thawing.

The results revealed large variations between lean and fat fish species measured as cold shortening or drip loss. Among the lean species saithe proved to be the most sensitive species to cold shortening with muscle contractions up to 40% of initial length after thawing. The mean value for cold shortening after thawing in saithe was 26%, with a drip loss of 16%. Cod gave intermediate results with a mean value for cold shortening of 15% and a drip loss of 9%. The lowest values were obtained for haddock with a shortening of 6% and a drip loss of 8%. Among the fat species the muscle shortening was low compared to the lean species with 3% shortening for halibut, 9% for salmon and 7% for trout.

More surprising was the impact of slaughtering temperature on the degree of cold shortening and drip losses. In cod, cold shortening increased from an average of 8% to 22% when the temperature of the environment for the fish before slaughter was raised from 9°C to 13°C. This was not as well documented for the other species, except for salmon, where the drip loss increased from an average of 1% to 7% when the temperature was raised from 9°C to 15°C, but there were no changes in the degree of muscle shortening. This phenomenon could also be observed in the control group, where the degree of rigor contraction and drip loss increased significantly in fish reared at higher temperatures. This shows that the fish temperature at slaughtering is not only important for the development of *rigor mortis*, but also for the degree of muscle shortening during storage and thawing of the fish.

The explanation behind the impact of pre-slaughter temperature on muscle shortening might be leakage of Ca^{2+} from the sarcoplasmic reticulum (SR) accelerating the Mg^{2+} Ca^{2+} ATP-ase within the myofibrils. What mechanism causes the Ca^{2+} leakage is unclear, but can be explained by the fact that the membrane flexibility is acclimatised to the environment of the fish before slaughter. Any change in temperature would then cause the membrane surrounding the SR to alter its function. We therefore recommend more systematic investigations on the effect rearing temperature would have on fish muscle physiology and *rigor mortis*.

A Mechanical Compression Test to Assess Development of *Rigor Mortis* in Fish; Atlantic Salmon, Atlantic Halibut and Plaice

N.K. Sørensen, T. Tobbiassen, S. Joensen, K. Midling and L. Akse

Norwegian Institute of Fisheries and Aquaculture Ltd, N-9291 Tromsø, Norway

The development of *rigor mortis* in fish muscle has gained increasing interest related to processing of farmed fish because it is important to avoid handling and packaging of fish in *rigor mortis*. Such handling may cause gaping of fillets and should therefore be kept at a minimum. In the industry, onset of *rigor mortis* starts usually 3–5 hours post-mortem, which leaves limited time for the necessary handling when producing boxed iced fish. It is of interest to understand the factors influencing onset and strength of *rigor mortis* in fish. These factors are related to species, degree of handling, temperature and catching or killing method, i.e. the energy status of the fish.

Rigor mortis is usually measured by different indexes describing the degree of bending the tail of the fish when hung over the edge of a table. Repeated measurements will be biased by the weight of the fish. It is, therefore, of interest to develop methods for assessing the development of *rigor mortis* in fish that are non-destructive. A mechanised and standardised method resembling the sensory 'finger test' has been developed.

The *rigor mortis* measurements using the mechanical test machine give promising data when assessing the development of *rigor mortis* in several fish species.

Slaughter Methods Affecting Adenosine Triphosphate and Derivatives in Chilled Stored Gilthead Seabream (*Sparus auratus*)

M. Tejada, A. Huidobro and A. Pastor

Instituto del Frío (CSIC), Ciudad Universitaria s/n. 28040 Madrid, Spain

Animal welfare is becoming an increasingly important part of consumer perception of quality; however, different fish slaughter procedures can affect the final quality of the fish. Given the diversity of slaughtering methods used in farmed fish, it is essential to assess how these methods affect fish quality during chilled storage. Breakdown of adenosine-5'-triphosphate (ATP) and derivatives and the ratio between them (as K value) are early indicators of changes in post-mortem fish; they are widely used to set safe consumption limits for raw fish and as indices of freshness of chilled fish or raw material for gels. Our aim was to determine whether gutting immediately after death and different methods of slaughter alter the evolution of these compounds or their ratio during chilled storage in gilthead seabream (*Sparus auratus*) killed by immersion in ice-water slurry, asphyxia in air, anaesthesia (AQUI-S^{TM}) followed by a blow on the head, and just a blow on the head. The fish was stored with ice flakes whole (*w*) or gutted (*g*) for a maximum time of 29 days. Forty kg (approximately 130 fish/lot) was used for each slaughter method and post-mortem treatment.

In all lots ATP rapidly degraded to inosine monophosphate (IMP) during chilling storage, leaving no appreciable amounts of adenosine-5'-diphosphate (ADP) or adenosine-5'-monophosphate (AMP). Dephosphorylation of IMP was slow and progressive. Inosine (Ino) and hypoxanthine (Hx) increased gradually over storage in all lots with no significant differences ($p < 0.05$), but Ino tended to accumulate and Hx tended to stabilise by the end of storage. The molar ratio Hx : Ino was $< 5{:}1$ throughout the period, and therefore this species was classified as intermediate.

None of the lots attained K values $> 20\%$ before seven days in chilled storage, which means that sashimi grade (raw fish) for this species was longer than for other commercial species. Maximum K values were established at 50–60% at the end of the storage period, well past the sensory limit, and around 35% (*w*) and 25% (*g*) of lots when the sensory evaluation was still within the limits.

This was financed by the EU project FAIR CT97-3127 (FAQUWEL).

Effect of Storage and Slaughter Temperature on Fillet Colour of Atlantic Salmon (*Salmo salar*)

O.J. Torrissen, B. Roth and E. Slinde

Institute of Marine Research, P.O. Box 1870-Nordnes, N-5817 Bergen, Norway

The aim of this study was to investigate the effects of pre- and post-harvest temperatures on the colour of Atlantic salmon (*Salmo salar*) fillets. The experiments were carried out by acclimatising salmon for two days at five different water temperatures; $2°$, $6°$, $10°$, $14°$ and $18°C$. After acclimatisation, five fish from each group were killed by a blow to the head, filleted, placed in plastic bags and stored at the temperatures indicated above. This created a 5×5 matrix with 25 combinations of pre- and post-harvest temperatures. After 24 hours of storage the colour of the fillets was measured by a Hunter Miniscan XE instrument using the L^*, a^*, b^* and chroma scale.

The results showed that lightness (L^*) was positively correlated to both the pre- and post-harvest temperatures. Darkest fish were obtained at a post-harvest temperature of $6°C$. There was significant interaction between the post- and pre-harvest temperatures on the fillet lightness. The redness (a^*) value was highest in fish kept at a pre-harvest temperature of $2°C$. However, post-harvest temperature also had a positive effect on fillet redness. The palest fish was obtained in the group kept at $2°C$ and a post-harvest temperature at $18°C$. Lowest chroma values were observed for the group both acclimatised and stored at $2°C$.

The results showed that the pre- and post-harvest temperature of salmon are important factors for the colour appearance of salmon fillets. However, further work is necessary in order to recommend harvest conditions for maximum colour yields.

Texture and Technological Properties of Fish

O.J. Torrissen[1], D. Sigurgisladottir[2] and E. Slind[1]

[1] *Institute of Marine Research, P.O. Box 1870-Nordnes, N-5817 Bergen, Norway;*
[2] *IceTec, Keldnaholt, IS0112 Reykjavik, Iceland*

The main quality parameters for fresh salmon are fat, colour, texture and freshness. Other quality parameters commonly cited are white stripes (myocommata), blood stains, marbling and melanin (Koteng 1992; Sigurgisladottir *et al.* 1997).

Atlantic salmon is often smoked or marinated and cut in thin slices and consumed without any further heat treatment. The firmness of the raw processed muscle and the visual appearance are critical parameters in determining the acceptability of the product. Gaping and soft mushy texture are not accepted by the processing industry or by consumers. Consumers attach negative associations to fat content, and apparently are associating loose texture with high fat content.

The classical method for evaluating fish texture is the finger method. A finger is pressed on the skin or the fillet and the firmness of the fish is evaluated as the hardness and the flexibility of the flesh. Texture of raw salmon fillets may be measured objectively by different methods using mechanical food testing equipment. The main techniques applied for fish may be classified into puncture, compression, shear and tensile techniques. Double compression makes it possible to perform a texture profile analysis (TPA) from a plot of force-time curves (Boune 1978).

The post-mortem tenderisation of fish muscle has been demonstrated in several microscopy studies to be closely related to the degradation of collagens, fibrils of the endomysium and perimysium. In severe collagen breakdown, the myotomes are separated from the myocommata and gaping occurs. Although little evidence exists on the exact mechanism of this degradation, some studies suggest that endogenous proteinases are involved. The *rigor mortis* tension and lipid infiltration of the myocommata are also possible factors involved.

Quality Changes in Rainbow Trout Fillets Under Different Storage Conditions

H. Wedekind and M. Griese

Institute for Inland Fisheries Potsdam-Sacrow, Jaegerhof, D-14476 Gross Glienicke, Germany

Flesh quality traits are known to be affected by storage conditions, e.g. temperature, wrapping, duration of storage. Especially for small scale processing and marketing in fish-farms it is essential to know about the significance of quality changes post mortem. The present investigation was carried out to investigate the development of several flesh quality parameters under the conditions occuring in fish farms and during distribution of fish fillets.

A random sample of 20 rainbow trout (*Oncorhynchus mykiss*) was slaughtered in a standard procedure. Immediately after slaughtering the post mortem values of flesh pH and electric conductivity were measured. After filleting, further parameters of flesh quality, e.g. colour, lightness, dry matter content, water-binding-capacity, lactate and glucose level, as well as

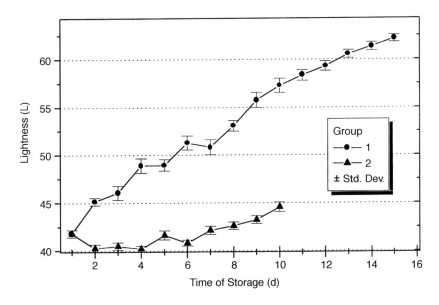

Fig. 1 Flesh-lightness during storage at 2°C (group 1) and 8°C (group 2).

the microbial population, were documented. The fish were divided into two groups of 20 fillets, each stored under different conditions: group 1was stored for 15 days on melting ice around +2°C; group 2 was stored for 10 days in containers covered with nylon foil at +8°C. At daily intervals the flesh quality was measured according to the physical and chemical parameters mentioned above.

The results indicate continuous quality changes in the fillets. pH values, water-binding-capacity, and several physiological parameters showed a pronounced degradation during storage. Moreover flesh colour and lightness changed significantly (Fig. 1). The comparison of the applied storage conditions resulted in a faster degradation of flesh quality at higher storage temperature. As expected the sensorial characters were conserved best during storage on ice. Whereas group 1 was in good condition after 10 days of storage, an unacceptable softening and spoilt odour was observed in group 2 after 7 days.

Effect of Different Extruded Diets on Flesh Quality in Rainbow Trout (*Oncorhynchus mykiss*)

H. Wedekind and N. Hjermitslev

Institute for Inland Fisheries Potsdam-Sacrow, Jaegerhof, D-14476 Gross Glienicke, Germany

A feeding trial was carried out with rainbow trout (*Oncorhynchus mykiss*) to compare three commercially available extruded diets. A full-sib-group of 100 individuals per group was raised in net-cages. Besides the growth rate and the feed conversion rate, the body composition (e.g. percentage of fillet, head, entrails, liver, gonads) and the flesh quality (dry matter content, pH, electric conductivity, L-a-b values, weight loss during cold storage, freezing, cooking and smoking) were evaluated.

Table 1 Chemical composition of the experimental diets.

	Diet I	Diet II	Diet III
Gross energy (MJ/kg)	22.4	23.5	23.2
Dry matter (%)	93.8	93.3	91.0
Crude protein (% dm)	45.0	44.3	43.6
Crude fat (% dm)	22.5	26.3	20.3
Crude fibre (% dm)	2.4	1.5	0.9
Crude ash (% dm)	7.19	6.77	9.64
Total P (% dm)	0.99	0.92	1.23
NFE (% dm)	23.9	15.2	18.0

The results indicated only small growth differences between the groups. The specific growth rates were around 1.0 and the feed conversion ratio ranged from 0.92 to 0.98. However, the fish showed clear differences in their body composition (corpulence, carcass-percentage). Group II had lower yield-percentage due to an advanced gonad development. Diet I resulted in the highest liver-percentages. Moreover, the latter group showed a significantly reduced intramuscular fat content and the highest pH-values 5 minutes post-mortem and 24 hours post-mortem. The physical flesh quality parameters, e.g. water-holding-capacity, colour and lightness, weight-loss in several stages of storage and processing showed significant differences between the experimental groups (Table 2).

Table 2 Average values and significance of some parameters of flesh quality.

Parameter	Diet I	Diet II	Diet III
Fat content (%)	2.53^a	4.69^b	4.03^b
pH 5 min. post-mortem	6.81^a	6.76^b	6.74^b
pH 24 h post-mortem	6.38^a	6.33^b	6.34^b
LF 24 h post-mortem (μS/cm^2)	1342^a	1597^b	1591^b
Cooking loss (%)	15.3^a	13.9^b	13.5^b

(same superscript letters = not significant; different superscript letters = significant, $p \leq 0.05$)

Index